D1574438

Probabilistic Modeling and Analysis in Science and Engineering

Probabilistic Modeling and Analysis in Science and Engineering

T. T. Soong

Faculty of Engineering and Applied Sciences
State University of New York at Buffalo
Buffalo, New York

John Wiley & Sons
NEW YORK   CHICHESTER   BRISBANE   TORONTO

Copyright © 1981 by John Wiley & Sons, Inc.

All rights reserved. Published simultaneously in Canada.

Reproduction or translation of any part of
this work beyond that permitted by Sections
107 and 108 of the 1976 United States Copyright
Act without the permission of the copyright
owner is unlawful. Requests for permission
or further information should be addressed to
the Permissions Department, John Wiley & Sons.

Library of Congress Cataloging in Publication Data

Soong, T. T.
   Probabilistic modeling and analysis in science
and engineering.

   Includes bibliographical references and index.
   1. Probabilities.   2. Mathematical statistics.
I. Title.

TA340. S66   519.2   80-39804
ISBN 0-471-08061-6

Printed in the United States of America

10 9 8 7 6 5 4 3 2 1

To the Memory of My Mother

Preface

This book was written for an introductory one-semester or two-quarter course in probability and statistics for students in engineering and applied sciences. No previous knowledge of probability or statistics is presumed but a course in calculus is a necessary prerequisite for the material.

The development of this book was guided by a number of considerations observed over many years of teaching courses in this subject area, including the following:

• As an introductory course, a sound and rigorous treatment of the basic principles is imperative for a proper understanding of the subject matter and for confidence in applying these principles to practical problem solving. A student, depending upon his or her major field of study, will no doubt pursue advanced work in this area in one or more of the many possible directions. How well is he or she prepared to do this strongly depends on his or her mastery of the fundamentals.

• It is important that the student develop an early appreciation for applications. Demonstrations of the utility of this material in nonsuperficial applications not only sustain student interest but also provide the student with stimulation to delve more deeply into the fundamentals.

• Most of the students in engineering and applied sciences can only devote one semester or two quarters to a course of this nature in their programs. Recognizing that the coverage is time limited, it is important that the material be self-contained, representing a reasonably complete and applicable body of knowledge.

The choice of the contents for this book is in line with the foregoing observations. The major objective is to give a careful presentation of the fundamentals in probability and statistics, the concept of probabilistic modeling, and the process of model selection, verification, and analysis. In this text, definitions and theorems are carefully stated and topics rigorously treated but care is taken not to become entangled in excessive mathematical details. Practical examples are emphasized; they are purposely selected from many different fields and not slanted toward any particular applied area. The same objective is observed in making up the exercises at the back of each chapter.

Because of the self-imposed criterion of writing a comprehensive text and presenting it within a limited time frame, there is a tight continuity from one topic to the next. Some flexibility exists in Chapters VI and VII that include discussions on more specialized distributions used in practice. For example, extreme-value distributions may be bypassed, if it is deemed necessary, without serious loss of continuity. Also, Chapter XI on linear models may be deferred to a follow-up course if time does not allow its full coverage.

It is a pleasure to acknowledge the substantial help I received from students in my courses over the past 10 years and from my colleagues and friends. Their constructive comments on preliminary versions of the manuscript led to many improvements. As in all my undertakings, my wife, Dottie, cared about this project and gave me her loving support for which I am deeply grateful. My thanks also go to my children, Karen, Stephen, and Susan, for their support and understanding throughout this writing period.

T.T.S.
Buffalo, New York

Contents

I. Introduction   1

   I.1 Organization of Text   1

PART A: PROBABILITY AND RANDOM VARIABLES

II. Basic Probability Concepts   7

   II.1. Elements of Set Theory   8
      II.1.1. Set Operations   10
      II.1.2. Borel Field   12
   II.2. Sample Space and Probability Measure   13
      II.2.1. Axioms of Probability   15
      II.2.2. Assignment of Probability   18
   II.3. Statistical Independence   18
   II.4. Conditional Probability   22
   References and Comments   30
   Problems   30

III.	**Random Variables and Probability Distributions**	37
	III.1. Random Variables	38
	III.2. Probability Distributions	39
	III.2.1. Probability Distribution Function (PDF)	39
	III.2.2. Probability Mass Function for Discrete Random Variables (pmf)	42
	III.2.3. Probability Density Function for Continuous Random Variables (pdf)	43
	III.2.4. Mixed-type Distribution	46
	III.3. Two or More Random Variables	47
	III.3.1. Joint Probability Distribution Function (JPDF)	47
	III.3.2. Joint Probability Mass Function (jpmf)	49
	III.3.3. Joint Probability Density Function (jpdf)	53
	III.4. Conditional Distributions and Independence	58
	References and Comments	64
	Problems	64
IV.	**Expectations and Moments**	71
	IV.1. Moments of a Single Random Variable	72
	IV.1.1. Mean, Median, and Mode	73
	IV.1.2. Central Moments, Variance, and Standard Deviation	75
	IV.1.3. Conditional Expectation	79
	IV.2. Chebyshev Inequality	81
	IV.3. Moments of Two or More Random Variables	83
	IV.3.1. Covariance and Correlation Coefficient	84
	IV.3.2. Schwarz Inequality	87
	IV.3.3. The Case of Three or More Random Variables	88
	IV.4. Moments of Sums of Random Variables	89
	IV.5. Characteristic Functions	93
	IV.5.1. Generation of Moments	94
	IV.5.2. Inversion Formulas	97
	IV.5.3. Joint Characteristic Functions	104
	References and Comments	107
	Problems	108
V.	**Functions of Random Variables**	113
	V.1. Functions of One Random Variable	114
	V.1.1. Probability Distributions	114
	V.1.2. Moments	129
	V.2. Functions of Two or More Random Variables	131
	V.2.1. Sums of Random Variables	139
	V.3. n Functions of n Random Variables	142

CONTENTS xi

		Reference	148
		Problems	148
VI.	**Some Important Discrete Distributions**		**156**
	VI.1.	Bernoulli Trials	156
		VI.1.1. The Binomial Distribution	*157*
		VI.1.2. The Geometric Distribution	*162*
		VI.1.3. The Negative Binomial Distribution	*164*
	VI.2.	The Multinomial Distribution	167
	VI.3.	The Poisson Distribution	168
		VI.3.1. Spatial Distributions	*176*
		VI.3.2. Poisson Approximation to the Binomial Distribution	*176*
	VI.4.	Summary	179
		References and Comments	179
		Problems	180
VII.	**Some Important Continuous Distributions**		**185**
	VII.1.	The Uniform Distribution	185
		VII.1.1. A Bivariate Uniform Distribution	*187*
	VII.2.	The Gaussian or Normal Distribution	190
		VII.2.1. The Central Limit Theorem	*192*
		VII.2.2. Probability Tabulations	*194*
		VII.2.3. The Multivariate Normal Distribution	*197*
		VII.2.4. Sums of Normal Random Variables	*201*
	VII.3.	The Lognormal Distribution	202
		VII.3.1. Probability Tabulations	*205*
	VII.4.	Gamma and Related Distributions	206
		VII.4.1. The Exponential Distribution	*209*
		VII.4.2. The Chi-square (χ^2) Distribution	*213*
	VII.5.	Beta and Related Distributions	214
		VII.5.1. Probability Tabulations	*217*
		VII.5.2. Generalized Beta Distribution	*219*
	VII.6.	Extreme Value Distributions	219
		VII.6.1 Type I Asymptotic Distributions of Extreme Values	*221*
		VII.6.2. Type II Asymptotic Distributions of Extreme Values	*226*
		VII.6.3. Type III Asymptotic Distributions of Extreme Values	*228*
	VII.7.	Summary	229
		References and Comments	231
		Problems	231

PART B: STATISTICAL INFERENCE, PARAMETER ESTIMATION, AND MODEL VERIFICATION

VIII. Observed Data and Graphical Representation — 241

    VIII.1. Histograms — 242
    References and Comments — 247
    Problems — 247

IX. Parameter Estimation — 254

    IX.1. Samples and Statistics — 254
        IX.1.1. Sample Mean — 256
        IX.1.2. Sample Variance — 257
        IX.1.3. Sample Moments — 259
        IX.1.4. Order Statistics — 259
    IX.2. Quality Criteria for Estimates — 260
        IX.2.1. Unbiasedness — 260
        IX.2.2. Minimum Variance — 261
        IX.2.3. Consistency — 269
        IX.2.4. Sufficiency — 270
    IX.3. Methods of Estimation — 272
    IX.4. Point Estimation — 273
        IX.4.1. The Method of Moments — 273
        IX.4.2. The Method of Maximum Likelihood — 283
    IX.5. Interval Estimation — 289
        IX.5.1. Confidence Interval for m in $N(m, \sigma^2)$ with Known σ^2 — 292
        IX.5.2. Confidence Interval for m in $N(m, \sigma^2)$ with Unknown σ^2 — 294
        IX.5.3. Confidence Interval for σ^2 in $N(m, \sigma^2)$ — 297
        IX.5.4. Confidence Interval for a Proportion — 299
    References and Comments — 301
    Problems — 302

X. Model Verification — 310

    X.1. Preliminaries — 311
        X.1.1. Type I and Type II Errors — 311
    X.2. Chi-Square Goodness-of-Fit Test — 312
        X.2.1. The Case of Known Parameters — 312
        X.2.2. The Case of Estimated Parameters — 318
    X.3. Kolmogorov–Smirnov Test — 322
    References and Comments — 325
    Problems — 326

XI. Linear Models and Linear Regression — 329

    XI.1. Simple Linear Regression — 329
        XI.1.1. Least Squares Method of Estimation — 330
        XI.1.2. Properties of Least-Square Estimators — 336
        XI.1.3. An Unbiased Estimator for σ^2 — 339
        XI.1.4. Confidence Intervals for Regression Coefficients — 341
        XI.1.5. Significance Tests — 345
    XI.2. Multiple Linear Regression — 348
        XI.2.1. Least Squares Method of Estimation — 348
    XI.3. Other Regression Models — 351
    References and Comments — 353
    Problems — 353

Appendix A: Tables — 359

    A.1. Binomial Mass Function — 360
    A.2. Poisson Mass Function — 363
    A.3. Standardized Normal Distribution Function — 366
    A.4. Student's t Distribution with n Degrees of Freedom — 367
    A.5. Chi Square Distribution with n Degrees of Freedom — 368
    A.6. D_2 Distribution with Sample Size n — 369
    References — 369

Appendix B: Answers to Selected Problems — 370

Index — 379

1
Introduction

As recent as 15 years ago, it would be difficult to find a course in probability in more than a handful of undergraduate programs in engineering and applied sciences. At present, almost all undergraduate curricula at major universities contain at least one basic course in this area. The recognition of this need of introducing the ideas of probability theory in a wide variety of scientific fields today reflects in part some of the profound changes in science and engineering education over the last decade.

One of the most significant is the greater emphasis that has been placed upon complexity and precision. A scientist now recognizes the importance of studying scientific phenomena having complex interrelations among their components; these components are often not only mechanical or electrical parts but also "soft-science" in nature, such as those stemming from behavioral and social sciences. The design of a comprehensive transportation system, for example, requires a good understanding of technological aspects of the problem as well as behavior patterns of the user, land use regulations, environmental requirements, pricing policies, and so on.

Moreover, precision is stressed—precision in describing interrelationships among factors involved in a scientific phenomenon and precision in predicting its behavior. This, coupled with increasing complexity in problems we face, leads to the recognition that a great deal of uncertainties and variabilities are inevitably present in problem formulations and one of the mathematical tools that is effective in dealing with them is probability and statistics.

Probabilistic ideas are used in a wide variety of scientific investigations involving randomness. Randomness is an empirical phenomenon characterized by the property that the quantities we are interested in do not have a predictable outcome under a given set of circumstances, but instead there is a statistical regularity associated with different possible outcomes. Loosely speaking, statistical regularity means that, in observing outcomes of an experiment a large number of times (say n), the ratio m/n, where m is the number of observed occurrences of a specific outcome, tends to a unique limit as n becomes large. For example, the outcome of flipping a coin is not predictable but there is statistical regularity in that the ratio m/n approaches $\frac{1}{2}$ for both heads and tails. Random phemonena in scientific areas abound: noise in radio signals, intensity of wind gusts, mechanical vibration due to atmospheric disturbances, Brownian motion of particles in a liquid, number of telephone calls made by a given population, length of queues at a ticket counter, choice of transportation modes by a group of individuals, and countless others. It is not inaccurate to say that randomness is present in any realistic conceptual model of a real world phenomenon.

I.1 Organization of Text

This book is concerned with the development of basic principles in constructing probability models and the subsequent analysis of these models. As in other scientific modeling procedures, the basic cycle of this undertaking consists of a number of fundamental steps; these are schematically presented in Figure 1.1. A basic understanding of probability theory and random variables is central to the whole modeling process as they provide needed mathematical machinery with which the modeling process is carried out and consequences deduced. The step from B to C in Figure 1.1 is the induction step by which the structure of the model is formed from factual observations of the scientific phenomenon under study. Model verification and parameter estimation (E) on the basis of observed data (D) fall within the framework of statistical inference. A model may be rejected at this stage due to inadequate inductive reasoning or to insufficient or deficient data. A reexamination of factual observations or additional data may be required here. Finally, model analysis and deduction are made to yield desired answers upon model substantiation.

In line with this outline of the basic steps, the book is divided into two parts. Part A (Chapters II–VII) addresses probability fundamentals involved in steps $A \to C$, $B \to C$, and $E \to F$. Chapters II through V provide these fundamentals that constitute the foundation of all subsequent development. Some important probability distributions are introduced in Chapters VI and VII. The nature and

INTRODUCTION

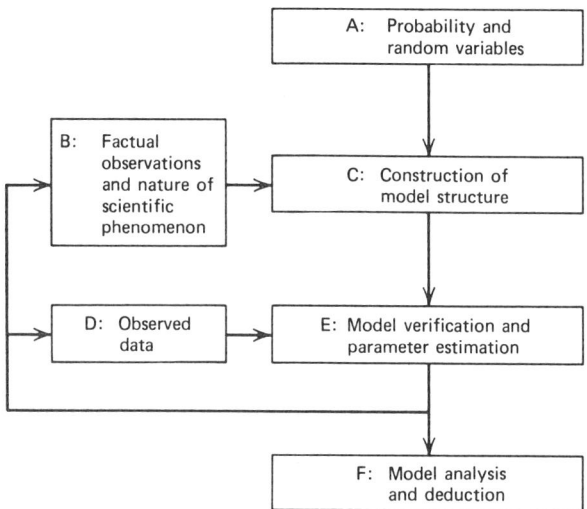

Figure 1.1 Basic cycle of probabilistic modeling and analysis.

applications of these distributions are discussed. An understanding of the situations in which these distributions arise enables us to choose an appropriate distribution for a scientific phenomenon.

Part B (Chapters VIII–XI) is concerned principally with step $D \to E$, the statistical inference portion of the text. Starting with data and data representation in Chapter VIII, parameter estimation techniques are carefully developed in Chapter IX, followed by a detailed discussion in Chapter X of a number of selected statistical tests that are useful for the purpose of model verification. In Chapter XI, the tools developed in Chapters IX and X for parameter estimation and model verification are applied to the study of linear regression models, a very useful class of models encountered in science and engineering.

The topics covered in Part B are somewhat selective but much of the foundation in statistical inference is laid. And this foundation should help the reader to pursue further studies in related and more advanced areas.

Part A

Probability and Random Variables

II
Basic Probability Concepts

The mathematical theory of probability gives us the basic tools for constructing and analyzing mathematical models for random phenomena. In studying a random phenomenon, we are dealing with an experiment whose outcome is not predictable in advance. Experiments of this type that immediately come to mind are those arising in games of chance. In fact, the earliest development in probability theory in the fifteenth and sixteenth centuries was motivated by problems of this type [see, for example, I. Todhunter (1949)].

In science and engineering, random phenomena describe a wide variety of situations. By and large, they can be grouped into two broad classes. The first class deals with physical or natural phenomena involving uncertainties. Uncertainty enters into problem formulation through complexity, through our lack of understanding of all the causes and effects, and through lack of information. Consider, for example, weather prediction. Information obtained from satellite tracking and other meteorological information simply is not sufficient to permit a reliable prediction of what weather condition will prevail in days ahead. It is therefore easily understandable that weather reports on radio and television are made in probabilistic terms.

The second class of problems widely studied by means of probabilistic models concerns those exhibiting variability. Consider, for example, a problem in traffic flow where an engineer wishes to know the number of vehicles crossing a certain point on a roadway within a specified interval of time. This number varies unpredictably from one interval to another, and this variability reflects variable driver behavior and is inherent in the problem. This property forces us to adopt a probabilistic point of view and probability theory provides a powerful tool for analyzing problems of this type.

It is safe to say that uncertainty and variability are present in our modeling of all real phenomena, and it is only natural to see that probabilistic modeling and analysis occupy a central place in the study of a wide variety of topics in science and engineering. There is no doubt that we will see an increasing reliance on the use of probabilistic formulation in most scientific disciplines in the future.

II.1 Elements of Set Theory

Our interest in the study of a random phenomenon is in the statements we can make concerning the events that can occur. Events and combinations of events thus play a central role in probability theory. The mathematics of events is closely tied to the theory of sets, and we give in this section some of its basic concepts and algebraic operations.

A *set* is a collection of objects possessing some common properties. These objects are called *elements* of the set and they can be of any kind with any specified properties. We may consider, for example, a set of numbers, a set of mathematical functions, a set of persons, or a set of a mixture of things. Capital letters A, B, C, Φ, Ω, ... shall be used to denote sets and lower case letters a, b, c, ϕ, ω, ... to denote their elements. A set is thus described by its elements. Notationally, we can write, for example,

$$A = \{1, 2, 3, 4, 5, 6\}$$

which means that the set A has as its elements integers 1 through 6. If the set B contains two elements, success and failure, it can be described by

$$B = \{s, f\}$$

where s and f are chosen to represent success and failure, respectively. For a set consisting of all nonnegative real numbers, a convenient description is

$$C = \{x : x \geq 0\}$$

We shall use the symbol

$$a \in A$$

to mean the element a belonging to set A.

A set containing no elements is called an *empty set* and is denoted by ∅. We distinguish between sets containing a finite number of elements and those having an infinite number. They are called, respectively, *finite sets* and *infinite sets*. An infinite set is called *enumerable* or *countable* if all of its elements can be arranged in such a way that there is a one-to-one correspondence between them and all positive integers; thus, a set containing all positive integers 1, 2, ... is a simple example of an enumerable set. A *nonenumerable* or *noncountable* set is one where the above-mentioned one-to-one correspondence cannot be established. A simple example of a nonenumerable set is the set C described above.

If every element of a set A is also an element of a set B, the set A is called a *subset* of B and this is represented symbolically by

$$A \subset B \quad \text{or} \quad B \supset A$$

Example 2.1. Let $A = \{2, 4\}$ and $B = \{1, 2, 3, 4\}$. Then $A \subset B$ since every element of A is also an element of B. This relationship can also be presented graphically using a Venn diagram as shown in Figure 2.1. The set B occupies the interior of the larger circle and A the shaded area in the figure.

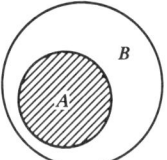

Figure 2.1 Venn diagram for $A \subset B$.

It is clear that an empty set is a subset of any set. When both $A \subset B$ and $B \subset A$ hold, the set A is then *equal* to B, and we write

$$A = B \tag{2.1}$$

We now give meaning to a particular set we shall call *space*. In our development we consider only sets that are subsets of a fixed (nonempty) set. This "largest" set containing all elements of all the sets under consideration is called *space*, and it is denoted by the symbol S.

Consider a subset A in S. The set of all elements in S which are not elements of A is called the *complement* of A, and we denote it by $\bar{A}$. A Venn diagram showing

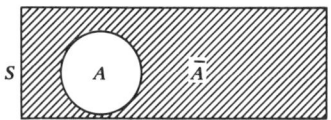

Figure 2.2 A and $\bar{A}$.

A and $\bar{A}$ is given in Figure 2.2 in which space S is shown as a rectangle and $\bar{A}$ is the shaded area. We note here that the following relations clearly hold.

$$\bar{S} = \varnothing, \qquad \bar{\varnothing} = S, \qquad \bar{\bar{A}} = A$$

II.1.1 SET OPERATIONS

Let us now consider some algebraic operations of sets $A, B, C, \ldots$ which are subsets of space S.

The *union* or *sum* of A and B, denoted by $A \cup B$, is the set of all elements belong to A or B or both.

The *intersection* or *product* of A and B, written as $A \cap B$ or simply AB, is the set of all elements that are common to A and B.

In terms of Venn diagrams, results of the above operations are shown in Figure 2.3 as sets having shaded areas.

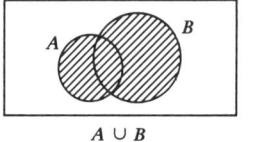
$A \cup B$

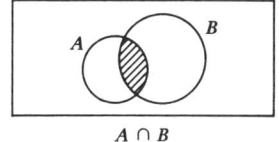
$A \cap B$

Figure 2.3 Union and intersection of sets A and B.

If $AB = \varnothing$, the sets A and B contain no common elements, and we call A and B *disjoint*. The symbol " $+$ " shall be reserved to denote the union of two disjoint sets when it is advantageous to do so.

Example 2.2. Let A be the set of all men and B consists of all men and women above 18 years of age. Then the set $A \cup B$ consists of all men as well as all women above 18 years of age. The elements of $A \cap B$ are all men above 18 years of age.

Example 2.3. Let S be the space consisting of a real line segment from 0 to 10 and A and B be sets of the real line segments, respectively, from 1 to 7 and 3 to 9.

BASIC PROBABILITY CONCEPTS

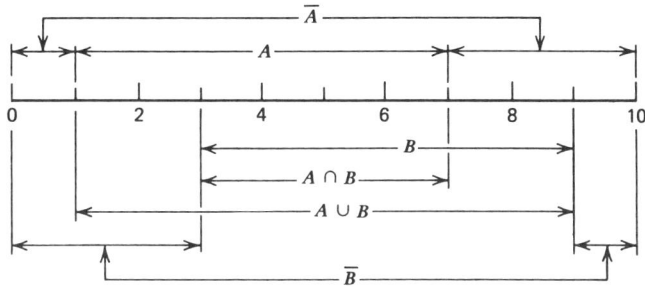

Figure 2.4 Sets defined in Example 2.3.

Line segments belonging to $A \cup B$, $A \cap B$, $\bar{A}$ and $\bar{B}$ are indicated in Figure 2.4. Let us note here that, by definition, a set and its complement are always disjoint.

The definitions of union and intersection can be directly generalized to those involving any arbitrary number (finite or countably infinite) of sets. Thus, the set

$$A_1 \cup A_2 \cdots \cup A_n = \bigcup_{j=1}^{n} A_j \qquad (2.2)$$

stands for the set of all elements belonging to one or more of the sets A_j, $j = 1, 2, \ldots, n$. The intersection

$$A_1 A_2 \cdots A_n = \bigcap_{j=1}^{n} A_j \qquad (2.3)$$

is the set of all elements common to *all* $A_j, j = 1, 2, \ldots, n$. The sets $A_j, j = 1, 2, \ldots, n$, are disjoint if

$$A_i A_j = \emptyset \qquad \text{for every } i, j \ (i \neq j) \qquad (2.4)$$

Using Venn diagrams or more formal procedures, it is easy to verify that union and intersection operations are associative, commutative, and distributive, that is,

$$\begin{aligned}
(A \cup B) \cup C &= A \cup (B \cup C) = A \cup B \cup C \\
A \cup B &= B \cup A \\
(AB)C &= A(BC) = ABC \\
AB &= BA \\
A(B \cup C) &= (AB) \cup (AC)
\end{aligned} \qquad (2.5)$$

Clearly, we also have

$$A \cup A = AA = A$$
$$A \cup \emptyset = A$$
$$A\emptyset = \emptyset$$
$$A \cup S = S \qquad (2.6)$$
$$AS = A$$
$$A \cup \bar{A} = S$$
$$A\bar{A} = \emptyset$$

Moreover, the following useful relations hold; all of which can be easily verified using Venn diagrams.

$$A \cup (BC) = (A \cup B)(A \cup C)$$
$$A \cup B = A \cup (\bar{A}B) = A + (\bar{A}B)$$
$$\overline{(A \cup B)} = \bar{A}\bar{B}$$
$$\overline{(AB)} = \bar{A} \cup \bar{B} \qquad (2.7)$$
$$\overline{\left(\bigcup_{j=1}^{n} A_j\right)} = \bigcap_{j=1}^{n} \bar{A}_j$$
$$\overline{\left(\bigcap_{j=1}^{n} A_j\right)} = \bigcup_{j=1}^{n} \bar{A}_j$$

The second relation gives the union of two sets in terms of the union of two disjoint sets. As we will see, this representation is useful in probability calculations. The last two relations in Equations 2.7 are referred to as *De Morgan's laws*.

Finally, we define the *difference* between two sets. The difference $A - B$ is the set of all elements belonging to A but not to B. From the definition we have the simple relations

$$A - \emptyset = A \qquad S - A = \bar{A} \qquad A - B = A\bar{B} \qquad (2.8)$$

II.1.2 BOREL FIELD

Given space S, a *Borel field* or a σ-field β of subsets of S is a class of, in general noncountable, subsets $A_j, j = 1, 2, \ldots$, having the following properties:

1. $S \in \beta$.
2. If $A_i \in \beta$, then $\bar{A}_i \in \beta$.
3. If $A_j \in \beta, j = 1, 2, \ldots$, then $\bigcup_{j=1}^{\infty} A_j \in \beta$.

BASIC PROBABILITY CONCEPTS

The first two properties imply that

$$\varnothing = \bar{S} \in \beta \tag{2.9}$$

and, with the aid of De Morgan's laws, the second and third properties lead to

$$\bigcap_{j=1}^{\infty} A_j = \overline{\left(\bigcup_{j=1}^{\infty} \bar{A}_j\right)} \in \beta \tag{2.10}$$

A Borel field is thus a class of sets, including the empty set $\varnothing$ and space S, which is closed under all countable unions and intersections of its sets. It is clear, of course, that a class of all subsets of S is a Borel field. However, in the development of the basic concepts of probability, this particular Borel field is too large and impractical. In general we consider the *smallest* class of subsets of S that is a Borel field and it contains all sets and elements under consideration.

II.2 Sample Space and Probability Measure

In probability theory we are concerned with an experiment with an outcome depending on chance, which is called a *random experiment*. It is assumed that all possible distinct outcomes of a random experiment are known and they are elements of a fundamental set known as the *sample space*. Each possible outcome is called a *sample point*, and an *event* is generally referred to as a subset of the sample space having one or more sample points as its elements.

It is important to point out that, for a given random experiment, its associated sample space is not unique and its construction depends upon the point of view adopted as well as the questions to be answered. For example, 100-ohm resistors are being manufactured by an industrial firm. Their values, due to inherent inaccuracies in the manufacturing and measurement processes, may range from 99 to 101 ohms. A measurement taken of a resistor is a random experiment whose possible outcomes can be defined in a variety of ways depending upon the purpose for performing such an experiment. If, for a given user, a resistor with resistance range of 99.9–100.1 ohms is considered acceptable and unacceptable otherwise, it is adequate to define the sample space as one consisting of two elements: acceptable and unacceptable. On the other hand, from the viewpoint of another user, possible outcomes may be reading ranges 99–99.5 ohms, 99.5–100 ohms, 100–100.5 ohms, and 100.5–101 ohms. The sample space in this case has four sample points. Finally, if each possible reading is a possible outcome, the sample space is now a real line from 99 to 101 on the ohm scale; there are noncountably infinite number of sample points and the sample space is a nonenumerable set.

To illustrate that a sample space is not fixed by the action of performing the experiment but by the point of view adopted by the observer, consider an energy negotiation between the United States and another country. From the point of view of the U.S. government, success and failure may be looked on as the only possible outcomes. To the consumer, on the other hand, a set of more direct possible outcomes may consist of price increase and decrease in gasoline purchases.

The description of sample space, sample points, and events shows that they fit nicely into the framework of set theory, a framework within which the analysis of outcomes of a random experiment can be performed. All relations between outcomes or events in probability theory can be described by sets and set operations. Consider a space S of elements $a, b, c, \ldots$, and with subsets $A, B, C, \ldots$. Some of these corresponding set and probability meanings are given in Table 2.1. As this table shows, the empty set $\emptyset$ is considered an impossible event since no possible outcome is an element of the empty set. Also, by occurrence of an event we mean that the observed outcome is an element of that set. For example, the event $A \cup B$ is said to occur if and only if the observed outcome is an element of A or B or both.

Table 2.1. *Corresponding Statements in Set Theory and Probability*

Set Theory	Probability Theory
Space S	Sample space, sure event
Empty set $\emptyset$	Impossible event
Elements $a, b, \ldots$	Sample points (or simple events)
Sets $A, B, \ldots$	Events
A	Event A occurs
$\bar{A}$	Event A does not occur
$A \cup B$	At least one of A and B occurs
AB	Both A and B occur
$A - B$	Only A of A and B occurs
$A \subset B$	A is a subevent of B (i.e., the occurrence of A necessarily implies the occurrence of B)
$AB = \emptyset$	A and B are mutually exclusive (i.e., they cannot occur simultaneously)

Example 2.4. Consider an experiment of counting the number of left-turn cars at an intersection in a group of 100 cars. The possible outcomes (possible numbers of left-turn cars) are 0, 1, 2, ..., 100. Then, the sample space S is $S = \{0, 1, 2, ..., 100\}$. Each element of S is a sample point or a possible outcome. The subset $A = \{0, 1, 2, ..., 50\}$ is the event that there are 50 or less cars turning left. The subset $B = \{40, 41, ..., 60\}$ is the event that between 40 and 60 (inclusive) cars take left turns. The set $A \cup B$ is the event of 60 or less cars turning left. The set $A \cap B$ is the event that the number of left-turn cars is between 40 and 50 (inclusive). Let $C = \{80, 81, ..., 100\}$. The events A and C are called *mutually exclusive* since they cannot occur simultaneously. Hence, disjoint sets are mutually exclusive events in probability theory.

In our subsequent discussion of probability theory, we require that the collection of all events associated with a random experiment constitutes a Borel field β, which implies that all sets formed through countable unions and intersections of events are events, and they are contained in β. This requirement, as we shall see, is reasonable and is always met in practice.

II.2.1 AXIOMS OF PROBABILITY

We now introduce the notion of a *probability function*. Given a random experiment, a finite number $P(A)$ is assigned to every event A in the Borel field β of all possible events. The number $P(A)$ is a function of the set A and is assumed to be defined for all sets in β. It is thus a set function and $P(A)$ is called the *probability measure* of A or simply the *probability* of A. It is assumed to have the following properties (axioms of probability):

1. $P(A) \geq 0$ (nonnegative).
2. $P(S) = 1$ (normed).
3. For a countable collection of mutually exclusive events $A_1, A_2, ...$ in β,

$$P(A_1 \cup A_2 \cup \cdots) = P\left(\sum_j A_j\right) = \sum_j P(A_j) \quad \text{(additive).} \quad (2.11)$$

These three axioms define a countably additive and nonnegative set function $P(A)$, $A \in \beta$. As we shall see, they constitute a sufficient set of postulates from which all useful properties of the probability function can be derived. Let us give some of these important properties.

1. $P(\emptyset) = 0$. Since S and $\emptyset$ are disjoint, we see from Axiom 3 that

$$P(S) = P(S + \emptyset) = P(S) + P(\emptyset)$$

It then follows from Axiom 2 that

$$1 = 1 + P(\emptyset)$$

or $P(\emptyset) = 0$.

2. *If* $A \subset C$, *then* $P(A) \leq P(C)$. Since $A \subset C$, the set $B = C - A$ is a subset of C. Moreover, A and B are disjoint and

$$A + B = C$$

Axiom 3 then gives

$$P(C) = P(A + B) = P(A) + P(B)$$

Since $P(B) \geq 0$ as required by Axiom 1, we have the desired result.

3. *Given two arbitrary events* A *and* B, *we have*

$$\boxed{P(A \cup B) = P(A) + P(B) - P(AB)} \qquad (2.12)$$

In order to show this, let us write $A \cup B$ in terms of the union of two mutually exclusive events. Following the second relation in Equations 2.7, we write

$$A \cup B = A + \bar{A}B$$

Hence, using Axiom 3,

$$P(A \cup B) = P(A + \bar{A}B) = P(A) + P(\bar{A}B) \qquad (2.13)$$

Furthermore, we note

$$\bar{A}B + AB = B$$

Hence, again using Axiom 3,

$$P(\bar{A}B) + P(AB) = P(B)$$

or

$$P(\bar{A}B) = P(B) - P(AB).$$

Substituting this equation into Equation 2.13 yields Equation 2.12.

BASIC PROBABILITY CONCEPTS   17

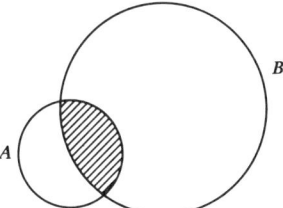

Figure 2.5 Venn diagram in derivation of Equation 2.12.

Equation 2.12 can also be verified by inspecting the Venn diagram above. The sum $P(A) + P(B)$ counts twice the events belonging to the shaded area AB. Hence, in computing $P(A \cup B)$, the probability associated with one AB must be subtracted from $P(A) + P(B)$, giving Equation 2.12 (see Figure 2.5).

The important result given by Equation 2.12 can be immediately generalized to the union of three or more events. Using the same procedure, we can show that, for arbitrary events A, B, and C,

$$P(A \cup B \cup C) = P(A) + P(B) + P(C) - P(AB) - P(AC) - P(BC) + P(ABC) \tag{2.14}$$

and more generally,

$$P\left(\bigcup_{j=1}^{n} A_j\right) = \sum_{j=1}^{n} P(A_j) - \sum_{\substack{i=1 \\ i<j}}^{n} \sum_{j=2}^{n} P(A_i A_j) + \sum_{\substack{i=1 \\ i<j<k}}^{n} \sum_{j=2}^{n} \sum_{k=3}^{n} P(A_i A_j A_k) - \cdots + (-1)^{n-1} P(A_1 A_2 \cdots A_n) \tag{2.15}$$

where $A_j, j = 1, 2, \ldots, n$, are arbitrary events.

Example 2.5.   Let us go back to Example 2.4 and assume that the probabilities $P(A)$, $P(B)$, and $P(C)$ are known. We wish to compute $P(A \cup B)$ and $P(A \cup C)$.

The probability $P(A \cup C)$, probability of having either 50 or less cars turning left or between 80 and 100 cars turning left, is simply $P(A) + P(C)$. This follows from Axiom 3 since A and C are mutually exclusive. On the other hand, $P(A \cup B)$, the probability of having 60 or less cars turning left is found from

$$P(A \cup B) = P(A) + P(B) - P(AB)$$

The information given above is thus not sufficient to determine this probability and we need the additional information $P(AB)$, the probability of having between 40 and 50 cars turning left.

With the statement of three axioms of probability, we have completed the mathematical description of a random experiment. It consists of three fundamental constituents: a sample space S, a Borel field β of events, and the probability function P. These three quantities constitute a *probability space* associated with a random experiment and it is denoted by (S, β, P).

II.2.2 ASSIGNMENT OF PROBABILITY

The axioms of probability define the properties of a probability measure, which are consistent with our intuitive notions. However, they do not guide us in assigning probabilities to various events. For problems in applied sciences, a natural way to assign the probability of an event is through the observation of *relative frequency*. Assuming that a random experiment is performed a large number of times, say n, then for any event A let n_A be the number of occurrences of A in the n trials and define the ratio n_A/n as the relative frequency of A. Under stable or statistical regularity conditions, it is expected that this ratio will tend to a unique limit as n becomes large. This limiting value of the relative frequency clearly possesses the properties required of the probability measure and is a natural candidate for the probability of A. This interpretation is used, for example, in saying that the probability of "heads" in flipping a coin is $\frac{1}{2}$. The relative frequency approach to probability assignment is objective and consistent with the axioms stated above and is one commonly adopted in science and engineering.

Another common but more subjective approach to probability assignment is that of *relative likelihood*. When it is not feasible or impossible to perform an experiment a large number of times, the probability of an event may be assigned as a result of subjective judgment. The statement "there is a 40% probability of rain tomorrow" is an example in this interpretation where the number 0.4 is assigned on the basis of available information and professional judgment.

In most problems considered in this book, probabilities of some simple but basic events are generally assigned using either of the two approaches. Other probabilities of interest are then derived through the theory of probability. Example 2.5 gives a simple illustration of this procedure where the probabilities of interest, $P(A \cup B)$ and $P(A \cup C)$, are derived upon assigning probabilities to simple events A, B, and C.

II.3 Statistical Independence

Let us pose the following question: Given the individual probabilities $P(A)$ and $P(B)$ of two events A and B, what is $P(AB)$, the probability that both A and B will occur? Upon little reflection, it is not difficult to see that the knowledge of $P(A)$ and $P(B)$ is not sufficient to determine $P(AB)$ in general. This is so because

BASIC PROBABILITY CONCEPTS

$P(AB)$ deals with the joint behavior of the two events whereas $P(A)$ and $P(B)$ are probabilities associated with individual events and do not yield information on their joint behavior. Let us then consider a special case in which the occurrence or nonoccurrence of one does not affect the occurrence or nonoccurrence of the other. In this situation events A and B are called *statistically independent* or simply *independent* and it is formalized with the following definition.

Definition. Two events A and B are said to be *independent* if and only if

$$\boxed{P(AB) = P(A)P(B)} \tag{2.16}$$

To show that this definition is consistent with our intuitive notion of independence, consider the following example.

Example 2.6. In a large number n of trials of a random experiment, let n_A and n_B be, respectively, the numbers of occurrences of two outcomes A and B and let n_{AB} be the number of times both A and B occur. Using the relative frequency interpretation, the ratios n_A/n and n_B/n tend to $P(A)$ and $P(B)$, respectively, as n becomes large. Similarly, n_{AB}/n tends to $P(AB)$. Let us now confine our attention to only those outcomes in which A is realized. If A and B are independent, we expect that the ratio n_{AB}/n_A also tends to $P(B)$ as n_A becomes large. The independence assumption then leads to the observation that

$$\frac{n_{AB}}{n_A} \cong P(B) \cong \frac{n_B}{n}$$

This then gives

$$\frac{n_{AB}}{n} \cong \left(\frac{n_A}{n}\right)\left(\frac{n_B}{n}\right)$$

or, in the limit as n becomes large,

$$P(AB) = P(A)P(B)$$

which is the definition of independence introduced above.

Example 2.7. In launching a satellite, the probability of an unsuccessful launch is q. What is the probability that two successive launches are unsuccessful? Assuming that satellite launchings are independent events, the answer to the above question is simply q^2. One can argue that these two events are not really completely independent, since they are manufactured using similar processes and launched by the same launcher. It is thus likely that the failures of both are attributable to the same source. However, we accept this answer as reasonable because, on the one hand, independence assumption is acceptable since there are a great deal of unknowns involved, any of which can be made accountable for the failure of a launch. On the other hand, the simplicity of computing the joint probability makes independence assumption attractive. In physical problems, therefore, independence assumption is often made whenever it is considered to be reasonable.

Care should be exercised in extending the concept of independence to more than two events. In the case of three events A_1, A_2, and A_3, for example, they are mutually independent if and only if

$$P(A_j A_k) = P(A_j)P(A_k) \qquad j \neq k \quad \text{and} \quad j, k = 1, 2, 3 \qquad (2.17)$$

and

$$P(A_1 A_2 A_3) = P(A_1)P(A_2)P(A_3) \qquad (2.18)$$

Equation 2.18 is required because pairwise independence does not generally lead to mutual independence. Consider, for example, three events A_1, A_2, and A_3 defined by

$$A_1 = B_1 \cup B_2 \qquad A_2 = B_1 \cup B_3 \qquad A_3 = B_2 \cup B_3$$

where B_1, B_2, and B_3 are mutually exclusive, each occurring with probability $\frac{1}{4}$. It is easy to calculate the following:

$$P(A_1) = P(B_1 \cup B_2) = P(B_1) + P(B_2) = \tfrac{1}{2}$$
$$P(A_2) = P(A_3) = \tfrac{1}{2}$$
$$P(A_1 A_2) = P[(B_1 \cup B_2) \cap (B_1 \cup B_3)] = P(B_1) = \tfrac{1}{4}$$
$$P(A_1 A_3) = P(A_2 A_3) = \tfrac{1}{4}$$
$$P(A_1 A_2 A_3) = P[(B_1 \cup B_2) \cap (B_1 \cup B_3) \cap (B_2 \cup B_3)] = P(\varnothing) = 0$$

We see that Equation 2.17 is satisfied for every j and k in this case but Equation 2.18 is not. In other words, events A_1, A_2, and A_3 are pairwise independent but they are not mutually independent.

In general, therefore, we have the following definition for mutual independence of n events.

BASIC PROBABILITY CONCEPTS

Definition. The events $A_1, A_2, \ldots, A_n$ are mutually independent if and only if, with $k_1, k_2, \ldots, k_m$ being any set of integers such that $1 \leq k_1 < k_2 \cdots < k_m \leq n$ and $m = 2, 3, \ldots, n$,

$$\boxed{P(A_{k_1} A_{k_2} \cdots A_{k_m}) = P(A_{k_1}) P(A_{k_2}) \cdots P(A_{k_m})} \qquad (2.19)$$

The total number of equations defined by Equation 2.19 is $2^n - n - 1$.

Example 2.8. A system consisting of five components is in working order only when each component is functioning (good). Let S_i, $i = 1, \ldots, 5$, be the event that the ith component is good and assume $P(S_i) = p_i$. What is the probability Q that the system fails?

Assuming that the five components perform in an independent manner, it is easier to determine Q through finding the probability of system success R. We have from the statement of the problem

$$R = P(S_1 S_2 S_3 S_4 S_5)$$

Equation 2.19 thus gives

$$R = P(S_1) P(S_2) \cdots P(S_5) = p_1 p_2 p_3 p_4 p_5 \qquad (2.20)$$

Hence

$$Q = 1 - R = 1 - p_1 p_2 p_3 p_4 p_5 \qquad (2.21)$$

An expression for Q may also be obtained by noting that the system fails if any of the five components fails or

$$Q = P(\bar{S}_1 \cup \bar{S}_2 \cup \bar{S}_3 \cup \bar{S}_4 \cup \bar{S}_5) \qquad (2.22)$$

where $\bar{S}_i$ is the complement of S_i and represents a bad ith component. Clearly, $P(\bar{S}_i) = 1 - p_i$. Since events $\bar{S}_i$, $i = 1, \ldots, 5$, are not mutually exclusive, the calculation of Q using Equation 2.22 requires the use of Equation 2.15. Another approach is to write the unions in Equation 2.22 in terms of unions of mutually exclusive events so that Axiom 3 can be directly utilized. The result is, upon applying the second relation in Equations 2.7,

$$\bar{S}_1 \cup \bar{S}_2 \cup \bar{S}_3 \cup \bar{S}_4 \cup \bar{S}_5 = \bar{S}_1 + S_1 \bar{S}_2 + S_1 S_2 \bar{S}_3 + S_1 S_2 S_3 \bar{S}_4 + S_1 S_2 S_3 S_4 \bar{S}_5$$

where the "∪" signs are replaced by " + " signs on the right-hand side to indicate mutually exclusive events. Axiom 3 then leads to

$$Q = P(\bar{S}_1) + P(S_1\bar{S}_2) + P(S_1S_2\bar{S}_3) + P(S_1S_2S_3\bar{S}_4) + P(S_1S_2S_3S_4\bar{S}_5)$$

and, using statistical independence,

$$Q = (1 - p_1) + p_1(1 - p_2) + p_1p_2(1 - p_3) \\ + p_1p_2p_3(1 - p_4) + p_1p_2p_3p_4(1 - p_5) \quad (2.23)$$

Some simple algebra will show that this result reduces to Equation 2.21.

Let us mention here that the probability R is called *reliability* of the system in systems literature.

II.4 Conditional Probability

The concept of conditional probability is a very useful one. Given two events A and B associated with a random experiment, the probability $P(A|B)$ is defined as the conditional probability of A, given that B has occurred. Intuitively, this probability can be interpreted by means of relative frequencies described in Example 2.6 except that the events A and B are no longer assumed to be independent. The number of outcomes where both A and B occur is n_{AB}. Hence, given that event B has occurred, the relative frequency of A is then n_{AB}/n_B. Thus we have, in the limit as n_B becomes large,

$$P(A|B) \cong \frac{n_{AB}}{n_B} = \frac{n_{AB}/n}{n_B/n} \cong \frac{P(AB)}{P(B)}$$

This relationship suggests the following definition.

Definition. The conditional probability of A given that B has occurred is given by

$$\boxed{P(A|B) = \frac{P(AB)}{P(B)} \quad P(B) \neq 0} \quad (2.24)$$

This definition is meaningless if $P(B) = 0$.

BASIC PROBABILITY CONCEPTS   23

It is noted that, in the discussion of conditional probabilities, we are dealing with a contracted sample space in which B is known to have occurred. In other words, B replaces S as the sample space and the conditional probability $P(A|B)$ is found as the probability of A with respect to this new sample space.

In the event that A and B are independent, it implies that the occurrence of B has no effect on the occurrence or nonoccurrence of A. We thus expect $P(A|B) = P(A)$ and Equation 2.24 gives

$$P(A) = \frac{P(AB)}{P(B)} \quad \text{or} \quad P(AB) = P(A)P(B)$$

which is precisely the definition of independence.

It is also important to point out that conditional probabilities are probabilities (i.e., they satisfy the three axioms). Using Equation 2.24, we see that the first axiom is automatically satisfied. For the second axiom we need to show that

$$P(S|B) = 1$$

This is certainly true since

$$P(S|B) = \frac{P(SB)}{P(B)} = \frac{P(B)}{P(B)} = 1$$

As for the third axiom, if $A_1, A_2, \ldots$ are mutually exclusive, then $A_1 B, A_2 B, \ldots$ are also mutually exclusive. Hence,

$$\begin{aligned} P[(A_1 \cup A_2 \cup \cdots)|B] &= \frac{P[(A_1 \cup A_2 \cup \cdots)B]}{P(B)} \\ &= \frac{P[A_1 B \cup A_2 B \cup \cdots]}{P(B)} \\ &= \frac{P(A_1 B)}{P(B)} + \frac{P(A_2 B)}{P(B)} + \cdots \\ &= P(A_1|B) + P(A_2|B) + \cdots \end{aligned}$$

and the third axiom holds.

The definition of conditional probability given by Equation 2.24 can be used not only to compute conditional probabilities but also to compute joint probabilities as the following examples show.

Example 2.9. Let us reconsider Example 2.8 and ask the following question: What is the conditional probability that the first two components are good given that (a) the first component is good and (b) at least one of the two is good?

The event $S_1 S_2$ means both are good components and $S_1 \cup S_2$ is the event that at least one of the two is good. Thus, for part (a) and in view of Equation 2.24,

$$P(S_1 S_2 | S_1) = \frac{P(S_1 S_2 S_1)}{P(S_1)} = \frac{P(S_1 S_2)}{P(S_1)} = \frac{p_1 p_2}{p_1} = p_2$$

This result is expected since S_1 and S_2 are independent. Intuitively, we see that this question is equivalent to one of computing $P(S_2)$.

For the (b) part, we have

$$P(S_1 S_2 | S_1 \cup S_2) = \frac{P[S_1 S_2 (S_1 \cup S_2)]}{P(S_1 \cup S_2)}$$

Now, $S_1 S_2 (S_1 \cup S_2) = S_1 S_2$. Hence,

$$P(S_1 S_2 | S_1 \cup S_2) = \frac{P(S_1 S_2)}{P(S_1 \cup S_2)} = \frac{P(S_1 S_2)}{P(S_1) + P(S_2) - P(S_1 S_2)}$$

$$= \frac{p_1 p_2}{p_1 + p_2 - p_1 p_2}$$

Example 2.10. In a game of cards, determine the probability of drawing, without replacement, two aces in succession.

Let A_1 be the event that the first card drawn is an ace and similarly for A_2. We wish to compute $P(A_1 A_2)$. From Equation 2.24 we write

$$P(A_1 A_2) = P(A_2 | A_1) P(A_1) \tag{2.25}$$

Now, $P(A_1) = \frac{4}{52}$ and $P(A_2 | A_1) = \frac{3}{51}$ (there are 51 cards left and three of them are aces). Therefore,

$$P(A_1 A_2) = \frac{3}{51}\left(\frac{4}{52}\right) = \frac{1}{221}$$

Equation 2.25 is seen to be useful for finding joint probabilities. Its extension to more than two events has the form

$$P(A_1 A_2 \cdots A_n) = P(A_1) P(A_2 | A_1) P(A_3 | A_1 A_2) \cdots P(A_n | A_1 A_2 \cdots A_{n-1}) \tag{2.26}$$

where $P(A_i) > 0$ for all i. This can be verified by successive applications of Equation 2.24.

BASIC PROBABILITY CONCEPTS

In another direction, let us state a useful theorem relating probability of an event to conditional probabilities.

Theorem of Total Probability
Suppose that events $B_1, B_2, \ldots,$ and B_n are mutually exclusive and exhaustive (i.e., $S = B_1 + B_2 + \cdots + B_n$). Then, for an arbitrary event A,

$$P(A) = P(A|B_1)P(B_1) + P(A|B_2)P(B_2) + \cdots + P(A|B_n)P(B_n)$$
$$= \sum_{j=1}^{n} P(A|B_j)P(B_j) \qquad (2.27)$$

Proof
Referring to the Venn diagram in Figure 2.6, we can clearly write A as the union of mutually exclusive events $AB_1, AB_2, \ldots, AB_n$ (i.e., $A = AB_1 + AB_2 + \cdots + AB_n$). Hence,

$$P(A) = P(AB_1) + P(AB_2) + \cdots + P(AB_n)$$

which gives Equation 2.27 on application of the definition of conditional probability.

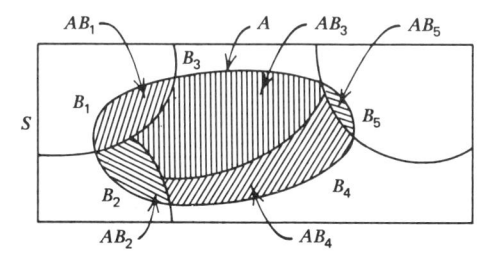

Figure 2.6 Venn diagram associated with total probability theorem.

The utility of this result rests with the fact that the probabilities in the sum in Equation 2.27 are often more readily obtainable than the probability of A itself.

Example 2.11. Our interest is in determining the probability that critical level of peak flow rate is reached during storms in a storm-sewer system on the basis of separate meteorological and hydrological measurements.

Let $B_i, i = 1, 2, 3$, be the different levels (low, medium, high) of precipitation caused by a storm and let $A_j, j = 1, 2$, denote, respectively, critical and noncritical levels of peak flow rate. Then the probabilties $P(B_i)$ can be estimated from meteorological records and $P(A_j|B_i)$ can be estimated from runoff analysis. Since $B_1, B_2,$ and B_3 constitute a set of mutually exclusive and exhaustive events, the desired probability $P(A_1)$ can be found from

$$P(A_1) = P(A_1|B_1)P(B_1) + P(A_1|B_2)P(B_2) + P(A_1|B_3)P(B_3)$$

Assuming the following information is available:

$$P(B_1) = 0.5 \qquad P(B_2) = 0.3 \qquad P(B_3) = 0.2$$

and $P(A_j|B_i)$ are tabulated as follows.

	B_1	B_2	B_3
A_1	0	0.2	0.6
A_2	1.0	0.8	0.4

The numerical value of $P(A_1)$ is

$$P(A_1) = 0(0.5) + 0.2(0.3) + 0.6(0.2) = 0.18$$

Let us observe that, in the foregoing conditional probability table, the sum of the probabilities in each column is one by virtue of conservation of probability. There is, however, no such requirement for the sum of each row.

A useful result generally referred to as Bayes' theorem can be derived based on the definition of conditional probability. Equation 2.24 permits us to write

$$P(AB) = P(A|B)P(B) \qquad \text{and} \qquad P(BA) = P(B|A)P(A)$$

Since $P(AB) = P(BA)$, we have the following theorem.

Bayes' Theorem

Let A and B be two arbitrary events with $P(A) \neq 0$ and $P(B) \neq 0$. Then

$$\boxed{P(B|A) = \frac{P(A|B)P(B)}{P(A)}} \qquad (2.28)$$

BASIC PROBABILITY CONCEPTS

Combining this result with the total probability theorem, we have a useful consequence

$$P(B_i | A) = \frac{P(A|B_i)P(B_i)}{\sum_{j=1}^{n} P(A|B_j)P(B_j)} \qquad (2.29)$$

for any i where the events B_j represent a set of mutually exclusive and exhaustive events.

The simple result given by Equation 2.28 is called Bayes' theorem after the English philosopher Thomas Bayes and is useful in the sense that it permits us to evaluate the *a posteriori* probability $P(B|A)$ in terms of *a priori* information $P(B)$ and $P(A|B)$ as the following examples illustrate.

Example 2.12. A simple binary communication channel carries messages using only two signals, say 0 and 1. We assume that, for a given binary channel, 40% of the time a 1 is transmitted, the probability that a transmitted 0 is correctly received is 0.90, and the probability that a transmitted 1 is correctly received is 0.95. Determine (a) the probability of a 1 being received and (b) given a 1 is received, the probability that 1 was transmitted.

Let

A = event that 1 is transmitted
$\bar{A}$ = event that 0 is transmitted
B = event that 1 is received
$\bar{B}$ = event that 0 is received

The information given in the problem statement gives us

$P(A) = 0.4 \qquad P(\bar{A}) = 0.6$
$P(B|A) = 0.95 \qquad P(\bar{B}|A) = 0.05$
$P(\bar{B}|\bar{A}) = 0.90 \qquad P(B|\bar{A}) = 0.1$

and they are represented diagrammatically in Figure 2.7.

For part (a) we wish to find $P(B)$. Since A and $\bar{A}$ are mutually exclusive and exhaustive, it follows from the theorem of total probability that

$$P(B) = P(B|A)P(A) + P(B|\bar{A})P(\bar{A}) = 0.95(0.4) + 0.1(0.6) = 0.44$$

The probability of interest in part (b) is $P(A|B)$ and this can be found using Bayes' theorem (Equation 2.28). It is

$$P(A|B) = \frac{P(B|A)P(A)}{P(B)} = \frac{0.95(0.4)}{0.44} = 0.863$$

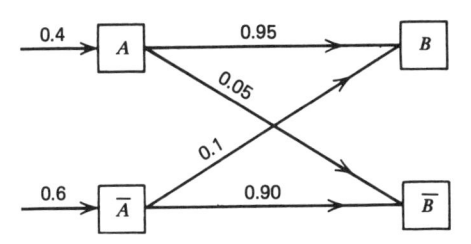

Figure 2.7 Probabilities associated with a binary channel.

It is worth mentioning that $P(B)$ in the calculation above is found by means of the total probability theorem. Hence, Equation 2.29 is the one actually used here in finding $P(A|B)$. In fact, the probability $P(A)$ in Equation 2.28 is often more conveniently found using the total probability theorem.

Example 2.13. In Example 2.11, determine $P(B_2|A_2)$, the probability that a noncritical level of peak flow rate will be caused by a medium-level storm.
From Equations 2.28 and 2.29 we have

$$P(B_2|A_2) = \frac{P(A_2|B_2)P(B_2)}{P(A_2)}$$

$$= \frac{P(A_2|B_2)P(B_2)}{P(A_2|B_1)P(B_1) + P(A_2|B_2)P(B_2) + P(A_2|B_3)P(B_3)}$$

$$= \frac{0.8(0.3)}{1.0(0.4) + 0.8(0.3) + 0.4(0.2)} = \frac{1}{3}$$

In closing, let us introduce the use of tree diagrams for dealing with more complicated experiments with "limited memory." Consider again Example 2.12 by adding a second stage to the communication channel with Figure 2.8 showing

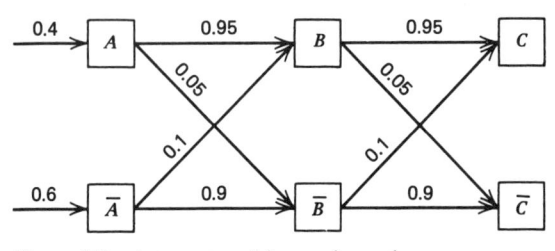

Figure 2.8 A two-stage binary channel.

BASIC PROBABILITY CONCEPTS

all the associated probabilities. We wish to determine $P(C)$, the probability of receiving a 1 at the second stage.

Tree diagrams are useful for determining the behavior of this system when the system has a "one-stage" memory, that is, the outcome at the second stage is only dependent on what has happened at the first stage and not on outcomes at stages prior to the first. Mathematically, it follows from this property that

$$P(C|BA) = P(C|B) \qquad P(\bar{C}|\bar{B}A) = P(\bar{C}|\bar{B}), \text{ etc.} \qquad (2.30)$$

Properties described above are commonly referred to as *Markovian* properties. Markov processes represent an important class of probabilistic processes that are studied at a more advanced level.

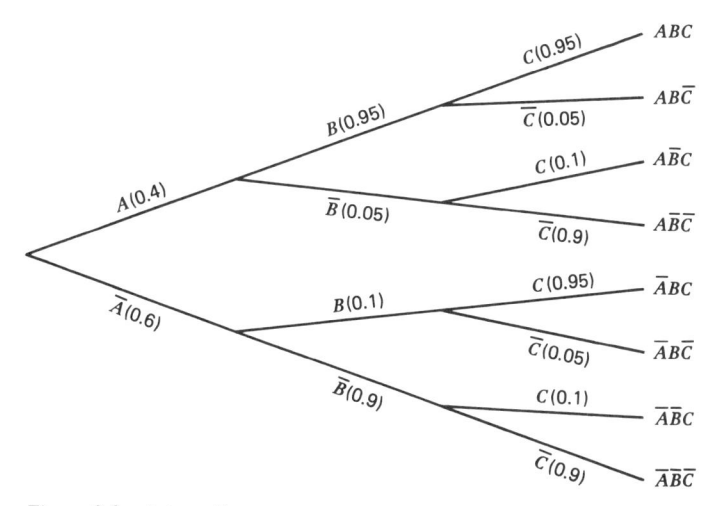

Figure 2.9 A tree diagram.

Suppose that Equations 2.30 hold for the system described in Figure 2.8. The tree diagram gives the flow of conditional probabilities originating from the source. Starting from the transmitter, the tree diagram for this problem has the appearance as shown in Figure 2.9. The top branch, for example, indicates the probability of the occurrence of event ABC which is, according to Equations 2.26 and 2.30,

$$\begin{aligned} P(ABC) &= P(A)P(B|A)P(C|BA) \\ &= P(A)P(B|A)P(C|B) \\ &= 0.4(0.95)(0.95) = 0.361 \end{aligned}$$

The probability of C is then found by summing probabilities of all events that end with a C. Thus,

$$P(C) = P(ABC) + P(A\bar{B}C) + P(\bar{A}BC) + P(\bar{A}\bar{B}C)$$
$$= 0.95(0.95)(0.4) + 0.1(0.05)(0.4) + 0.95(0.1)(0.6) + 0.1(0.9)(0.6)$$
$$= 0.472$$

REFERENCES AND COMMENTS

1. I. Todhunter, *A History of the Mathematical Theory of Probability from the Time of Pascal to Laplace*, Chelsea, New York, 1949.

More accounts of early development of probability theory related to gambling can be found in:

2. F. N. David, *Games, Gods, and Gambling*, Hafner, New York, 1962.

PROBLEMS

2.1. Let A, B, C be arbitrary sets. Determine which of the following relations are correct and which are incorrect.
   (a) $ABC = AB(C \cup B)$
   (b) $\overline{AB} = \bar{A} \cup \bar{B}$
   (c) $(A \cup B) - C = A \cup (B - C)$
   (d) $\overline{A \cup B} = \bar{A}\bar{B}$
   (e) $\overline{(A \cup B)C} = \bar{A}\bar{B}C$
   (f) $A \cup B \cup C = A \cup (B - AB) \cup (C - AC)$
   (g) $A\bar{B} \subset A \cup B$
   (h) $(AB)(\bar{A}C) = \emptyset$

2.2. The second relation in Equations 2.7 expresses the union of two sets as the union of two disjoint sets (i.e., $A \cup B = A + \bar{A}B$). Express $A \cup B \cup C$ in terms of unions of disjoint sets where A, B, and C are arbitrary sets.

2.3. Verify De Morgan's laws given as the last two equations of Equations 2.7.

2.4. Let $S = \{1, 2, \ldots, 10\}$, $A = \{1, 3, 5\}$, $B = \{1, 4, 6\}$ and $C = \{2, 5, 7\}$. Determine the elements of the following sets.
- (a) $S \cup C$
- (b) $A \cup B$
- (c) $\bar{A}C$
- (d) $\bar{A} \cup (BC)$
- (e) $\overline{ABC}$
- (f) $\overline{\bar{A}\bar{B}}$
- (g) $(AB) \cup (BC) \cup (CA)$

2.5. Repeat Problem 2.4 if $S = \{x\colon 0 \le x \le 10\}$, $A = \{x\colon 1 \le x \le 5\}$, $B = \{x\colon 1 \le x \le 6\}$, and $C = \{x\colon 2 \le x \le 7\}$.

2.6. Let A, B, C be arbitrary events. Find expressions for the events that of A, B, C:
- (a) None occurs.
- (b) Only A occurs.
- (c) Only one occurs.
- (d) At least one occurs.
- (e) A occurs and either B or C occurs but not both.
- (f) B and C, but not A, occur.
- (g) Two or more occur.
- (h) At most two occur.
- (i) All three occur.

2.7. Verify Equation 2.14.

2.8. Show that, for arbitrary events $A_1, A_2, \ldots, A_n$,

$$P(A_1 \cup A_2 \cup \cdots \cup A_n) \le P(A_1) + P(A_2) + \cdots + P(A_n)$$

This is known as Boole's inequality.

2.9. A box contains 20 parts of which 5 are defective. Two parts are drawn at random from the box. What is the probability that:
- (a) Both are good?
- (b) Both are defective?
- (c) One is good and one is defective?

2.10. An automobile braking device consists of three subsystems, all of which must work for the device to work. These systems are an electronic system, a hydraulic system, and a mechanical activator. In braking, the reliabilities (probabilities of success) of these units are 0.96, 0.95, and 0.95, respectively. Estimate the system reliability assuming that these subsystems function independently.

Comment: Systems of this type can be graphically represented as follows in which the subsystems A (electronic system), B (hydraulic system), and C (mechanical activator) are arranged in series. Consider the path a-b as the "path to success." A breakdown of any or all of A, B, and C will block the path from a to b.

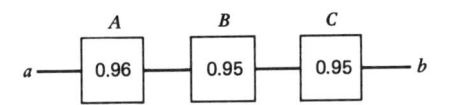

```
         A          B          C
a ──┤ 0.96 ├──┤ 0.95 ├──┤ 0.95 ├── b
```

2.11. A spacecraft has 1000 components in series. If the required reliability of the spacecraft is 0.9 and if all components have the same reliability, what is the required reliability of each component?

2.12. Automobiles are now equipped with redundant braking circuits; their brakes fail only when *all* circuits fail. Consider one with two redundant or parallel braking circuits, each having a reliability of 0.95. Determine the system reliability assuming that these circuits act independently.

Comment: Systems of this type are graphically represented as follows in which the circuits (A and B) have a parallel arrangement. The path to success is broken only when breakdowns of both A and B occur.

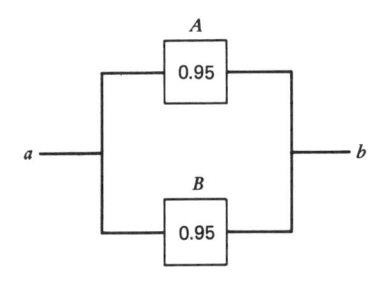

```
            A
        ┌─────────┐
        │  0.95   │
        ├─────────┤
a ──────┤         ├────── b
        ├─────────┤
        │  0.95   │
        └─────────┘
            B
```

2.13. On the basis of definitions given in Problems 2.10 and 2.12 for series and parallel arrangements of system components, determine reliabilities of the systems described by the block diagrams given as follows.

(a)

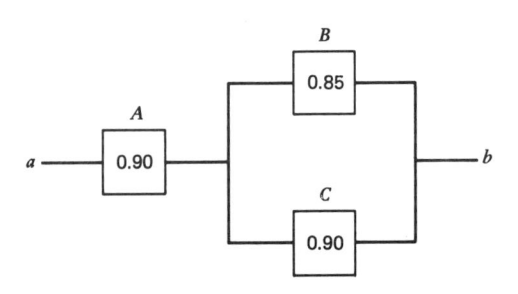

```
                         B
                    ┌─────────┐
                    │  0.85   │
            A       ├─────────┤
        ┌─────────┐ │         │
a ──────┤  0.90   ├─┤         ├────── b
        └─────────┘ ├─────────┤
                    │  0.90   │
                    └─────────┘
                         C
```

(b)

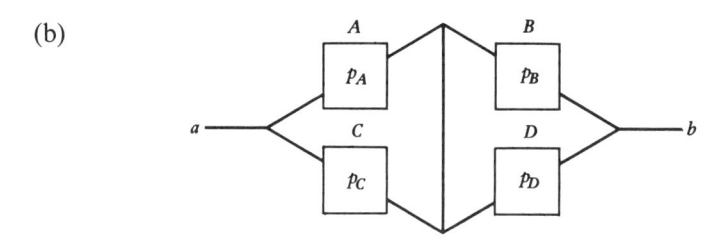

2.14. A rifle is fired at a target. Assuming that the probability of scoring a hit is 0.9 for each shot and that the shots are independent, compute the probability that, in order to score a hit:
   (a) It takes more than two shots.
   (b) The number of shots required is between four and six (inclusive).

2.15. The events A and B are mutually exclusive. Can they also be independent? Explain.

2.16. Let $P(A) = 0.4$ and $P(A \cup B) = 0.7$. What is $P(B)$ if:
   (a) A and B are independent?
   (b) A and B are mutually exclusive?

2.17. Let $P(A \cup B) = 0.75$ and $P(AB) = 0.25$. Is it possible to determine $P(A)$ and $P(B)$? Answer the same question if, in addition:
   (a) A and B are independent.
   (b) A and B are mutually exclusive.

2.18. The events A and B are mutually exclusive. Determine which of the following relations are true and which are false.
   (a) $P(A|B) = P(A)$.
   (b) $P(A \cup B|C) = P(A|C) + P(B|C)$.
   (c) $P(A) = 0$, $P(B) = 0$, or both.
   (d) $\dfrac{P(A|B)}{P(B)} = \dfrac{P(B|A)}{P(A)}$.
   (e) $P(AB) = P(A)P(B)$.

   Repeat the above if the events A and B are independent.

2.19. On a stretch of highway, the probability of an accident due to human error in any given minute is 10^{-5} and the probability of an accident due to mechanical breakdown in any given minute is 10^{-7}. Assuming that these two causes are independent:
   (a) Find the probability of the occurrence of an accident on this stretch of highway during any minute.

(b) In this case, can the above answer be approximated by P(accident due to human error) + P(accident due to mechanical failure)? Explain.

(c) If the events in succeeding minutes are mutually independent, what is the probability that there will be no accident at this location in a year?

2.20. Rapid transit trains arrive at a given station every five minutes and depart after stopping at the station for one minute to pick up passengers. Assuming trains arrive every hour on the hour, what is the probability that a passenger will be able to board a train immediately if he or she arrives at the station at a random instant between 7:54 A.M. and 8:06 A.M.?

2.21. A telephone call occurs at random in the interval $(0, t)$. Let T be its time of occurrence. Determine:
 (a) $P(t_0 \leq T \leq t_1)$.
 (b) $P(t_0 \leq T \leq t_1 | T \geq t_0)$
 where $0 \leq t_0 \leq t_1 \leq t$.

2.22. For a storm-sewer system, estimates of annual maximum flow rates (AMFR) and their likelihood of occurrence [assuming that a maximum of 12 cfs (cubic feet per second) is possible] are given as follows:

 Event A = (5 to 10 cfs) $P(A) = 0.6$
 Event B = (8 to 12 cfs) $P(B) = 0.6$
 Event $C = A \cup B$ $P(C) = 0.7$

 Determine:
 (a) $P(8 \leq \text{AMFR} \leq 10)$, probability that AMFR is between 8 and 10 cfs.
 (b) $P(5 \leq \text{AMFR} \leq 12)$.
 (c) $P(10 \leq \text{AMFR} \leq 12)$.
 (d) $P(8 \leq \text{AMFR} \leq 10 | 5 \leq \text{AMFR} \leq 10)$.
 (e) $P(5 \leq \text{AMFR} \leq 10 | \text{AMFR} \geq 5)$.

2.23. At a major- and minor-street intersection, one finds that out of every 100 gaps on the major street, 65 are acceptable, that is, large enough for a car arriving on the minor street to cross. When a vehicle arrives on the minor street:
 (a) What is the probability that the first gap is not an acceptable one?
 (b) What is the probability that the first two gaps are both unacceptable?
 (c) The first car has crossed the intersection. What is the probability that the second will be able to cross at the very next gap?

2.24. A machine part may be selected from any of three manufacturers with probabilities $p_1 = 0.25$, $p_2 = 0.50$, and $p_3 = 0.25$. The probabilities that it will function properly during a specified period of time are 0.2, 0.3, and 0.4, respectively, for the three manufacturers. Determine the probability that a randomly chosen machine part will function properly for the specified time period.

2.25. Consider the possible failure of a transportation system to meet demand during rush hour.
   (a) Determine the probability that the system will fail if the probabilities shown in the following table are known.

Demand Level	P(level)	$P\left(\dfrac{system\ failure}{} \middle\| level\right)$
Low	0.6	0
Medium	0.3	0.1
High	0.1	0.5

   (b) If system failure was observed, find the probability that a "medium" demand level was its cause.

2.26. A cancer diagnostic test is 95% accurate both on those who have cancer and on those who do not. If 0.005 of the population actually does have cancer, compute the probability that a particular individual has cancer, given that the test indicates he or she has cancer.

2.27. A quality control record of transistors gives the following results when classified by manufacturer and quality.

| Manufacturer | Quality | | | |
	Acceptable	Marginal	Unacceptable	Total
A	128	10	2	140
B	97	5	3	105
C	110	5	5	120

Let one transistor be selected at random. What is the probability of it being:
   (a) From manufacturer A and with acceptable quality?
   (b) Acceptable given that it is from manufacturer C?
   (c) From manufacturer B given that it is marginal?

2.28. Verify Equation 2.26 for three events.

2.29. In an elementary study of synchronized traffic lights, consider a simple four-light system. Suppose that each light is red for 30 seconds of a 50-second cycle and suppose

$$P(S_{j+1}|S_j) = 0.15 \quad \text{and} \quad P(S_{j+1}|\bar{S}_j) = 0.40$$

for $j = 1, 2, 3$, where S_j is the event that a driver is stopped by the jth light. We assume a "one-light" memory for the system.

By means of the tree diagram, determine the probability that a driver:
(a) Will be delayed by all four lights.
(b) Will not be delayed by any of the four lights.
(c) Will be delayed by at most one light.

III
Random Variables and Probability Distributions

We have mentioned that our interest in the study of a random phenomenon is in the statements we can make concerning the events that can occur, and these statements are made based on probabilities assigned to simple outcomes. While basic concepts have been developed in Chapter II, a systematic and unified procedure is needed to facilitate making these statements, which can be quite complex. One of the immediate steps that can be taken in this unifying attempt is to require that each of the possible outcomes of a random experiment be represented by a real number. In this way, when the experiment is performed, its outcome is identified by its assigned real number rather than by its physical description. For example, when the possible outcomes of a random experiment consist of success and failure, we arbitrarily assign the number one to the event success and the number zero to the event failure. The associated sample space has now $\{1, 0\}$ as its sample points instead of success and failure, and the statement "the outcome is 1" means "the outcome is success."

This procedure not only permits us to replace a sample space of arbitrary elements by a new sample space having real numbers as its elements but also enables us to use arithmetic means for probability calculations. Furthermore, most problems in science and engineering deal with quantitative measures. Consequently, sample spaces associated with many random experiments of interest are already themselves sets of real numbers. The real-number assignment procedure is thus a natural unifying agent. On this basis, we may introduce a variable X, which is used to denote a real number whose value is determined by the outcome of a random experiment. This leads to the notion of a random variable that is defined more precisely as follows.

III.1. Random Variables

Consider a random experiment whose outcomes s are elements of S in the underlying probability space (S, β, P). In order to construct a model for a random variable, we assume that it is possible to assign a real number $X(s)$ for each s following a certain set of rules. We see that the "number" $X(s)$ is really a real-valued *point function* defined over the domain of the basic probability space.

Definition. The point function $X(s)$ is called a *random variable* (r.v.) if (a) it is a finite real-valued function defined on the sample space S of a random experiment for which probability function is defined on the Borel field β of events and (b) for every real number x, the set $\{s: X(s) \leq x\}$ is an event in β. The relation $X = X(s)$ takes every element s in S of the probability space on to a point X on the real line $R^1 = (-\infty, \infty)$.

Notationally, the dependence of a r.v. $X(s)$ on s will be omitted for convenience.

The second condition stated in the definition is a so-called "measurability condition." It insures us that it is meaningful to consider the probability of the event $X \leq x$ for every x or, more generally, the probability of any finite or countably infinite combination of such events.

To see more clearly the role a random variable plays in the study of a random phenomenon, consider again the simple example where the possible outcomes of a random experiment are success and failure. Let us again assign the number one to the event success and the number zero to failure. If X is the random variable associated with this experiment, then X takes on two possible values: 1 and 0. Moreover, the following statements are equivalent:

1. The outcome is success.
2. The outcome is 1.
3. $X = 1$.

A r.v. X is called a *discrete* random variable if it is defined over a sample space having finite or countably infinite number of sample points. In this case, the r.v. X takes on discrete values and it is possible to enumerate all the values it may assume. In the case of a sample space having noncountably infinite number of sample points, the associated random variable is called a *continuous* random variable, with its values distributed over a continuous interval on the real line. We make this distinction because they require different probability assignment considerations. Both types of random variables are important in science and engineering and we shall see ample evidence of this in the subsequent chapters.

In the sequel, all random variables will be written in capital letters, $X, Y, Z, \ldots$. The values that a r.v. X can assume will be denoted by corresponding lowercase x or $x_1, x_2, \ldots$.

We will have many occasions to consider a sequence of r.v.'s $X_j, j = 1, 2, \ldots, n$. In these cases we assume that they are defined on the same probability space. The r.v.'s $X_1, X_2, \ldots, X_n$ will then map every element s of S in the probability space onto a point in the n-dimensional Euclidian space R^n. We note here that an analysis involving n random variables is equivalent to considering a *random vector* having the n random variables as components. The notion of a random vector will be used frequently in what follows, and we will denote them by boldface capital letters **X, Y, Z**, ...

III.2 Probability Distributions

The behavior of a random variable is characterized by its probability distribution, that is, the way probabilities are distributed over the values it assumes. Probability distribution function and probability mass function are two ways to characterize this distribution for a discrete random variable. They are equivalent in the sense that the knowledge of either one completely specifies the random variable. The corresponding functions for a continuous random variable are the probability distribution function, defined in the same way as in the discrete random variable case, and the probability density function. The definitions of these functions now follow.

III.2.1 PROBABILITY DISTRIBUTION FUNCTION (DPF)

Given a random experiment with its associated r.v. X and given a real number x, let us consider the probability of the event $\{s: X(s) \leq x\}$, or simply $P(X \leq x)$. This probability is clearly dependent on the assigned value x. The function

$$\boxed{F_X(x) = P(X \leq x)} \tag{3.1}$$

is defined as the *probability distribution function* or simply *distribution function* of the r.v. X. In the above, the subscript X identifies the random variable. This subscript is sometimes omitted when there is no risk of confusion. Let us repeat that $F_X(x)$ is simply $P(A)$, the probability of an event A occurring, the event being $X \leq x$. This observation ties what we do here with the development in Chapter II.

The probability distribution function or PDF is thus the probability that X will assume a value lying in a subset of S, the subset being the point x and all points lying to the left of x. As x increases, the subset covers more of the real line and the value of PDF increases until it reaches 1. The PDF of a random variable thus accumulates probability as x increases and the name *cumulative distribution function* is also used for this function.

In view of the definition and the discussion above, we give below some of the important properties possessed by a PDF.

1. It exists for both discrete and continuous random variables and it has values between 0 and 1.
2. It is a nonnegative, continuous to the left, and nondecreasing function of the real variable x. Moreover, we have

$$F_X(-\infty) = 0 \quad \text{and} \quad F_X(+\infty) = 1 \tag{3.2}$$

3. If a and b are two real numbers such that $a < b$, then

$$P(a < X \leq b) = F_X(b) - F_X(a) \tag{3.3}$$

This relation is a direct result of the identity

$$P(X \leq b) = P(X \leq a) + P(a < X \leq b)$$

We see from Equation 3.3 that the probability of X having a value in an arbitrary interval can be represented by a difference between two values of the PDF. Generalizing, probabilities associated with any sets of intervals are derivable from the probability distribution function.

Example 3.1. Let a discrete r.v. X assume values $-1, 1, 2,$ and 3 with probabilities $\frac{1}{4}, \frac{1}{8}, \frac{1}{8},$ and $\frac{1}{2}$, respectively. We then have

$$\begin{aligned}
F_X(x) &= 0 & x &< -1 \\
&= \tfrac{1}{4} & -1 &\leq x < 1 \\
&= \tfrac{3}{8} & 1 &\leq x < 2 \\
&= \tfrac{1}{2} & 2 &\leq x < 3 \\
&= 1 & x &\geq 3
\end{aligned}$$

RANDOM VARIABLES AND PROBABILITY DISTRIBUTIONS   41

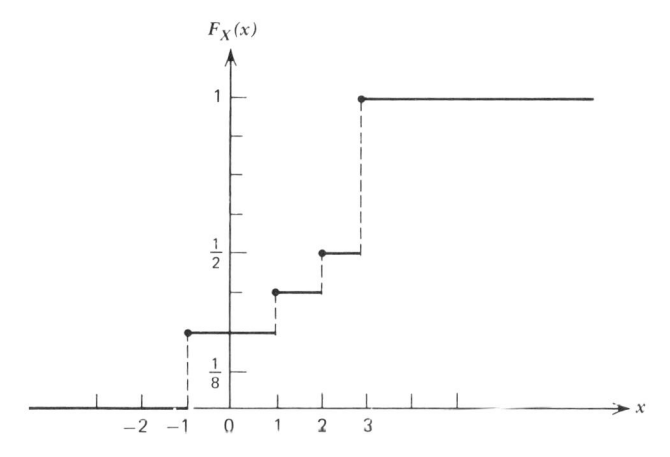

Figure 3.1 PDF of X in Example 3.1.

This function is plotted in Figure 3.1. It is typical of PDF's associated with discrete random variables, increasing from 0 to 1 in a staircase fashion.

A continuous random variable assumes a nonenumerable number of values over the real line. Hence, the probability of a continuous r.v. assuming any particular value is zero and therefore no discrete jumps are possible for its PDF. A typical PDF for continuous r.v.'s is shown in Figure 3.2. It is smooth as opposed to having a staircase shape. The probability of X having a value in a given interval is found using Equation 3.3, and it makes sense to speak only of this kind of probabilities for continuous random variables. For example, $P(-1 < X \leq 1) = F_X(1) - F_X(-1) = 0.8 - 0.4 = 0.4$. Clearly, $P(X = a) = 0$ for any a.

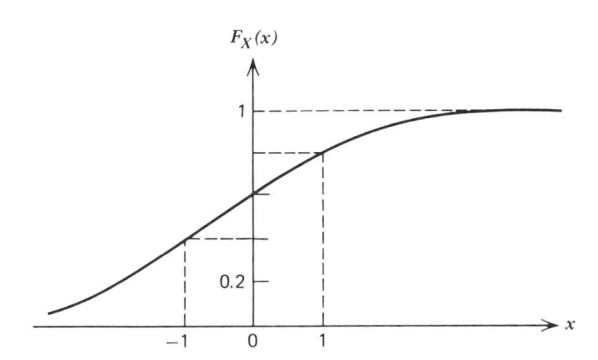

Figure 3.2 PDF of a continuous r.v.

III.2.2 PROBABILITY MASS FUNCTION FOR DISCRETE RANDOM VARIABLES (pmf)

Let X be a discrete random variable that assumes at most a countably infinite number of values $x_1, x_2, \ldots$ with nonzero probabilities. If we denote $P(X = x_i) = p(x_i)$, $i = 1, 2, \ldots$, then clearly

$$0 < p(x_i) \leq 1 \text{ for all } i$$
$$\sum_i p(x_i) = 1 \tag{3.4}$$

Definition. The function

$$\boxed{p_X(x) = P(X = x)} \tag{3.5}$$

is defined as the *probability mass function* (pmf) of X. Again, the subscript X is used to identify the associated random variable. The pmf is zero everywhere except at x_i, $i = 1, 2, \ldots$, and it has the appearance shown in Figure 3.3 for the random variable defined in Example 3.1.

This is a typical shape of pmf's for discrete random variables. Since $P(X = x) = 0$ for continuous random variables, it does not exist in the continuous r.v. case. We also observe that, like $F_X(x)$, the specification of $p_X(x)$ completely characterizes the r.v. X; furthermore, these two functions are simply related by (assuming $x_1 < x_2 < \cdots$)

$$p_X(x_i) = F_X(x_i) - F_X(x_{i-1}) \tag{3.6}$$

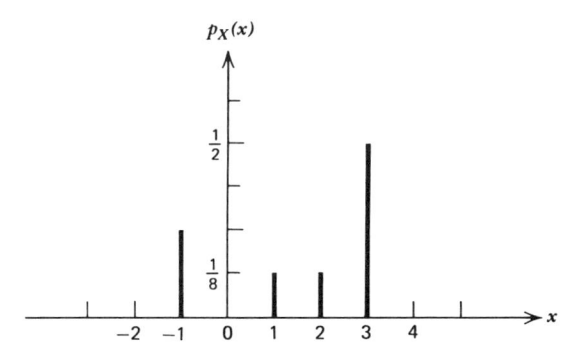

$$F_X(x) = \sum_{\substack{i=1 \\ i: \, x_i \leq x}} p_X(x_i) \tag{3.7}$$

Figure 3.3 pmf of X in Example 3.1.

where the upper limit for the sum in Equation 3.7 means that the sum is taken over all i satisfying $x_i \leq x$. Hence, we see that the PDF and pmf of a discrete random variable contain the same information, each one is recoverable from the other.

One can also give PDF and pmf a useful physical interpretation. In terms of the distribution of one unit of mass over the real line $-\infty < x < \infty$, the PDF of a random variable at x, $F_X(x)$, can be interpreted as the total mass associated with the point x and all points lying to the left of x. The pmf, on the other hand, shows the distribution of this unit of mass over the real line; it is distributed at discrete points x_i with the amount of mass equal to $p_X(x_i)$ at x_i, $i = 1, 2, \ldots$.

Example 3.2. A discrete distribution arising in a large number of physical models is the *binomial distribution*. Much more will be said of this important distribution but, at present, let us use it as an illustration for graphing PDF and pmf of a random variable.

A discrete r.v. X has a binomial distribution when

$$p_X(k) = \binom{n}{k} p^k (1-p)^{n-k} \qquad k = 0, 1, 2, \ldots, n \qquad (3.8)$$

where n and p are two parameters of the distribution, n being a positive integer and $0 < p < 1$. The binomial coefficient $\binom{n}{k}$ is defined by

$$\binom{n}{k} = \frac{n!}{k!(n-k)!} \qquad (3.9)$$

The pmf and PDF of X for $n = 10$ and $p = 0.2$ are given in Figure 3.4. The numerical values of $p_X(k)$ for this case are found from

$$p_X(k) = \binom{10}{k} (0.2)^k (0.8)^{10-k} \qquad k = 0, 1, 2, \ldots, 10$$

III.2.3 PROBABILITY DENSITY FUNCTION FOR CONTINUOUS RANDOM VARIABLES (pdf)

For a continuous r.v. X, its PDF, $F_X(x)$, is a continuous function of x and the derivative

$$\boxed{f_X(x) = \frac{dF_X(x)}{dx}} \qquad (3.10)$$

exists for all x. The function $f_X(x)$ is called the *probability density function* (pdf) or simply *density function* of X.

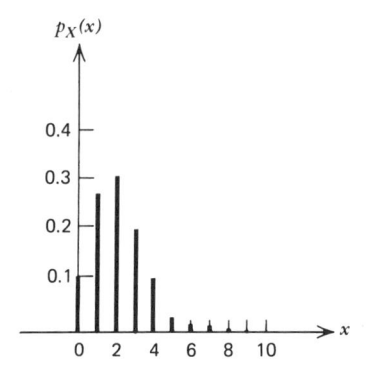

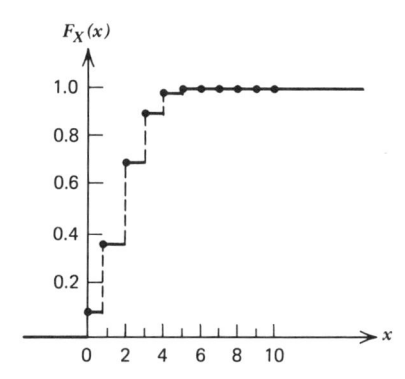

Figure 3.4 pmf and PDF of X in Example 3.2.

Since $F_X(x)$ is monotone nondecreasing, we clearly have

$$f_X(x) \geq 0 \tag{3.11}$$

Additional properties of $f_X(x)$ can be derived easily from the definition (3.10); these include

$$\boxed{F_X(x) = \int_{-\infty}^{x} f_X(u)du} \tag{3.12}$$

and

$$\int_{-\infty}^{\infty} f_X(x)dx = 1$$

$$P(a < X \leq b) = F_X(b) - F_X(a) = \int_{a}^{b} f_X(x)dx \tag{3.13}$$

A typical probability density function has the shape shown in Figure 3.5. As indicated by Equations 3.13, the total area under the curve is unity and the shaded area from a to b gives the probability $P(a < X \leq b)$. We again observe that the knowledge of either pdf or PDF completely characterizes a continuous r.v. The probability density function does not exist for a discrete r.v. since its associated probability distribution function has discrete jumps and is not differentiable at these points of discontinuity.

RANDOM VARIABLES AND PROBABILITY DISTRIBUTIONS

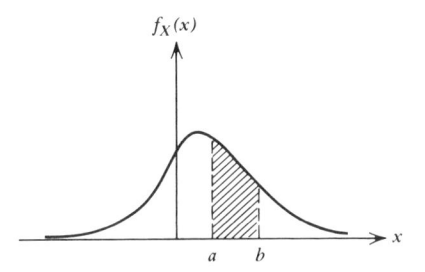

Figure 3.5 A probability density function.

Using the mass distribution analogy, the pdf of a continuous r.v. plays exactly the same role as the pmf of a discrete r.v. The function $f_X(x)$ can be interpreted as the mass per unit length along the real line, or the mass density. There are no masses attached to discrete points as in the discrete r.v. case. The use of the term *density function* is therefore appropriate here for $f_X(x)$.

Example 3.3. A r.v. X whose density function has the form ($a > 0$)

$$f_X(x) = ae^{-ax} \quad x \geq 0$$
$$= 0 \quad \text{elsewhere} \quad (3.14)$$

is said to be *exponentially distributed*. We can easily check that all the conditions given by Equation 3.11 and Equations 3.13 are satisfied. It is graphically presented in Figure 3.6 together with its associated probability distribution function. The functional form of the PDF as obtained from Equation 3.12 is

$$F_X(x) = \int_{-\infty}^{x} f_X(u)du = 0 \quad x < 0$$
$$= 1 - e^{-ax} \quad x \geq 0 \quad (3.15)$$

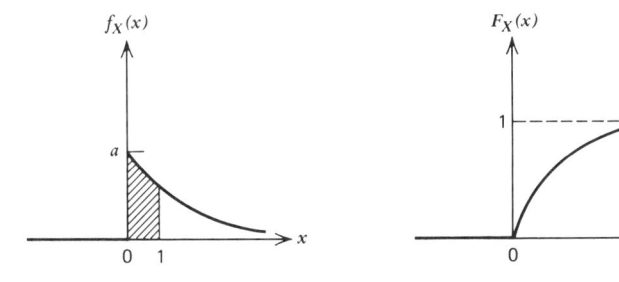

Figure 3.6 pdf and PDF of X in Example 3.3.

Let us compute some of the probabilities using $f_X(x)$. The probability $P(0 < X \leq 1)$ is numerically equal to the area under $f_X(x)$ from $x = 0$ to $x = 1$. It is

$$P(0 < X \leq 1) = \int_0^1 f_X(x)dx = 1 - e^{-a}$$

The probability $P(X > 3)$ is obtained by computing the area under $f_X(x)$ to the right of $x = 3$. Hence,

$$P(X > 3) = \int_3^\infty f_X(x)dx = e^{-3a}$$

The same probabilities can be obtained from $F_X(x)$ by taking appropriate differences. We have

$$P(0 < X \leq 1) = F_X(1) - F_X(0) = (1 - e^{-a}) - 0 = 1 - e^{-a}$$
$$P(X > 3) = F_X(\infty) - F_X(3) = 1 - (1 - e^{-3a}) = e^{-3a}$$

Let us note that there is no numerical difference between $P(0 < X \leq 1)$ and $P(0 \leq X \leq 1)$ for continuous random variables since $P(X = 0) = 0$.

III.2.4 MIXED-TYPE DISTRIBUTION

There are situations in which one encounters a random variable that is partially discrete and partially continuous. The PDF given in Figure 3.7 represents such a case in which the r.v. X is continuously distributed over the real line except at $X = 0$ where $P(X = 0)$ is a positive quantity. This situation may arise when, for example, the r.v. X represents the waiting time of a customer at a ticket counter. Let X be the time interval from time of arrival at the ticket counter to the time being served. It is reasonable to expect that X will assume values over the interval $X \geq 0$. At

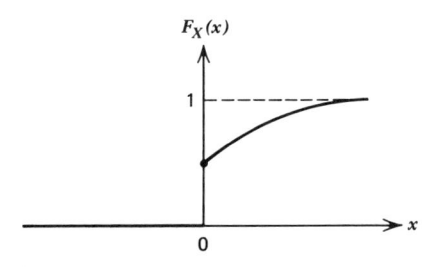

Figure 3.7 A mixed-type PDF.

$X = 0$, however, there is a finite probability for not having to wait at all, giving rise to the situation depicted in Figure 3.7.

Strictly speaking, neither pmf nor pdf exists for a random variable of the mixed type. We can, however, still use them separately for different portions of the distribution for computational purposes. Let $f_X(x)$ be the pdf for the *continuous* portion of the distribution, it can be used for calculating probabilities in the positive range of x-values for this example. We observe that the total area under the pdf curve is no longer 1 but is equal to $1 - P(X = 0)$.

III.3 Two or More Random Variables

In many cases it is more natural to describe the outcome of a random experiment by two or more numerical numbers simultaneously. For example, the characterization of both weight and height in a given population, the study of temperature and pressure variations in a physical experiment, and the distribution of monthly temperature readings in a given region over a given year. In these situations, two or more random variables are considered jointly and the description of their joint behavior is our concern.

Let us first consider the case of two r.v.'s X and Y. We proceed analogously to the single r.v. case in defining their joint probability distributions. We note that the r.v.'s X and Y can also be considered as components of a two-dimensional random vector, say **Z**. Joint probability distributions associated with two random variables are sometimes called *bivariate distributions*.

As we shall see, extensions to cases of more than two random variables, or *multivariate distributions*, are straightforward.

III.3.1 JOINT PROBABILITY DISTRIBUTION FUNCTION (JPDF)

The joint probability distribution function of r.v.'s X and Y, denoted by $F_{XY}(x, y)$, is defined by

$$F_{XY}(x, y) = P(X \leq x \cap Y \leq y) \qquad (3.16)$$

for all x and y. It is the probability of the intersection of two events; the r.v.'s X and Y thus induce a probability distribution over a two-dimensional Euclidean plane.

Using again the analogy to a mass distribution, let one unit of mass be distributed over the (x, y)-plane in such a way that the mass in any given region R be equal to the probability that X and Y take values in R. Then the JPDF $F_{XY}(x, y)$

represents the total mass in the quadrant to the left and below the point (x, y), inclusive of the boundaries. In the case where both X and Y are discrete, all the mass is concentrated at a finite or countably infinite number of places in the (x, y)-plane as point masses. When both are continuous, the mass is distributed continuously over the (x, y)-plane.

It is clear from the definition that $F_{XY}(x, y)$ is nonnegative, nondecreasing in x and y, and continuous to the left with respect to x and y. The following properties are also direct consequences of the definition.

$$\begin{aligned} F_{XY}(-\infty, -\infty) &= F_{XY}(-\infty, y) = F_{XY}(x, -\infty) = 0 \\ F_{XY}(+\infty, +\infty) &= 1 \\ F_{XY}(x, +\infty) &= F_X(x) \\ F_{XY}(+\infty, y) &= F_Y(y) \end{aligned} \quad (3.17)$$

For example, the third relation follows from the fact that the joint event $X \leq x \cap Y \leq +\infty$ is the same as the event $X \leq x$ since $Y \leq +\infty$ is a sure event. Hence,

$$F_{XY}(x, +\infty) = P(X \leq x \cap Y \leq +\infty) = P(X \leq x) = F_X(x)$$

Similarly, we can show that, for any $x_1, x_2, y_1,$ and y_2 such that $x_1 < x_2$ and $y_1 < y_2$, the probability $P(x_1 < X \leq x_2 \cap y_1 < Y \leq y_2)$ is given in terms of $F_{XY}(x, y)$ by

$$\begin{aligned} P(x_1 < X \leq x_2 \cap y_1 < Y \leq y_2) = &F_{XY}(x_2, y_2) - F_{XY}(x_1, y_2) \\ &- F_{XY}(x_2, y_1) + F_{XY}(x_1, y_1) \end{aligned} \quad (3.18)$$

which shows that all probability calculations involving r.v.'s X and Y can be made with the knowledge of their JPDF.

Finally, we note that the last two equations in Equations 3.17 show that distribution functions of individual random variables are directly derivable from their joint distribution function. The converse, of course, is not true. In the context of several random variables, these individual distribution functions are called *marginal distribution functions*. For example, $F_X(x)$ is the marginal distribution function of X.

The general shape of $F_{XY}(x, y)$ can be visualized from the properties given in Equations 3.17. In the case where X and Y are discrete, it has the appearance of a corner of an irregular staircase, something like that shown in Figure 3.8. It rises from zero to the height of one in the direction from the third quadrant to the first quadrant. When both X and Y are continuous, $F_{XY}(x, y)$ becomes a smooth surface with the same features. It is a staircase type in one direction and smooth in the other if one of the random variables is discrete and the other continuous.

RANDOM VARIABLES AND PROBABILITY DISTRIBUTIONS 49

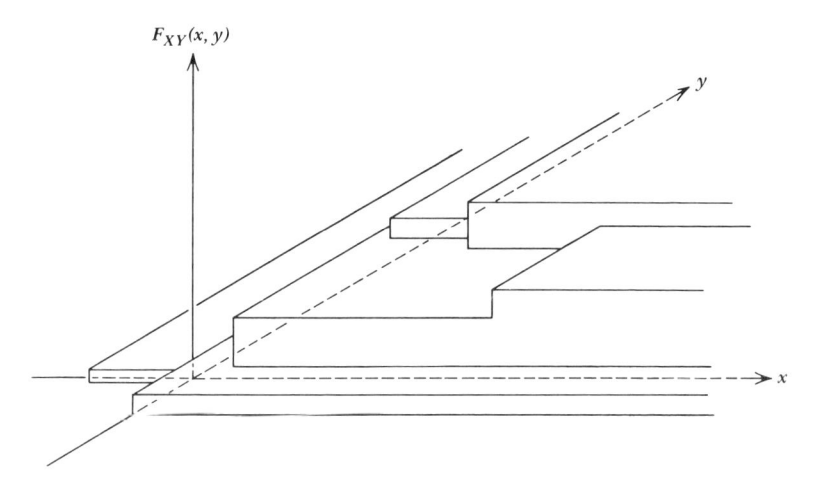

Figure 3.8 A joint probability distribution function of X and Y.

The joint probability distribution function of more than two random variables is defined in a similar fashion. Consider n r.v.'s $X_1, X_2, \ldots, X_n$. Their JPDF is defined by

$$F_{X_1 X_2 \cdots X_n}(x_1, x_2, \ldots, x_n) = P(X_1 \leq x_1 \cap X_2 \leq x_2 \cap \cdots \cap X_n \leq x_n) \quad (3.19)$$

These random variables induce a probability distribution in an n-dimensional Euclidean space. One can deduce immediately its properties in parallel with those noted in Equations 3.17 and 3.18 for the two-r.v. case.

As we have mentioned previously, the finite sequence $\{X_j, j = 1, 2, \ldots, n\}$ may be regarded as the components of an n-dimensional random vector **X**. The JPDF of **X** is identical to that given above but it can be written in a more compact form, namely, $F_{\mathbf{X}}(\mathbf{x})$, where **x** is a vector with components $x_1, x_2, \ldots, x_n$.

III.3.2 JOINT PROBABILITY MASS FUNCTION (jpmf)

The joint probability mass function is another, and more direct, characterization of the joint behavior of two or more random variables when they are discrete. Let X and Y be two discrete random variables that assume at most a countably infinite number of value pairs (x_i, y_j), $i, j = 1, 2, \ldots$, with nonzero probabilities. The jpmf of X and Y is defined by

$$p_{XY}(x, y) = P(X = x \cap Y = y) \quad (3.20)$$

for all x and y. It is zero everywhere except at the points (x_i, y_j), $i, j = 1, 2, \ldots$, where it takes values equal to the joint probability $P(X = x_i \cap Y = y_j)$. We observe the following properties, which are direct extensions of those noted in Equations 3.4, 3.6, and 3.7 for the single r.v. case.

$$0 \leq p_{XY}(x, y) \leq 1$$

$$\sum_i \sum_j p_{XY}(x_i, y_j) = 1 \qquad (3.21)$$

$$\sum_i p_{XY}(x_i, y) = p_Y(y) \qquad \sum_j p_{XY}(x, y_j) = p_X(x)$$

where $p_X(x)$ and $p_Y(y)$ are now called *marginal probability mass functions*. We also have

$$F_{XY}(x, y) = \sum_{i=1}^{i:\, x_i \leq x} \sum_{j=1}^{j:\, y_j \leq y} p_{XY}(x_i, y_j) \qquad (3.22)$$

Example 3.4. Consider a simplified version of a two-dimensional "random walk" problem. We imagine a particle that moves in a plane in unit steps starting from the origin. Each step is one unit in the positive direction with probability p along the x-axis and probability q ($p + q = 1$) along the y-axis. We further assume that each step is taken independent of the others. What is the probability distribution of the position of this particle after five steps?

Since the position is conveniently represented by two coordinates, we wish to establish $p_{XY}(x, y)$ where the r.v. X represents the x-coordinate of the position after five steps and Y represents the y-coordinate. It is clear that the jpmf $p_{XY}(x, y)$ is zero everywhere except at those points satisfying $x + y = 5$ and $x, y \geq 0$. Invoking independence of events of taking successive steps, it follows from Section II.3 that $p_{XY}(5, 0)$, the probability of the particle being at $(5, 0)$ after five steps, is the product of probabilities of taking five successive steps in the positive x-direction. Hence,

$$p_{XY}(5, 0) = p^5$$

For $p_{XY}(4, 1)$, there are five distinct ways of reaching that position (4 steps in x-direction and 1 in y; 3 in x, 1 in y, and 1 in x; etc.), each with a probability of $p^4 q$. We thus have

$$p_{XY}(4, 1) = 5p^4 q$$

RANDOM VARIABLES AND PROBABILITY DISTRIBUTIONS 51

Similarly, other nonvanishing values of $p_{XY}(x, y)$ are easily calculated to be

$$\begin{aligned} p_{XY}(x, y) &= 10p^3q^2 & (x, y) &= (3, 2) \\ &= 10p^2q^3 & &= (2, 3) \\ &= 5pq^4 & &= (1, 4) \\ &= q^5 & &= (0, 5) \end{aligned}$$

where $p_X(x)$ and $p_Y(y)$ are now called *marginal probability mass functions*. We discussed in Example 3.2.

The jpmf $p_{XY}(x, y)$ is graphically presented in Figure 3.9 for $p = 0.4$ and $q = 0.6$. It is easy to check that the sum of $p_{XY}(x, y)$ over all x and y is one as required by the second of Equations 3.21.

The joint probability distribution function $F_{XY}(x, y)$ can also be constructed using Equation 3.22. Rather than showing it in three-dimensional form, Figure 3.10 gives this function by indicating its value in each of the dividing regions. One

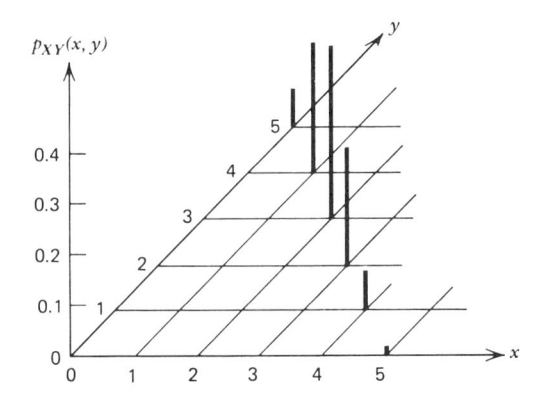

$$\begin{aligned} p_{XY}(x, y) &= 0.01024 & (x, y) &= (5, 0) \\ &= 0.0768 & &= (4, 1) \\ &= 0.2304 & &= (3, 2) \\ &= 0.3456 & &= (2, 3) \\ &= 0.2592 & &= (1, 4) \\ &= 0.07776 & &= (0, 5) \end{aligned}$$

Figure 3.9 $p_{XY}(x, y)$ in Example 3.4 with $p = 0.4$ and $q = 0.6$.

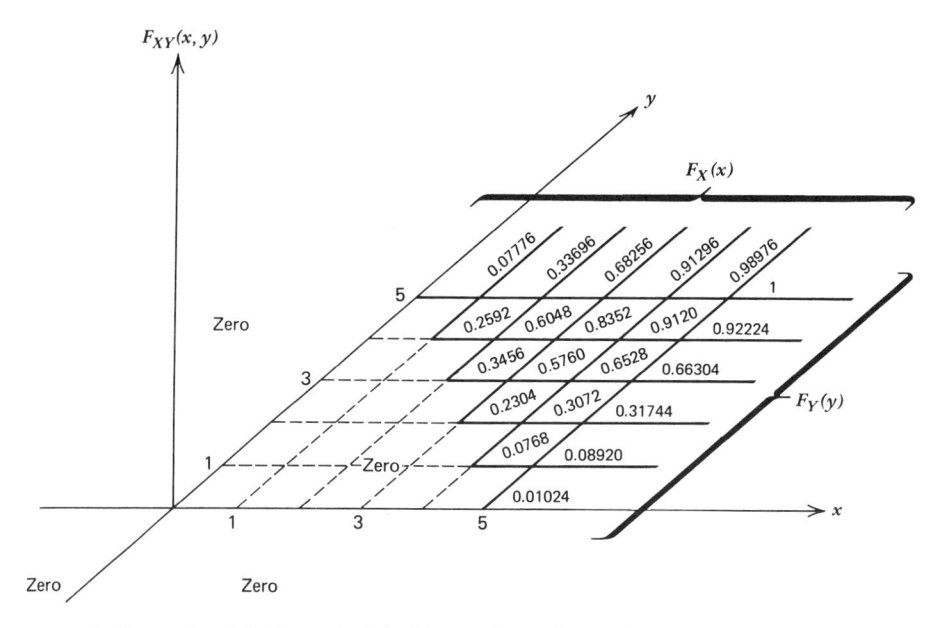

Figure 3.10 $F_{XY}(x, y)$ in Example 3.4 with $p = 0.4$ and $q = 0.6$.

also notes that the arrays of indicated numbers beyond $y = 5$ are values associated with the marginal distribution function $F_X(x)$. Similarly, $F_Y(y)$ takes those values situated beyond $x = 5$. These observations are also indicated on the graph.

The knowledge of the joint probability mass function permits us to make all probability calculations of interest. The probability of any event being realized involving X and Y is found by determining the pairs of values of X and Y that give rise to this event and then simply summing over the values of $p_{XY}(x, y)$ at all such pairs. In Example 3.4, suppose we wish to determine the probability of $X > Y$, it is given by

$$P(X > Y) = P(X = 5 \cap Y = 0) + P(X = 4 \cap Y = 1) + P(X = 3 \cap Y = 2)$$
$$= 0.01024 + 0.0768 + 0.2304 = 0.31744$$

Example 3.5. Let us discuss again Example 2.11 in the context of random variables. Let X be the r.v. representing precipitation levels with values 1, 2, and 3 indicating low, medium, and high, respectively. The r.v. Y will be used for peak flow rate with the value 1 when it is critical and 2 when noncritical. The information given in Example 2.11 defines the jpmf $p_{XY}(x, y)$ whose values are tabulated as follows.

y \ x	1	2	3
1	0.0	0.06	0.12
2	0.5	0.24	0.08

In order to determine the probability of reaching critical level of peak flow rate, for example, we simply sum over all $p_{XY}(x, y)$ satisfying $y = 1$, regardless of x-values. Hence,

$$P(Y = 1) = p_{XY}(1, 1) + p_{XY}(2, 1) + p_{XY}(3, 1) = 0.18$$

The definition of joint probability mass function for more than two random variables is a direct extension of that for the two-r.v. case. Consider n r.v.'s $X_1, X_2, \ldots, X_n$. Their jpmf is defined by

$$p_{X_1 X_2 \cdots X_n}(x_1, x_2, \ldots, x_n) = P(X_1 = x_1 \cap X_2 = x_2 \cap \cdots \cap X_n = x_n) \quad (3.23)$$

which is the probability of the intersection of n events. Its properties and utilities follow directly from our discussion in the two-r.v. case. Again, a more compact form for the jpmf is $p_\mathbf{X}(\mathbf{x})$, where $\mathbf{X}$ is an n-dimensional random vector with components $X_1, X_2, \ldots, X_n$.

III.3.3 JOINT PROBABILITY DENSITY FUNCTION (jpdf)

As in the case of single random variables, probability density functions become appropriate when the random variables are continuous. The joint probability density function of r.v.'s X and Y is defined by the partial derivative

$$f_{XY}(x, y) = \frac{\partial^2 F_{XY}(x, y)}{\partial x \partial y} \quad (3.24)$$

Since $F_{XY}(x, y)$ is monotone nondecreasing in both x and y, $f_{XY}(x, y)$ is nonnegative for all x and y. We also see from Equation 3.24 that

$$F_{XY}(x, y) \equiv P(X \leq x \cap Y \leq y) = \int_{-\infty}^{y} \int_{-\infty}^{x} f_{XY}(u, v) du\, dv \quad (3.25)$$

Moreover, with $x_1 < x_2$ and $y_1 < y_2$,

$$P(x_1 < X \leq x_2 \cap y_1 < Y \leq y_2) = \int_{y_1}^{y_2} \int_{x_1}^{x_2} f_{XY}(x, y)dx\, dy \qquad (3.26)$$

The jpdf $f_{XY}(x, y)$ defines a surface over the (x, y)-plane. As indicated by Equation 3.26, the probability that the r.v.'s X and Y fall within a certain region R is equal to the volume under the surface $f_{XY}(x, y)$ and bounded by that region. This is illustrated in Figure 3.11.

We also note the following important properties.

$$\int_{-\infty}^{\infty} \int_{-\infty}^{\infty} f_{XY}(x, y)dx\, dy = 1 \qquad (3.27)$$

$$\int_{-\infty}^{\infty} f_{XY}(x, y)dy = f_X(x) \qquad (3.28)$$

$$\int_{-\infty}^{\infty} f_{XY}(x, y)dx = f_Y(y) \qquad (3.29)$$

Equation 3.27 follows from Equation 3.25 by letting $x, y \to +\infty$, and this shows that the total volume under the $f_{XY}(x, y)$-surface is unity. To give a derivation of Equation 3.28, we know that

$$F_X(x) = F_{XY}(x, \infty) = \int_{-\infty}^{\infty} \int_{-\infty}^{x} f_{XY}(u, y)du\, dy$$

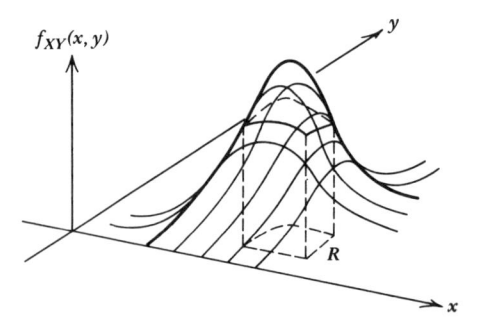

Figure 3.11 A joint probability density function.

RANDOM VARIABLES AND PROBABILITY DISTRIBUTIONS

Differentiating the above with respect to x gives the desired result immediately. The density functions $f_X(x)$ and $f_Y(y)$ in Equations 3.28 and 3.29 are now called the *marginal density functions* of X and Y, respectively.

Example 3.6. A boy and a girl plan to meet at a certain place between 9 A.M. and 10 A.M., each not waiting more than 10 minutes for the other. If all times of arrival within the hour are equally likely for each person, and if their times of arrival are independent, find the probability that they will meet.

For a single continuous random variable X that takes all values over an interval a to b with equal likelihood, the distribution is called a *uniform* distribution and its density function $f_X(x)$ has the form

$$f_X(x) = 1/(b - a) \quad a \leq x \leq b$$
$$= 0 \quad \text{elsewhere} \tag{3.30}$$

The height of $f_X(x)$ over the interval (a, b) must be $1/(b - a)$ in order that the area is 1 below the curve (see Figure 3.12). For a two-dimensional case as described in this example, a uniform distribution is one whose joint density function is a flat surface within prescribed bounds. The volume under the surface is unity.

Let the boy arrive at X minutes past 9 A.M. and the girl arrive at Y minutes past 9 A.M. The jpdf $f_{XY}(x, y)$ thus takes the form as shown in Figure 3.13 and is given by

$$f_{XY}(x, y) = \tfrac{1}{3600} \quad 0 \leq x \leq 60 \quad 0 \leq y \leq 60$$
$$= 0 \quad \text{elsewhere}$$

The probability we are seeking is thus the volume under this surface over an appropriate region R. For this problem, the region R is given by

$$R: |X - Y| \leq 10$$

and is shown in Figure 3.14 in the (x, y)-plane.

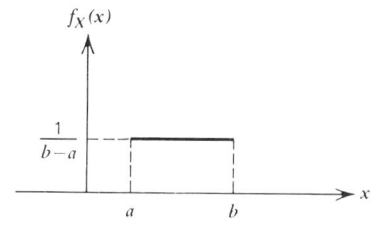

Figure 3.12 A uniform density function.

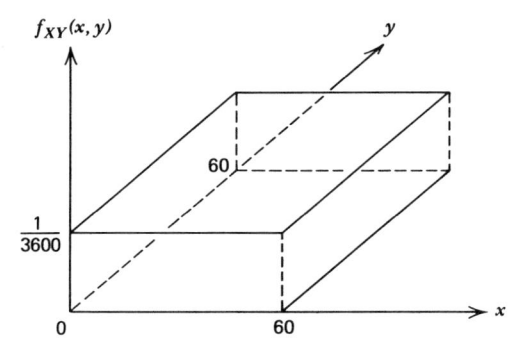

Figure 3.13 $f_{XY}(x, y)$ in Example 3.6.

The volume of interest can be found by inspection in this simple case. Dividing R into three regions as shown, we have

$$P(\text{they will meet}) = P(|X - Y| \le 10)$$
$$= [2(5)(10) + 10\sqrt{2}(50\sqrt{2})]/3600 = \tfrac{11}{36}$$

We note that, for a more complicated jpdf, one needs to carry out the volume integral $\iint_R f_{XY}(x, y)dx\, dy$ for volume calculations.

As an exercise, let us determine the joint probability distribution function and the marginal density functions of the r.v.'s X and Y defined in Example 3.6.

The JPDF of X and Y is obtained from Equation 3.25. It is clear that

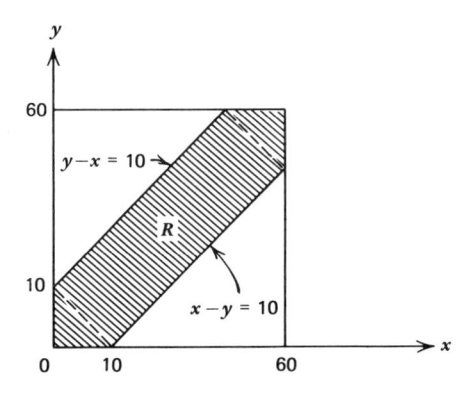

$$F_{XY}(x, y) = 0 \quad (x, y) < (0, 0)$$
$$= 1 \quad (x, y) > (60, 60)$$

Figure 3.14 Region R in Example 3.6.

RANDOM VARIABLES AND PROBABILITY DISTRIBUTIONS

Within the region $(0, 0) \leq (x, y) \leq (60, 60)$, we have

$$F_{XY}(x, y) = \int_0^y \int_0^x (\tfrac{1}{3600}) dx\, dy = \frac{xy}{3600}$$

For marginal density functions, Equations 3.28 and 3.29 give us

$$f_X(x) = \int_0^{60} (\tfrac{1}{3600}) dy = \tfrac{1}{60} \qquad 0 \leq x \leq 60$$
$$= 0 \qquad \text{elsewhere}$$

Similarly,

$$f_Y(y) = \tfrac{1}{60} \qquad 0 \leq y \leq 60$$
$$= 0 \qquad \text{elsewhere}$$

Both random variables are thus uniformly distributed over the interval $(0, 60)$.

Example 3.7. In structural reliability studies, the resistance Y of a structural element and the force X applied to it are generally regarded as random variables. The probability of failure, p_f, is defined by $P(Y \leq X)$. Suppose that the jpdf of X and Y is specified to be

$$f_{XY}(x, y) = abe^{-(ax+by)} \qquad (x, y) > 0$$
$$= 0 \qquad (x, y) \leq 0$$

where a and b are known positive constants, we wish to determine p_f.

The probability p_f is determined from

$$p_f = \iint_R f_{XY}(x, y) dx\, dy$$

where R is the region satisfying $Y \leq X$. Since X and Y take only positive values, the region R is that shown in Figure 3.15. Hence,

$$p_f = \int_0^\infty \int_y^\infty abe^{-(ax+by)}\, dx\, dy = \frac{b}{a+b}$$

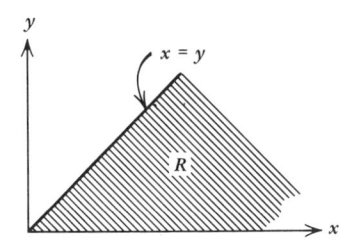

Figure 3.15 Region R in Example 3.7.

In closing this section, let us note that generalization to the case of many random variables is again straightforward. The joint distribution function of n r.v.'s $X_1, X_2, \ldots, X_n$, or $\mathbf{X}$, is given by Equation 3.19 as

$$F_{\mathbf{X}}(\mathbf{x}) = P(X_1 \leq x_1 \cap X_2 \leq x_2 \cdots \cap X_n \leq x_n) \tag{3.31}$$

The corresponding joint density function, denoted by $f_{\mathbf{X}}(\mathbf{x})$, is then

$$f_{\mathbf{X}}(\mathbf{x}) = \frac{\partial^n F_{\mathbf{X}}(\mathbf{x})}{\partial x_1 \partial x_2 \cdots \partial x_n} \tag{3.32}$$

if the indicated partial derivative exists. Various properties possessed by these functions can be readily inferred from those indicated for the two-r.v. case.

III.4 Conditional Distributions and Independence

The important concepts of conditional probability and independence introduced in Sections II.3 and II.4 play equally important roles in the context of random variables. The *conditional distribution function* of a r.v. X, given that another r.v. Y has taken a value y, is defined by

$$F_{XY}(x|y) = P(X \leq x | Y = y) \tag{3.33}$$

Similarly, when the r.v. X is discrete, the definition of *conditional mass function* of X given $Y = y$ is

$$p_{XY}(x|y) = P(X = x | Y = y) \tag{3.34}$$

RANDOM VARIABLES AND PROBABILITY DISTRIBUTIONS

Using the definition of conditional probability given by Equation 2.24, we have

$$p_{XY}(x|y) = P(X = x | Y = y) = \frac{P(X = x \cap Y = y)}{P(Y = y)}$$

or

$$p_{XY}(x|y) = \frac{p_{XY}(x, y)}{p_Y(y)} \quad \text{if } p_Y(y) \neq 0 \tag{3.35}$$

which is expected. It gives the relationship between the joint probability mass function and the conditional mass function. As we will see in Example 3.8, it is sometimes more convenient to derive joint mass functions using Equation 3.35 as conditional mass functions are more readily available.

If the r.v.'s X and Y are independent, then the definition of independence, Equation 2.16, implies

$$p_{XY}(x|y) = p_X(x) \tag{3.36}$$

and Equation 3.35 becomes

$$p_{XY}(x, y) = p_X(x) p_Y(y) \tag{3.37}$$

Thus, when, and only when, the r.v.'s X and Y are independent, their joint probability mass function is the product of the marginal mass functions.

Let X be a continuous random variable. A consistent definition of the *conditional density function* of X given $Y = y$, $f_{XY}(x|y)$, is the derivative of its corresponding conditional distribution function. Hence,

$$\boxed{f_{XY}(x|y) = \frac{dF_{XY}(x|y)}{dx}} \tag{3.38}$$

where $F_{XY}(x|y)$ is defined in Equation 3.33. To see what this definition leads to, let us consider

$$P(x_1 < X \leq x_2 | y_1 < Y \leq y_2) = \frac{P(x_1 < X \leq x_2 \cap y_1 < Y \leq y_2)}{P(y_1 < Y \leq y_2)} \tag{3.39}$$

In terms of jpdf $f_{XY}(x, y)$, it is given by

$$P(x_1 < X \le x_2 | y_1 < Y \le y_2) = \int_{y_1}^{y_2} \int_{x_1}^{x_2} f_{XY}(x, y)dx\, dy \Big/ \int_{y_1}^{y_2} \int_{-\infty}^{\infty} f_{XY}(x, y)dx\, dy$$

$$= \int_{y_1}^{y_2} \int_{x_1}^{x_2} f_{XY}(x, y)dx\, dy \Big/ \int_{y_1}^{y_2} f_Y(y)dy \quad (3.40)$$

By setting $x_1 = -\infty$, $x_2 = x$, $y_1 = y$, $y_2 = y + \Delta y$, and taking the limit $\Delta y \to 0$, Equation 3.40 reduces to

$$F_{XY}(x|y) = \frac{\int_{-\infty}^{x} f_{XY}(u, y)du}{f_Y(y)} \quad (3.41)$$

provided that $f_Y(y) > 0$.

Now we see that Equation 3.38 leads to

$$f_{XY}(x|y) = \frac{dF_{XY}(x|y)}{dx} = \frac{f_{XY}(x, y)}{f_Y(y)} \quad f_Y(y) \ne 0 \quad (3.42)$$

which is in a form identical to Equation 3.35 for the mass functions, a satisfying result. We should add here that this relationship between the conditional density function and the joint density function is obtained at the expense of the definition (3.33) for $F_{XY}(x|y)$. We say "at the expense of" because the definition given to $F_{XY}(x|y)$ does not lead to a convenient relationship between $F_{XY}(x|y)$ and $F_{XY}(x, y)$, that is,

$$F_{XY}(x|y) \ne \frac{F_{XY}(x, y)}{F_Y(y)} \quad (3.43)$$

This inconvenience, however, is not a severe penalty since we deal with density functions and mass functions more often.

When the r.v.'s X and Y are independent, $F_{XY}(x|y) = F_X(x)$ and, as seen from Equation 3.42,

$$f_{XY}(x|y) = f_X(x) \quad (3.44)$$

and

$$f_{XY}(x, y) = f_X(x)f_Y(y) \quad (3.45)$$

RANDOM VARIABLES AND PROBABILITY DISTRIBUTIONS

which shows again that the joint density function is equal to the product of the associated marginal density functions when X and Y are independent.

Finally, let us note that

$$F_{XY}(x|y) = \sum_{i=1}^{i:\, x_i \leq x} p_{XY}(x_i|y) \tag{3.46}$$

$$F_{XY}(x|y) = \int_{-\infty}^{x} f_{XY}(u|y)\,du \tag{3.47}$$

Comparing these equations with Equations 3.7 and 3.12, they are identical to those relating these functions for X alone.

Extensions of the above results to the case of more than two random variables are straightforward. Starting from (see Equation 2.26)

$$P(ABC) = P(A|BC)P(B|C)P(C)$$

for three events A, B, and C, we have, in the case of three r.v.'s X, Y, and Z,

$$\begin{aligned} p_{XYZ}(x, y, z) &= p_{XYZ}(x|y, z)p_{YZ}(y|z)p_Z(z) \\ f_{XYZ}(x, y, z) &= f_{XYZ}(x|y, z)f_{YZ}(y|z)f_Z(z) \end{aligned} \tag{3.48}$$

Hence, for the general case of n r.v.'s $X_1, X_2, \ldots, X_n$, or $\mathbf{X}$, we can write

$$\begin{aligned} p_{\mathbf{X}}(\mathbf{x}) &= p_{X_1 X_2 \cdots X_n}(x_1|x_2, \ldots, x_n)p_{X_2 \cdots X_n}(x_2|x_3, \ldots, x_n) \cdots \\ & \quad p_{X_{n-1} X_n}(x_{n-1}|x_n)p_{X_n}(x_n) \\ f_{\mathbf{X}}(\mathbf{x}) &= f_{X_1 X_2 \cdots X_n}(x_1|x_2, \ldots, x_n)f_{X_2 \cdots X_n}(x_2|x_3, \ldots, x_n) \cdots \\ & \quad f_{X_{n-1} X_n}(x_{n-1}|x_n)f_{X_n}(x_n) \end{aligned} \tag{3.49}$$

In the event that the random variables are mutually independent, Equations 3.49 become

$$\begin{aligned} p_{\mathbf{X}}(\mathbf{x}) &= p_{X_1}(x_1)p_{X_2}(x_2) \cdots p_{X_n}(x_n) \\ f_{\mathbf{X}}(\mathbf{x}) &= f_{X_1}(x_1)f_{X_2}(x_2) \cdots f_{X_n}(x_n) \end{aligned} \tag{3.50}$$

Example 3.8. To show that joint mass functions are sometimes more easily found by finding first the conditional mass functions, let us consider a traffic problem as described below.

A group of n cars enters an intersection from the south. Through prior observations, it is known that each car has the probability p of turning east, probability q of turning west, and probability r of going straight ($p + q + r = 1$).

Assume that drivers behave independently and let X be the number of cars turning east and Y the number turning west. Determine the jpmf $p_{XY}(x, y)$.

Since

$$p_{XY}(x, y) = p_{XY}(x|y)p_Y(y)$$

we proceed by determining $p_{XY}(x|y)$ and $p_Y(y)$. The marginal mass function $p_Y(y)$ is found in a way very similar to the random walk situation described in Example 3.4. Each car has two alternatives: turning west and not turning west. By enumeration, we can show that it has a binomial distribution (to be more fully justified in Chapter VI)

$$p_Y(y) = \binom{n}{y} q^y (1-q)^{n-y} \qquad y = 0, 1, 2, \ldots, n \qquad (3.51)$$

Consider now the conditional mass function $p_{XY}(x|y)$. With $Y = y$ having happened, the situation is again similar to that for determining $p_Y(y)$ except that the number of cars available for possible eastward turn is now $n - y$ and that the probabilities p and r need to be renormalized so that they sum to 1. Hence, $p_{XY}(x|y)$ takes the form

$$p_{XY}(x|y) = \binom{n-y}{x} \left(\frac{p}{r+p}\right)^x \left(1 - \frac{p}{r+p}\right)^{n-y-x} \qquad (3.52)$$

$$x = 0, 1, \ldots, n-y \qquad y = 0, 1, \ldots, n$$

Finally, we have $p_{XY}(x, y)$ as the product of the two expressions given by Equations 3.51 and 3.52. The ranges of values for x and y are $x = 0, 1, \ldots, n - y$ and $y = 0, 1, \ldots, n$.

We note that $p_{XY}(x, y)$ has a rather formidable expression that could not have come easily in a direct way. It also points out the need to exercise care in determining the limits of validity for x and y.

Example 3.9. Resistors are designed to have a resistance R of 50 ± 2 ohms. Due to inprecision in the manufacturing process, the actual density function of R has the form as shown in Figure 3.16 by the solid curve. Determine the density function of R after screening, that is, after all the resistors having resistances beyond the (48–52)-ohm range are rejected.

We are interested in the conditional density function $f_R(r|A)$ where A is the event $\{48 \leq R \leq 52\}$. This is not the usual definition of a conditional density function but it can be found from the basic definition of conditional probability.

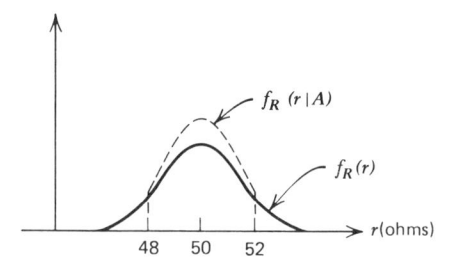

RANDOM VARIABLES AND PROBABILITY DISTRIBUTIONS

Figure 3.16 $f_R(r)$ and $f_R(r|A)$.

We start by considering

$$F_R(r|A) = P(R \leq r | 48 \leq R \leq 52) = \frac{P(R \leq r \cap 48 \leq R \leq 52)}{P(48 \leq R \leq 52)}$$

However,

$$\begin{aligned} R \leq r \cap 48 \leq R \leq 52 &= \emptyset & &\text{if } r < 48 \\ &= 48 \leq R \leq r & &\text{if } 48 \leq r \leq 52 \\ &= 48 \leq R \leq 52 & &\text{if } r > 52 \end{aligned}$$

Hence,

$$\begin{aligned} F_R(r|A) &= 0 & &r < 48 \\ &= \frac{P(48 \leq R \leq r)}{P(48 \leq R \leq 52)} = \frac{\int_{48}^{r} f_R(r) dr}{c} & &48 \leq r \leq 52 \\ &= 1 & &r > 52 \end{aligned}$$

where $c = \int_{48}^{52} f_R(r) dr$, a constant.

The desired $f_R(r|A)$ is then obtained from the above by differentiation. We obtain

$$\begin{aligned} f_R(r|A) = \frac{dF_R(r|A)}{dr} &= \frac{f_R(r)}{c} & &48 \leq r \leq 52 \\ &= 0 & &\text{elsewhere} \end{aligned}$$

It is seen that the effect of screening is essentially a truncation of the tails of the distribution beyond the allowable limits. This is accompanied by an adjustment within the limits by a multiplicative factor $1/c$ so that the area under the curve is again equal to 1. This conditional density function is indicated by the dashed curve in Figure 3.16.

REFERENCES AND COMMENTS

We have discussed in Section III.3 the determination of (unique) marginal distributions from the knowledge of joint distributions. It is noted here that the knowledge of marginal distributions does not in general lead to a unique joint distribution. The following reference shows that all joint distributions having a specified set of marginals can be obtained by repeated applications of the "so-called" θ-transformation to the product of the marginals.

1. P. W. Becker, "A Note on Joint Densities which have the Same Set of Marginal Densities," *Proc. International Symp. Information Theory*, The Netherlands, 1970.

PROBLEMS

3.1. For each of the functions given below, determine the constant a so that it possesses all the properties of a PDF. Determine, in each case, its associated pdf or pmf if it exists and sketch all functions.

(a) $F(x) = 0 \quad x < 5$
$ = a \quad x \geq 5$

(b) $F(x) = 0 \quad x < 5$
$ = \tfrac{1}{3} \quad 5 \leq x < 7$
$ = a \quad x \geq 7$

(c) $F(x) = 0 \quad x < 1$
$ = \sum_{j=1}^{k} 1/a^j \quad k \leq x < k+1 \quad k = 1, 2, 3, \ldots$

(d) $F(x) = 0 \quad x \leq 0$
$ = 1 - e^{-ax} \quad x > 0$

(e) $F(x) = 0 \quad x < 0$
$ = x^a \quad 0 \leq x \leq 1$
$ = 1 \quad x > 1$

(f) $F(x) = 0$ $x < 0$
       $= a \sin^{-1} \sqrt{x}$ $0 \le x \le 1$
       $= 1$ $x > 1$

(g) $F(x) = 0$ $x < 0$
       $= a(1 - e^{-x/2}) + \frac{1}{2}$ $x \ge 0$

3.2. For each part of Problem 3.1, determine
  (a) $P(X \le 6)$
  (b) $P(\frac{1}{2} < x \le 7)$

3.3. For each of the following, sketch roughly in scale the corresponding PDF $F_X(x)$ and show on all graphs the procedure for finding $P(2 < X < 4)$.

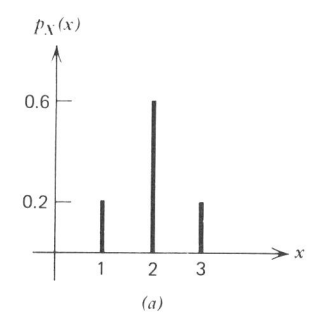

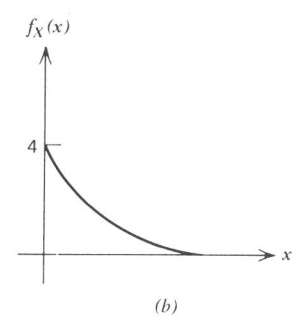

(a) Plot of $p_X(x)$: impulses at $x=1$ (height 0.2), $x=2$ (height 0.6), $x=3$ (height 0.2).

(b) Plot of $f_X(x)$: decreasing exponential-like curve starting at 4.

3.4. For each part, find the corresponding PDF for the r.v. X
  (a) $f_X(x) = 0.1$ $90 \le x < 100$
             $= 0$ elsewhere
  (b) $f_X(x) = 2(1 - x)$ $0 \le x < 1$
             $= 0$ elsewhere
  (c) $f_X(x) = \dfrac{1}{\pi(1 + x^2)}$ $-\infty < x < \infty$

3.5. The life X, in hours, of a certain kind of electronic component has a pdf given by

$$f_X(x) = 0 \qquad x < 100$$
$$= \frac{100}{x^2} \qquad x \ge 100$$

Determine the probability that a component will survive 150 hours of operation?

66 PROBABILITY AND RANDOM VARIABLES

3.6. Time, in minutes, required for a student to travel from home to a morning class is uniformly distributed between 20 and 25. If the student leaves home promptly at 7:38 A.M., what is the probability that the student will not be late for class at 8:00 A.M.?

3.7. In constructing a bridge as shown, an engineer is concerned with forces acting on the end supports caused by a randomly applied concentrated load P, the term "randomly applied" meaning that the probability of the load lying in any region is proportional only to the length of that region. Suppose that the bridge has a span $2b$. Determine the PDF and pdf of the r.v. X, which is the distance from the load to the nearest edge support. Sketch these functions.

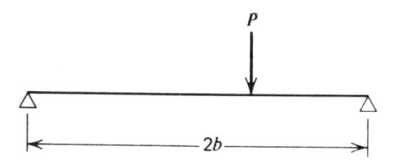

3.8. Fire can erupt at random at any point along a stretch of forest AB. The fire station is located as shown in the figure. Determine the PDF and pdf of X, representing the distance between the fire and the fire station. Sketch these functions.

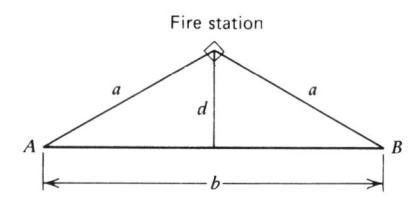

3.9. Pollutant concentrations caused by a pollution source can be modeled by the pdf ($a > 0$)

$$f_R(r) = 0 \qquad r < 0$$
$$ = ae^{-ar} \qquad r \geq 0$$

where R is the distance from the source. Determine the radius within which 95% of the pollutant is contained.

3.10. Since it is more economical to limit long distance telephone calls within three minutes, the PDF of X, duration in minutes of long distance calls, may be of the form

$$F_X(x) = 0 \qquad x < 0$$
$$= 1 - e^{-x/3} \qquad 0 \le x < 3$$
$$= 1 - \frac{e^{-x/3}}{2} \qquad x \ge 3$$

Determine the probability that X is (a) more than two minutes and (b) between two and six minutes.

3.11. As another example of a mixed probability distribution, consider the following problem. A particle is at rest at the origin ($x = 0$) at time $t = 0$.

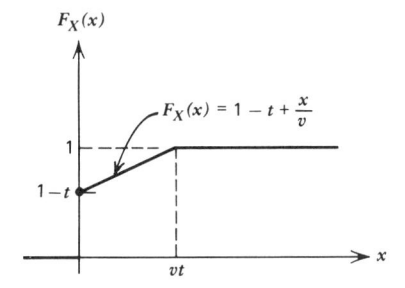

At a randomly selected time uniformly distributed over the interval $0 < t < 1$, the particle is suddenly given a velocity v in the positive x-direction.

(a) Show that X, the particle position at t ($0 < t < 1$), has the PDF as shown in the figure.
(b) Calculate the probability that the particle is at least $v/3$ away from the origin at $t = \frac{1}{2}$.

3.12. For each of the joint probability mass functions or joint probability density functions given below, determine:
(a) The marginal mass or density functions.
(b) Whether the random variables are independent.

(i) $\quad P_{XY}(x, y) = 0.5 \qquad (x, y) = (1, 1)$
$\qquad\qquad\qquad = 0.1 \qquad (x, y) = (1, 2)$
$\qquad\qquad\qquad = 0.1 \qquad (x, y) = (2, 1)$
$\qquad\qquad\qquad = 0.3 \qquad (x, y) = (2, 2)$

(ii) $f_{XY}(x, y) = a(x + y)$ $0 < x \leq 1$ $1 < y \leq 2$
       $= 0$ elsewhere

(iii) $f_{XY}(x, y) = e^{-(x+y)}$ $(x, y) > (0, 0)$
       $= 0$ elsewhere

(iv) $f_{XY}(x, y) = 4y(x - y)e^{-(x+y)}$ $0 < x < \infty$ $0 < y \leq x$
       $= 0$ elsewhere

3.13. Let X_1, X_2, and X_3 be independent random variables, each taking values ± 1 with probability $\frac{1}{2}$. Define r.v.'s Y_1, Y_2, and Y_3 by

$$Y_1 = X_1 X_2 \quad\quad Y_2 = X_1 X_3 \quad\quad Y_3 = X_2 X_3$$

Show that any two of these new random variables are independent but that Y_1, Y_2, and Y_3 are not independent.

3.14. The r.v.'s X and Y are distributed according to the jpdf given by Problem 3.12(ii). Determine
   (a) $P(X \leq 0.5 \cap Y > 1.0)$
   (b) $P(XY < \frac{1}{2})$
   (c) $P(X \leq 0.5 | Y = 1.5)$
   (d) $P(X \leq 0.5 | Y \leq 1.5)$

3.15. Let the r.v. X denote the time of failure in years of a system whose PDF is $F_X(x)$. In terms of $F_X(x)$, determine the probability

$$P(X \leq x | X \geq 100)$$

which is the conditional distribution function of X assuming that the system did not fail up to 100 years.

3.16. The pdf of r.v. X is

$$f_X(x) = 3x^2 \quad\quad -1 < x \leq 0$$
$$= 0 \quad\quad \text{elsewhere}$$

Determine $P(X > b | X < b/2)$ with $-1 < b < 0$.

3.17. Using the joint probability distribution given in Example 3.4 for r.v.'s X and Y, determine:
   (a) $P(X > 3)$
   (b) $P(0 \leq Y < 3)$
   (c) $P(X > 3 | Y \leq 2)$

RANDOM VARIABLES AND PROBABILITY DISTRIBUTIONS 69

3.18. A commuter is accustomed to leaving home between 7:30 and 8:00 A.M., the drive to the station taking between 20 and 30 minutes. It is assumed that departure time and time for trip are independent r.v.'s, uniformly distributed over their respective intervals. There are two trains the commuter can take; the first leaves at 8:05 A.M. and takes 30 minutes for the trip and the second leaves at 8:25 A.M. and takes 35 minutes. What is the probability that the commuter misses both trains?

3.19. The distance X (in miles) from a nuclear plant to the epicenter of potential earthquakes within 50 miles is distributed according to

$$f_X(x) = \frac{2x}{2500} \quad 0 \leq x \leq 50$$
$$= 0 \quad \text{elsewhere}$$

and the magnitude Y of potential earthquakes of scales 5 to 9 is distributed according to

$$f_Y(y) = \frac{3(9-y)^2}{64} \quad 5 \leq y \leq 9$$
$$= 0 \quad \text{elsewhere}$$

Assume that X and Y are independent. Determine $P(X \leq 25 \cap Y > 8)$, the probability that the next earthquake within 50 miles will have a magnitude greater than 8 and that its epicenter will lie within 25 miles of the nuclear plant.

3.20. Let r.v.'s X and Y be independent and uniformly distributed in the interval $(0, 0) < (X, Y) < (1, 1)$. Determine the probability that $XY < \frac{1}{2}$.

3.21. In splashdown maneuvers, spacecrafts often miss the target due to guidance inaccuracies, atmospheric disturbances, and other error sources. Taking the origin of coordinates as the designed point of impact, the X and Y coordinates of the actual impact point are random with marginal density functions

$$f_X(x) = \frac{1}{\sigma\sqrt{2\pi}} e^{-x^2/2\sigma^2} \quad -\infty < x < \infty$$

$$f_Y(y) = \frac{1}{\sigma\sqrt{2\pi}} e^{-y^2/2\sigma^2} \quad -\infty < y < \infty$$

Assume that the r.v.'s are independent. Show that the probability of a splashdown lying within a circle of radius a centered at the origin is $1 - e^{-a^2/2\sigma^2}$.

3.22. Let $X_1, X_2, \ldots, X_n$ be independent and identically distributed r.v.'s each with PDF $F_X(x)$. Show that:

$$P[\min(X_1, X_2, \ldots, X_n) \leq u] = 1 - [1 - F_X(u)]^n$$
$$P[\max(X_1, X_2, \ldots, X_n) \leq u] = [F_X(u)]^n$$

The above are examples of extreme value distributions. They are of considerable practical importance and will be discussed in Chapter VII.

3.23. In studies of social mobility, assume that social classes can be ordered from one (professional) to seven (unskilled). Let the r.v. X_k denote the class order of the kth generation. Then, for a given region, the following information is given:
(a) The pmf of X_0 is described by $p_{X_0}(1) = 0.00$, $p_{X_0}(2) = 0.00$, $p_{X_0}(3) = 0.04$, $p_{X_0}(4) = 0.06$, $p_{X_0}(5) = 0.11$, $p_{X_0}(6) = 0.28$, and $p_{X_0}(7) = 0.51$.
(b) The conditional mass functions $P(X_{k+1} = i | X_k = j)$ for $i, j = 1, 2, \ldots, 7$ and for every k are given in the following table:

Table of $P(X_{k+1} = i | X_k = j)$

i \ j	1	2	3	4	5	6	7
1	0.388	0.107	0.035	0.021	0.009	0.000	0.000
2	0.146	0.267	0.101	0.039	0.024	0.013	0.008
3	0.202	0.227	0.188	0.112	0.075	0.041	0.036
4	0.062	0.120	0.191	0.212	0.123	0.088	0.083
5	0.140	0.206	0.357	0.430	0.473	0.391	0.364
6	0.047	0.053	0.067	0.124	0.171	0.312	0.235
7	0.015	0.020	0.061	0.062	0.125	0.155	0.274

(c) The outcome at the $(k + 1)$th generation is only dependent on the class order at the kth generation and not on any generation prior to it, that is,

$$P(X_{k+1} = i | X_k = j \cap X_{k-1} = k \cap \cdots) = P(X_{k+1} = i | X_k = j)$$

Determine (a) the pmf of X_3 and (b) the jpmf of X_3 and X_4.

IV
Expectations and Moments

While a probability distribution ($F_X(x)$, $p_X(x)$, or $f_X(x)$) contains a complete description of a r.v. X, it is often of interest to seek a set of simple numbers that gives the random variable some of its dominant features. These numbers include moments of various orders associated with X. Let us first give a general definition.

Definition. Let $g(X)$ be a real-valued function of a r.v. X. The *mathematical expectation*, or simply *expectation*, of $g(X)$, denoted by $E\{g(X)\}$, is defined by

$$\boxed{E\{g(X)\} = \sum_i g(x_i) p_X(x_i)} \qquad (4.1)$$

if X is discrete. In the above, $x_1, x_2, \ldots$ are possible values assumed by X. When the range of i extends from 1 to infinity, the sum in Equation 4.1 exists if it converges absolutely, that is,

$$\sum_{i=1}^{\infty} |g(x_i)| p_X(x_i) < \infty$$

The symbol $E\{\ \}$ is regarded here and in the sequel as *expectation operator*.

If r.v. X is continuous, the expectation $E\{g(X)\}$ is defined by

$$E\{g(X)\} = \int_{-\infty}^{\infty} g(x) f_X(x) dx \qquad (4.2)$$

if the improper integral is absolutely convergent, that is,

$$\int_{-\infty}^{\infty} |g(x)| f_X(x) dx < \infty$$

Let us note some basic properties associated with the expectation operator. For any constant c and any functions $g(X)$ and $h(X)$ whose expectations exist, we have

$$\begin{aligned}
&E\{c\} = c \\
&E\{cg(X)\} = cE\{g(X)\} \\
&E\{g(X) + h(X)\} = E\{g(X)\} + E\{h(X)\} \\
&E\{g(X)\} \leq E\{h(X)\} \qquad \text{if } g(X) \leq h(X) \text{ for all values of } X
\end{aligned} \qquad (4.3)$$

These relations follow directly from the definition of $E\{g(X)\}$. For example,

$$\begin{aligned}
E\{g(X) + h(X)\} &= \int_{-\infty}^{\infty} [g(x) + h(x)] f_X(x) dx \\
&= \int_{-\infty}^{\infty} g(x) f_X(x) dx + \int_{-\infty}^{\infty} h(x) f_X(x) dx \\
&= E\{g(X)\} + E\{h(X)\}
\end{aligned}$$

The proof is similar when X is discrete.

IV.1 Moments of a Single Random Variable

Let $g(X) = X^n$, $n = 1, 2, \ldots$, the expectation $E\{X^n\}$, when exists, is called the *nth moment* of X. It is denoted by α_n and is given by

$$E\{X^n\} = \sum_i x_i^n p_X(x_i) \qquad (X \text{ discrete}) \qquad (4.4)$$

$$E\{X^n\} = \int_{-\infty}^{\infty} x^n f_X(x) dx \qquad (X \text{ continuous}) \qquad (4.5)$$

IV.1.1 MEAN, MEDIAN, AND MODE

One of the most important moments is α_1, the first moment. Using the mass analogy for the probability distribution, the first moment may be regarded as the center of mass of its distribution. It is thus the average value of r.v. X and certainly reveals one of the most important characteristics of its distribution. The first moment of X is synonymously called *mean, expectation,* or *average value* of X. A common notation for it is m_X or simply m.

Example 4.1. In Example 3.8, determine the average number of cars turning west in a group of n cars.

We wish to determine the mean of Y, $E\{Y\}$, whose mass function is

$$p_Y(k) = \binom{n}{k} q^k (1-q)^{n-k} \quad k = 0, 1, 2, \ldots, n$$

Equation 4.4 then gives

$$E\{Y\} = \sum_{k=0}^{n} k p_Y(k) = \sum_{k=0}^{n} k \binom{n}{k} q^k (1-q)^{n-k}$$

$$= \sum_{k=1}^{n} \frac{n!}{(k-1)!(n-k)!} q^k (1-q)^{n-k}$$

Let $k - 1 = m$. We have

$$E\{Y\} = nq \sum_{m=0}^{n-1} \binom{n-1}{m} q^m (1-q)^{n-1-m}$$

The sum in this expression is simply the sum of binomial probabilities and hence equals one. Therefore,

$$E\{Y\} = nq$$

which is a number since n and q are known constants.

Example 4.2. The waiting time X (in minutes) of a customer waiting to be served at a ticket counter has the density function

$$f_X(x) = 2e^{-2x} \quad x \geq 0$$
$$= 0 \quad \text{elsewhere}$$

Determine the average waiting time.

Referring to Equation 4.5 we have, using integration by parts,

$$E\{X\} = \int_0^\infty x(2e^{-2x})dx = \tfrac{1}{2} \text{ minute}$$

Example 4.3. In Example 3.9, find the average resistance of the resistors after screening.

The average value required in this example is a *conditional mean* of R given the event A. While there is no formal definition given, it should be clear that the desired average is obtained from

$$E\{R|A\} = \int_{48}^{52} r f_R(r|A) dr = \int_{48}^{52} \frac{r f_R(r)}{c} dr$$

This integral can be evaluated when $f_R(r)$ is specified.

Two other quantities in common usage that also give a measure of the center of a probability distribution are the *median* and the *mode*.

A *median* of X is any point that divides the mass of the distribution into two equal parts; that is, x_0 is a median of X if

$$P(X \le x_0) = \tfrac{1}{2}$$

While the mean of X may not exist, there exists at least one median.

In comparison with the mean, the median is sometimes preferred as a measure of central tendency when a distribution is skewed, particularly where there are a small number of extreme values in the distribution. For example, we speak of *median income* as a good central measure of personal income for a population. This is a better average because the median is not as sensitive to a small number of extremely high incomes or extremely low incomes as is the mean.

Example 4.4. Let T be the time between emissions of particles by a radioactive atom. It is well established that T is a random variable and it obeys an exponential distribution, that is,

$$\begin{aligned} f_T(t) &= \lambda e^{-\lambda t} & t \ge 0 \\ &= 0 & \text{elsewhere} \end{aligned}$$

where λ is a constant. The r.v. T is called the lifetime of the atom and a common

EXPECTATIONS AND MOMENTS

average measure of this lifetime is called half-life, which is defined as the median of T. Thus, if τ is half-life, it is found from

$$\int_0^\tau f_T(t) = \tfrac{1}{2}$$

or

$$\tau = \ln\frac{2}{\lambda}$$

Let us note that the mean life, $E\{T\}$, is

$$E\{T\} = \int_0^\infty t f_T(t)dt = \frac{1}{\lambda}$$

A point x_i such that

$$p_X(x_i) > p_X(x_{i+1}) \quad \text{and} \quad p_X(x_i) > p_X(x_{i-1}) \quad (X \text{ discrete})$$

$$f_X(x_i) > f_X(x_i + \varepsilon) \quad \text{and} \quad f_X(x_i) > f_X(x_i - \varepsilon) \quad (X \text{ continuous})$$

where ε is an arbitrarily small positive quantity, is called a *mode* of X. A mode is thus a value of X corresponding to a peak in its mass function or density function. The term *unimodal distribution* refers to a probability distribution possessing a unique mode.

To give a comparison of these three measures of central tendency of a distribution, Figure 4.1 shows their relative positions in three different situations. It is clear that the mean, the median, and the mode coincide when a unimodal distribution is symmetric.

IV.1.2 CENTRAL MOMENTS, VARIANCE, AND STANDARD DEVIATION

Besides the mean, the next most important moment is the *variance*, which measures the dispersion or spread of a r.v. X about its mean. Its definition will follow a general definition of central moments.

Definition. The *central moments* of a r.v. X are the moments of X with respect to its mean. Hence, the nth central moment of X, μ_n, is defined as

$$\mu_n = E\{(X - m)^n\} = \sum_i (x_i - m)^n p_X(x_i) \quad (X \text{ discrete}) \quad (4.6)$$

$$\mu_n = E\{(X - m)^n\} = \int_{-\infty}^\infty (x - m)^n f_X(x)dx \quad (X \text{ continuous}) \quad (4.7)$$

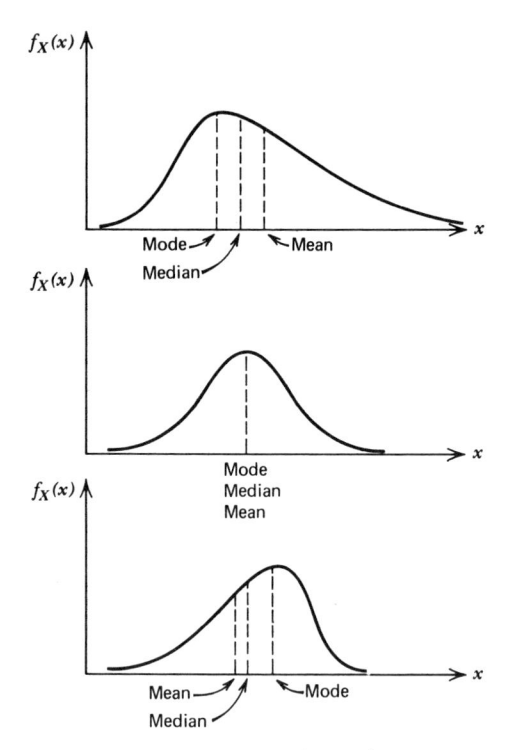

Figure 4.1 Relative positions of mean, median, and mode.

The *variance* of X is the second central moment μ_2, commonly denoted by σ_X^2 or simply σ^2 or var(X). It is the most common measure of dispersion of a distribution about its mean. Large values of σ_X^2 imply a large spread in the distribution of X about its mean. Conversely, small values imply a sharp concentration of the mass of distribution in the neighborhood of the mean. This is illustrated in Figure 4.2 in which two density functions are shown with the same mean but different variances. When $\sigma_X^2 = 0$, the whole mass of the distribution is concentrated at the mean. In this extreme case, $X = m_X$ with probability 1.

An important relation between the variance and simple moments is

$$\sigma^2 = \alpha_2 - m^2 \tag{4.8}$$

This can be shown by making use of Equations 4.3. We get

$$\sigma^2 = E\{(X - m)^2\} = E\{X^2 - 2mX + m^2\} = E\{X^2\} - 2mE\{X\} + m^2$$
$$= \alpha_2 - 2m^2 + m^2 = \alpha_2 - m^2$$

EXPECTATIONS AND MOMENTS

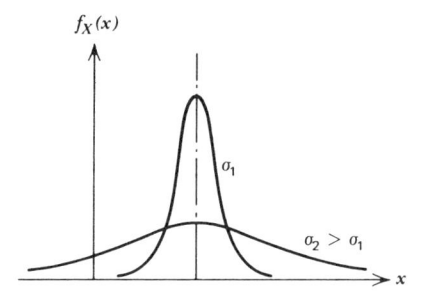

Figure 4.2 Density functions with different variances.

We note two other properties of the variance of a r.v. X which can be similarly verified. They are

$$\text{var}(X + c) = \text{var}(X)$$
$$\text{var}(cX) = c^2 \text{var}(X) \qquad (4.9)$$

where c is any constant.

It is further noted from Equations 4.6 and 4.7 that, since each term in the sum in Equation 4.6 and the integrand in Equation 4.7 are nonnegative, the variance of a random variable is always nonnegative. The positive square root

$$\sigma_X = +\sqrt{E\{(X - m)^2\}}$$

is called the *standard deviation* of X. An advantage of using σ_X rather than σ_X^2 as a measure of dispersion is that it has the same unit as the mean. It can therefore be compared with the mean on the same scale to gain some feeling for the degree of spread of the distribution. A dimensionless number that characterizes dispersion relative to the mean and also facilitates comparison among random variables of different units is the *coefficient of variation* v_X, defined by

$$\boxed{v_X = \frac{\sigma_X}{m_X}} \qquad (4.10)$$

Example 4.5. Let us determine the variance of Y defined in Example 4.1.

$$\sigma_Y^2 = E\{Y^2\} - m_Y^2 = E\{Y^2\} - n^2 q^2$$

Now,

$$E\{Y^2\} = \sum_{k=0}^{n} k^2 p_Y(k) = \sum_{k=0}^{n} k(k-1) p_Y(k) + \sum_{k=0}^{n} k p_Y(k)$$

and

$$\sum_{k=0}^{n} k p_Y(k) = nq$$

Proceeding as in Example 4.1,

$$\sum_{k=0}^{n} k(k-1) p_Y(k) = n(n-1)q^2 \sum_{k=2}^{n} \frac{(n-2)!}{(k-2)!(n-k)!} q^{k-2}(1-q)^{n-k}$$

$$= n(n-1)q^2 \sum_{j=0}^{m} \binom{m}{j} q^j (1-q)^{m-j}$$

$$= n(n-1)q^2$$

Thus,

$$E\{Y^2\} = n(n-1)q^2 + nq$$

and

$$\sigma_Y^2 = n(n-1)q^2 + nq - (nq)^2 = nq(1-q)$$

Example 4.6. We again use Equation 4.8 to determine the variance of X defined in Example 4.2. The second moment of X is, on integrating by parts,

$$E\{X^2\} = 2 \int_0^\infty x^2 e^{-2x} \, dx = \tfrac{1}{2}$$

Hence,

$$\sigma_X^2 = E\{X^2\} - m_X^2 = \tfrac{1}{2} - \tfrac{1}{4} = \tfrac{1}{4}$$

Example 4.7. Due to inherent manufacturing and scaling inaccuracies, the tape measures manufactured by a certain company have a standard deviation of 0.03 feet for 3-foot-long tape measures. What is a reasonable estimate of the standard deviation associated with 3-yard tape measures?

For this problem, it is reasonable to expect that errors introduced in the making of 3-foot-long tape measures again are accountable for inaccuracies in the 3-yard tape measures. It is thus reasonable to assume that the coefficient of variation $v = \sigma/m$ is constant for tape measures of all lengths manufactured by this company. Thus

$$v = \frac{0.03}{3} = 0.01$$

and the standard deviation for 3-yard tape measures is 0.01 (9 feet) = 0.09 feet.

This example illustrates the fact that the coefficient of variation is often used as a measure of quality for products of different sizes or different weights. In concrete industry, for example, the quality in terms of concrete strength is specified by a coefficient of variation, which is a constant for all mean strengths.

Central moments of higher orders reveal additional features of a distribution. The *coefficient of skewness*, defined by

$$\gamma_1 = \frac{\mu_3}{\sigma^3} \tag{4.11}$$

gives a measure of symmetry of a distribution. It is usually positive when a unimodal distribution has a dominant tail on the right. The opposite arrangement usually produces a negative γ_1. It is zero when a distribution is symmetrical about the mean. In fact, a symmetrical distribution about the mean implies that all odd-order central moments vanish.

The degree of flattening of a distribution near its peaks can be measured by the *coefficient of excess*

$$\gamma_2 = \frac{\mu_4}{\sigma^4} - 3 \tag{4.12}$$

A positive γ_2 usually implies a slim sharp peak in the neighborhood of a mode in a unimodal distribution, while a negative γ_2 implies, as a rule, a flattened peak.

IV.1.3 CONDITIONAL EXPECTATION

We conclude this section by introducing a useful relation involving conditional expectation. Let us denote by $E\{X|Y\}$ that function of the r.v. Y whose value at $Y = y_i$ is $E\{X|Y = y_i\}$. Hence, $E\{X|Y\}$ is itself a random variable and one of its very useful properties is that

$$\boxed{E\{X\} = E\{E\{X|Y\}\}} \tag{4.13}$$

If Y is a discrete random variable taking on values $y_1, y_2, \ldots$, the above states that

$$E\{X\} = \sum_i E\{X|Y = y_i\}P(Y = y_i) \tag{4.14}$$

and

$$E\{X\} = \int_{-\infty}^{\infty} E\{X|y\} f_Y(y) dy \tag{4.15}$$

if Y is continuous.

To establish the relation given by Equation 4.13, let us show that Equation 4.14 is true when both X and Y are discrete. Starting from the right-hand side of Equation 4.14, we have

$$\sum_i E\{X|Y = y_i\}P(Y = y_i) = \sum_i \sum_j x_j P(X = x_j|Y = y_i)P(Y = y_i)$$

Since

$$P(X = x_j|Y = y_i) = \frac{P(X = x_j \cap Y = y_i)}{P(Y = y_i)}$$

we have

$$\sum_i E\{X|Y = y_i\}P(Y = y_i) = \sum_i \sum_j x_j p_{XY}(x_j, y_i)$$

$$= \sum_j x_j \sum_i p_{XY}(x_j, y_i)$$

$$= \sum_j x_j p_X(x_j)$$

$$= E\{X\}$$

and the result is obtained.

The usefulness of Equation 4.13 is analogous to what we found in using the theorem of total probability discussed in Chapter II. It states that, in order to determine $E\{X\}$, it can be found by taking a weighted average of the conditional expectation of X given $Y = y_i$; each of these terms is weighted by the probability $P(Y = y_i)$.

Example 4.8. The survival of a motorist stranded in a snowstorm depends on which of the three directions the motorist chooses to walk. The first road leads to safety after one hour of travel, the second leads to safety after three hours of travel, but the third will circle back to the original spot after two hours. Determine the average time to safety if the motorist is equally likely to choose any one of the roads.

Let $Y = 1, 2, 3$ be the events that the motorist chooses the first, second, and third road, respectively. Then $P(Y = i) = \frac{1}{3}$ for $i = 1, 2, 3$. Let X be the time to safety in hours. We have

$$E\{X\} = \sum_{i=1}^{3} E\{X \mid Y = i\} P(Y = i)$$

$$= \frac{1}{3} \sum_{i=1}^{3} E\{X \mid Y = i\}$$

Now,

$$\begin{aligned} E\{X \mid Y = 1\} &= 1 \\ E\{X \mid Y = 2\} &= 3 \\ E\{X \mid Y = 3\} &= 2 + E\{X\} \end{aligned} \quad (4.16)$$

Hence

$$E\{X\} = \tfrac{1}{3}(1 + 3 + 2 + E\{X\})$$

or

$$E\{X\} = 3 \text{ hours}$$

Let us remark that the third relation in Equations 4.16 is obtained by noting that, if the motorist chooses the third road, then it takes two hours to find that he or she is back to the starting point and the problem is as before. Hence, the motorist's expected additional time to safety is just $E\{X\}$. The result is thus $2 + E\{X\}$. We further remark that problems of this type would require much more work using other approaches.

IV.2 Chebyshev Inequality

In the discussion of expectations and moments, there are two aspects to be considered in applications. The first is that of calculating moments of various orders of a random variable knowing its distribution, and the second is concerned with making statements about the behavior of a random variable when only some of its moments are available. The latter arises in numerous number of practical situations in which available information only leads to estimates of some simple moments of a random variable.

The knowledge of mean and variance of a random variable, while very useful, is not sufficient to determine its distribution, and therefore does not permit us to give answers to such questions as "what is $P(X \leq 5)$?" However, as is shown as follows, it is possible to establish some probability bounds knowing only the mean and the variance.

The *Chebyshev inequality* states that

$$P(|X - m_X| \geq k\sigma_X) \leq \frac{1}{k^2} \qquad (4.17)$$

for any $k > 0$.

Proof.

From the definition we have

$$\sigma_X^2 = \int_{-\infty}^{\infty} (x - m_X)^2 f_X(x)dx \geq \int_{|x-m_X| \geq k\sigma_X} (x - m_X)^2 f_X(x)dx$$

$$\geq k^2 \sigma_X^2 \int_{|x-m_X| \geq k\sigma_X} f_X(x)dx = k^2 \sigma_X^2 P(|X - m_X| \geq k\sigma_X)$$

and Equation 4.17 follows. The proof is similar when X is discrete.

Example 4.9. In Example 4.7 and for 3-foot tape measures, we can write

$$P(|X - 3| \geq 0.03k) \leq \frac{1}{k^2}$$

If $k = 2$,

$$P(|X - 3| \geq 0.06) \leq \tfrac{1}{4}$$

or

$$P(2.94 \leq X \leq 3.06) \geq \tfrac{3}{4}$$

In words, the probability of a 3-foot tape measure being in error less than or equal to ± 0.06 feet is at least 0.75. Various probability bounds can be found by assigning different values to k.

EXPECTATIONS AND MOMENTS

The complete generality with which the Chebyshev inequality is derived suggests that the bounds given by Equation 4.17 can be quite conservative. This is indeed true. Sharper bounds can be achieved if more is known about the distribution.

IV.3 Moments of two or More Random Variables

Let $g(X, Y)$ be a real-valued function of two r.v.'s X and Y. Its expectation is defined by

$$E\{g(X, Y)\} = \sum_i \sum_j g(x_i, y_j) p_{XY}(x_i, y_j) \qquad (X \text{ and } Y \text{ discrete}) \qquad (4.18)$$

$$E\{g(X, Y)\} = \int_{-\infty}^{\infty} \int_{-\infty}^{\infty} g(x, y) f_{XY}(x, y) dx\, dy \qquad (X \text{ and } Y \text{ continuous}) \qquad (4.19)$$

if the indicated sums or integrals exist.

In a completely analogous way, the *joint moments* α_{nm} of X and Y are given by, if they exist,

$$\alpha_{nm} = E\{X^n Y^m\} \qquad (4.20)$$

They are computed from Equation 4.18 or 4.19 by letting $g(X, Y) = X^n Y^m$.

Similarly, the *joint central moments* of X and Y, when they exist, are

$$\mu_{nm} = E\{(X - m_X)^n (Y - m_Y)^m\} \qquad (4.21)$$

They are computed from Equation 4.18 or 4.19 by letting $g(X, Y) = (X - m_X)^n \times (Y - m_Y)^m$.

Some of the most important moments in the two-r.v. case are clearly the individual means and variances of X and Y. In the notation used here, the means of X and Y are, respectively, α_{10} and α_{01}. Using Equation 4.19, for example,

$$\alpha_{10} = E\{X\} = \int_{-\infty}^{\infty} \int_{-\infty}^{\infty} x f_{XY}(x, y) dx\, dy = \int_{-\infty}^{\infty} x \int_{-\infty}^{\infty} f_{XY}(x, y) dy\, dx$$

$$= \int_{-\infty}^{\infty} x f_X(x) dx$$

where $f_X(x)$ is the marginal density function of X. We thus see that the result is identical to that in the single r.v. case.

This observation is, of course, also true for the individual variances. They are, respectively, μ_{20} and μ_{02}, and can be found from Equation 4.21 with appropriate substitutions for n and m. As in the single r.v. case, we also have

$$\begin{aligned} \mu_{20} &= \alpha_{20} - \alpha_{10}^2 \\ \mu_{02} &= \alpha_{02} - \alpha_{01}^2 \end{aligned} \quad \text{or} \quad \begin{aligned} \sigma_X^2 &= \alpha_{20} - m_X^2 \\ \sigma_Y^2 &= \alpha_{02} - m_Y^2 \end{aligned} \quad (4.22)$$

IV.3.1 COVARIANCE AND CORRELATION COEFFICIENT

The first and simplest joint moment of X and Y that gives some measure of their interdependence is $\mu_{11} = E\{(X - m_X)(Y - m_Y)\}$. It is called the *covariance* of X and Y. Let us first note some of its properties.

1. It is related to α_{nm} by

$$\mu_{11} = \alpha_{11} - \alpha_{10}\alpha_{01} = \alpha_{11} - m_X m_Y \quad (4.23)$$

This is obtained by expanding $(X - m_X)(Y - m_Y)$ and then taking expectation of each term. We have

$$\begin{aligned} \mu_{11} &= E\{(X - m_X)(Y - m_Y)\} = E\{XY - m_Y X - m_X Y + m_X m_Y\} \\ &= E\{XY\} - m_Y E\{X\} - m_X E\{Y\} + m_X m_Y \\ &= \alpha_{11} - \alpha_{10}\alpha_{01} - \alpha_{10}\alpha_{01} + \alpha_{10}\alpha_{01} \\ &= \alpha_{11} - \alpha_{10}\alpha_{01} \end{aligned}$$

2. Let the *correlation coefficient* of X and Y be defined by

$$\boxed{\rho = \frac{\mu_{11}}{\sqrt{\mu_{20}\mu_{02}}} = \frac{\mu_{11}}{\sigma_X \sigma_Y}} \quad (4.24)$$

Then $|\rho| \leq 1$. To show this, let t and u be any real quantities and form

$$\begin{aligned} \phi(t, u) &= E\{[t(X - m_X) + u(Y - m_Y)]^2\} \\ &= \mu_{20} t^2 + 2\mu_{11} tu + \mu_{02} u^2 \end{aligned}$$

Since the expectation of a nonnegative function of X and Y must be nonnegative, $\phi(t, u)$ is a nonnegative quadratic form in t and u and we must have

$$\mu_{20}\mu_{02} - \mu_{11}^2 \geq 0 \quad (4.25)$$

which gives the desired result.

EXPECTATIONS AND MOMENTS

The normalization of the covariance through Equation 4.24 renders ρ a useful substitute for μ_{11}. Furthermore, the correlation coefficient is dimensionless and independent of the origin, that is, for any constants a_1, a_2, b_1, and b_2 with $a_1 > 0$ and $a_2 > 0$, we can easily verify that

$$\rho(a_1 X + b_1, a_2 Y + b_2) = \rho(X, Y) \tag{4.26}$$

3. If X and Y are independent, then

$$\mu_{11} = 0 \quad \text{and} \quad \rho = 0 \tag{4.27}$$

Proof.

Let X and Y be continuous, their joint moment α_{11} is found from

$$\alpha_{11} = E\{XY\} = \int_{-\infty}^{\infty} \int_{-\infty}^{\infty} xy f_{XY}(x, y) dx\, dy$$

If X and Y are independent, we see from Equation 3.45 that

$$f_{XY}(x, y) = f_X(x) f_Y(y)$$

and

$$\alpha_{11} = \int_{-\infty}^{\infty} \int_{-\infty}^{\infty} xy f_X(x) f_Y(y) dx\, dy = \int_{-\infty}^{\infty} x f_X(x) dx \int_{-\infty}^{\infty} y f_Y(y) dy$$

$$= m_X m_Y$$

Equations 4.23 and 4.24 then show that $\mu_{11} = 0$ and $\rho = 0$. A similar result can be obtained for two independent discrete random variables.

This result leads immediately to an important generalization. Consider a function of X and Y in the form $g(X)h(Y)$ whose expectation exists. Then, if X and Y are independent,

$$E\{g(X)h(Y)\} = E\{g(X)\}E\{h(Y)\} \tag{4.28}$$

When the correlation coefficient of two random variables vanishes, we say they are *uncorrelated*. It should be carefully pointed out that what we have shown is that independence implies zero correlation. The converse, however, is not true. This point is more fully discussed in what follows.

The covariance or the correlation coefficient is of great importance in the analysis of two random variables. It is a measure of their *linear* interdependence in the sense that its value is a measure of accuracy with which one random variable can be approximated by a linear function of the other. In order to see this, let us consider the problem of approximating a r.v. X by a linear function of a second r.v. Y, $aY + b$, where a and b are chosen so that the mean-square error e, defined by

$$e = E\{[X - (aY + b)]^2\} \tag{4.29}$$

is minimized. Upon taking partial derivatives of e with respect to a and b and setting them to zero, straightforward calculations show that this minimum is attained when

$$a = \frac{\sigma_X \rho}{\sigma_Y} \quad \text{and} \quad b = m_X - am_Y$$

Substituting these values into Equation 4.29 then gives $\sigma_X^2(1 - \rho^2)$ as the minimum mean-square error. We thus see that an exact fit in the mean-square sense is achieved when $|\rho| = 1$ and the linear approximation is the worst when $\rho = 0$. More specifically, when $\rho = +1$, the r.v.'s X and Y are said to be *positively perfectly correlated* in the sense that the values they assume fall on a straight line with positive slope; they are *negatively perfectly correlated* when $\rho = -1$ and their values form a straight line with negative slope. These two extreme cases are illustrated in Figure 4.3. The value of $|\rho|$ decreases as scatter about these lines increases.

Let us again stress the fact that the correlation coefficient measures only the linear interdependence between two random variables. It is by no means a general measure of dependence between X and Y. Thus, $\rho = 0$ does not necessarily imply

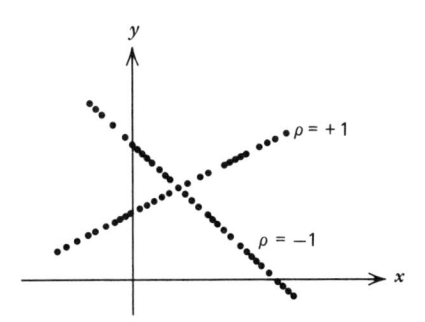

Figure 4.3 An illustration of perfect correlation.

EXPECTATIONS AND MOMENTS

independence of the random variables. In fact, as the example below shows, the correlation coefficient can vanish when the values of one random variable are completely determined by the values of another.

Example 4.10. Determine the correlation coefficient of r.v.'s X and Y when X takes the values ± 1 and ± 2, each with probability $\frac{1}{4}$, and $Y = X^2$.

Clearly, Y assumes the values 1 and 4, each with probability $\frac{1}{2}$, and their joint mass function is

$$\begin{aligned} p_{XY}(x, y) &= \tfrac{1}{4} & (x, y) &= (-1, 1) \\ &= \tfrac{1}{4} & &= (1, 1) \\ &= \tfrac{1}{4} & &= (-2, 4) \\ &= \tfrac{1}{4} & &= (2, 4) \end{aligned}$$

The means and the second moment α_{11} are given by

$$\begin{aligned} m_X &= (-2)(\tfrac{1}{4}) + (-1)(\tfrac{1}{4}) + (1)(\tfrac{1}{4}) + (2)(\tfrac{1}{4}) = 0 \\ m_Y &= (1)(\tfrac{1}{2}) + (4)(\tfrac{1}{2}) = 2.5 \\ \alpha_{11} &= (-1)(1)(\tfrac{1}{4}) + (1)(1)(\tfrac{1}{4}) + (-2)(4)(\tfrac{1}{4}) + (2)(4)(\tfrac{1}{4}) = 0 \end{aligned}$$

Hence,

$$\alpha_{11} - m_X m_Y = 0$$

and, from Equations 4.23 and 4.24,

$$\rho = 0$$

This is a simple example showing that X and Y are uncorrelated but they are completely dependent on each other in a nonlinear way.

IV.3.2 SCHWARZ INEQUALITY

In the preceding section, an inequality given by Equation 4.25 was established in the process of proving that $|\rho| \le 1$. It is

$$\mu_{11}^2 = |\mu_{11}|^2 \le \mu_{20}\mu_{02} \tag{4.30}$$

We can also show following a similar procedure that

$$\boxed{E^2\{XY\} = |E\{XY\}|^2 \le E\{X^2\}E\{Y^2\}} \tag{4.31}$$

Equations 4.30 and 4.31 are usually referred to as the *Schwarz inequality*. We point them out here because they are useful in a number of situations involving moments in subsequent chapters.

IV.3.3 THE CASE OF THREE OR MORE RANDOM VARIABLES

The expectation of a function $g(X_1, X_2, \ldots, X_n)$ of n r.v.'s $X_1, X_2, \ldots, X_n$ is defined in an analogous manner. Following Equations 4.18 and 4.19 for the two-r.v. case, we have

$$E\{g(X_1, \ldots, X_n)\} = \sum_{i_1} \cdots \sum_{i_n} g(x_{1i_1}, \ldots, x_{ni_n}) p_{X_1 \cdots X_n}(x_{1i_1}, \ldots, x_{ni_n})$$

$$(X_1, \ldots, X_n \text{ discrete}) \quad (4.32)$$

$$E\{g(X_1, \ldots, X_n)\} = \int_{-\infty}^{\infty} \cdots \int_{-\infty}^{\infty} g(x_1, \ldots, x_n) f_{X_1 \cdots X_n}(x_1, \ldots, x_n) dx_1 \cdots dx_n$$

$$(X_1, \ldots, X_n \text{ continuous}) \quad (4.33)$$

where $p_{X_1 \cdots X_n}$ and $f_{X_1 \cdots X_n}$ are, respectively, joint mass function and joint density function of the associated random variables.

The important moments associated with n random variables are still the individual means, individual variances, and pairwise covariances. Let **X** be the random column vector whose components are $X_1, \ldots, X_n$ and let the means of $X_1, \ldots, X_n$ be represented by the vector $\mathbf{m_X}$. A convenient representation of their variances and covariances is the *covariance matrix* Λ, defined by

$$\Lambda = E\{(\mathbf{X} - \mathbf{m_X})(\mathbf{X} - \mathbf{m_X})^T\} \quad (4.34)$$

where the superscript T denotes matrix transpose. The $n \times n$ matrix Λ has the structure whose diagonal elements are the variances and nondiagonal elements are covariances. Specifically, it is given by

$$\Lambda = \begin{bmatrix} \text{var}(X_1) & \text{cov}(X_1, X_2) & \cdots & \text{cov}(X_1, X_n) \\ \text{cov}(X_2, X_1) & \text{var}(X_2) & \cdots & \text{cov}(X_2, X_n) \\ \vdots & \vdots & & \vdots \\ \text{cov}(X_n, X_1) & \text{cov}(X_n, X_2) & \cdots & \text{var}(X_n) \end{bmatrix} \quad (4.35)$$

In the above, "var" reads "variance of" and "cov" reads "covariance of." Since $\text{cov}(X_i, X_j) = \text{cov}(X_j, X_i)$, the covariance matrix is always symmetrical.

In closing, let us state without proof an important result which is a direct extension of Equation 4.28.

Theorem.

If $X_1, X_2, \ldots, X_n$ are mutually independent, then

$$E\{g_1(X_1)g_2(X_2)\cdots g_n(X_n)\} = E\{g_1(X_1)\}E\{g_2(X_2)\}\cdots E\{g_n(X_n)\} \quad (4.36)$$

where $g_j(X_j)$ is an arbitrary function of X_j. It is assumed, of course, that all indicated expectations exist.

IV.4 Moments of Sums of Random Variables

Let $X_1, X_2, \ldots, X_n$ be n random variables. Their sum is also a random variable. In this section, we are interested in the moments of this sum in terms of those associated with X_j, $j = 1, 2, \ldots, n$. These relations find applications in a large number of derivations to follow and in a variety of physical situations.

Consider a r.v. Y formed by

$$Y = X_1 + X_2 + \cdots + X_n \quad (4.37)$$

Let m_j and σ_j^2 denote the respective mean and variance of X_j. The following are some of the important results concerning the mean and the variance of Y. Verifications of these results are carried out for the case where the r.v.'s $X_1, \ldots, X_n$ are continuous. Same procedures can be used when they are discrete.

1. The mean of the sum is the sum of the means, that is,

$$m_Y = m_1 + m_2 + \cdots + m_n \quad (4.38)$$

To establish this result, consider

$$m_Y = E\{Y\} = E\{X_1 + X_2 + \cdots + X_n\}$$

$$= \int_{-\infty}^{\infty} \cdots \int_{-\infty}^{\infty} (x_1 + \cdots + x_n) f_{X_1 \cdots X_n}(x_1, \ldots, x_n) dx_1 \cdots dx_n$$

$$= \int_{-\infty}^{\infty} \cdots \int_{-\infty}^{\infty} x_1 f_{X_1 \cdots X_n}(x_1, \ldots, x_n) dx_1 \cdots dx_n$$

$$+ \int_{-\infty}^{\infty} \cdots \int_{-\infty}^{\infty} x_2 f_{X_1 \cdots X_n}(x_1, \ldots, x_n) dx_1 \cdots dx_n$$

$$+ \cdots$$

$$+ \int_{-\infty}^{\infty} \cdots \int_{-\infty}^{\infty} x_n f_{X_1 \cdots X_n}(x_1, \ldots, x_n) dx_1 \cdots dx_n$$

The first integral in the final expression can be immediately integrated with respect to $x_2, x_3, \ldots, x_n$, yielding $f_{X_1}(x_1)$, the marginal density function of X_1. Similarly, the $(n-1)$-fold integration with respect to $x_1, x_3, \ldots, x_n$ in the second integral gives $f_{X_2}(x_2)$, and so on. Hence, the foregoing reduces to

$$m_Y = \int_{-\infty}^{\infty} x_1 f_{X_1}(x_1) dx_1 + \cdots + \int_{-\infty}^{\infty} x_n f_{X_n}(x_n) dx_n$$

$$= m_1 + m_2 + \cdots + m_n$$

Combining this result with some basic properties of the expectation will lead to some useful generalizations. For example, in view of the second of Equations 4.3, we have the following extension:

2. If

$$Z = a_1 X_1 + a_2 X_2 + \cdots + a_n X_n \qquad (4.39)$$

where $a_1, a_2, \ldots, a_n$ are constants, then

$$m_Z = a_1 m_1 + a_2 m_2 + \cdots + a_n m_n \qquad (4.40)$$

3. Let $X_1, \ldots, X_n$ be mutually independent random variables. Then the variance of the sum is the sum of the variances, that is,

$$\boxed{\sigma_Y^2 = \sigma_1^2 + \sigma_2^2 + \cdots + \sigma_n^2} \qquad (4.41)$$

Let us verify this relation for $n = 2$. The proof for the case of n random variables follows at once by mathematical induction.

Consider

$$Y = X_1 + X_2$$

We know from Equation 4.38 that

$$m_Y = m_1 + m_2$$

Subtracting m_Y from Y and $(m_1 + m_2)$ from $(X_1 + X_2)$ yields

$$Y - m_Y = (X_1 - m_1) + (X_2 - m_2)$$

EXPECTATIONS AND MOMENTS 91

and

$$\sigma_Y^2 = E\{(Y - m_Y)^2\} = E\{[(X_1 - m_1) + (X_2 - m_2)]^2\}$$
$$= E\{(X_1 - m_1)^2 + 2(X_1 - m_1)(X_2 - m_2) + (X_2 - m_2)^2\}$$
$$= E\{(X_1 - m_1)^2\} + 2E\{(X_1 - m_1)(X_2 - m_2)\} + E\{(X_2 - m_2)^2\}$$
$$= \sigma_1^2 + 2\operatorname{cov}(X_1, X_2) + \sigma_2^2$$

The covariance $\operatorname{cov}(X_1, X_2)$ vanishes since X_1 and X_2 are independent (see Equation 4.27), thus the desired result.

Again, many generalizations are possible. For example, if Z is given by Equation 4.39, we have, following the second of Equations 4.9,

$$\sigma_Z^2 = a_1^2 \sigma_1^2 + \cdots + a_n^2 \sigma_n^2 \qquad (4.42)$$

Let us again emphasize that, while Equation 4.38 is valid for any set of r.v.'s $X_1, \ldots, X_n$, Equation 4.41 pertaining to the variance holds only under the independence assumption. Removal of the condition of independence would, as seen from the proof, add covariance terms to the right-hand side of Equation 4.41. It would then have the form

$$\sigma_Y^2 = \sigma_1^2 + \sigma_2^2 + \cdots + \sigma_n^2 + 2\operatorname{cov}(X_1, X_2) + 2\operatorname{cov}(X_1, X_3) + \cdots$$
$$+ 2\operatorname{cov}(X_{n-1}, X_n)$$
$$= \sum_{j=1}^{n} \sigma_j^2 + 2 \sum_{i=1}^{n-1} \sum_{\substack{j=1 \\ i<j}}^{n} \operatorname{cov}(X_i, X_j) \qquad (4.43)$$

Example 4.11. An inspection is made on a group of n TV picture tubes. If each passes the inspection with probability p and fails with probability q ($p + q = 1$), we wish to calculate the average number of tubes in n tubes that pass the inspection.

This problem may be easily solved if we introduce a r.v. X_j to represent the outcome of the jth inspection and define

$$X_j = \begin{cases} 1 & \text{if } j\text{th tube passes inspection} \\ 0 & \text{if } j\text{th tube does not pass} \end{cases}$$

Then the r.v. Y defined by

$$Y = X_1 + X_2 + \cdots + X_n$$

has the desirable property that its value is the total number of tubes passing the inspection. The mean of X_j is

$$E\{X_j\} = 0(q) + 1(p) = p$$

Therefore, as seen from Equation 4.38, the desired average number is given by

$$m_Y = E\{X_1\} + \cdots + E\{X_n\} = np$$

We can also calculate the variance of Y if $X_1, \ldots, X_n$ are assumed to be independent. The variance of X_j is

$$\sigma_j^2 = E\{(X_j - p)^2\} = (0 - p)^2(q) + (1 - p)^2 p = pq$$

Equation 4.41 then gives

$$\sigma_Y^2 = \sigma_1^2 + \cdots + \sigma_n^2 = npq$$

Example 4.12. Let $X_1, \ldots, X_n$ be a set of mutually independent random variables with a common distribution, each having mean m. Show that, for every $\varepsilon > 0$ and as $n \to \infty$,

$$P\left(\left|\frac{Y}{n} - m\right| \geq \varepsilon\right) \to 0 \quad \text{where } Y = X_1 + \cdots + X_n \quad (4.44)$$

This is a statement of the *law of large numbers*. The r.v. Y/n can be interpreted as an average of n independently observed random variables from the same distribution. Equation 4.44 then states that the probability that this average will differ from the mean by less than an arbitrarily prescribed ε tends to one.

To proceed with the proof of Equation 4.44, we first note that, if σ^2 is the variance of each X_j, it follows from Equation 4.41 that

$$\sigma_Y^2 = n\sigma^2$$

According to the Chebyshev inequality given by Equation 4.17, for every $k > 0$ we have

$$P(|Y - nm| \geq k) \leq \frac{n\sigma^2}{k^2}$$

For $k > \varepsilon n$, the left-hand side is less than $\sigma^2/(\varepsilon^2 n)$, which tends to zero as $n \to \infty$. This establishes the proof.

EXPECTATIONS AND MOMENTS

Let us note that the proof given above requires the existence of σ^2. This is not necessary but more work is required without this restriction.

Among many of its uses, statistical sampling is an example in which the law of large numbers plays an important role. Suppose that in a group of m families there are m_j number of families with exactly j children ($j = 0, 1, \ldots$ and $m_0 + m_1 + \cdots = m$). For a family chosen at random, the number of children is a random variable that assumes the value r with probability $p_r = m_r/m$. A sample of n families among this group represents n observed independent random variables $X_1, \ldots, X_n$ with the same distribution. The quantity $(X_1 + \cdots + X_n)/n$ is the sample average and the law of large numbers then states that, for sufficiently large samples, the sample average is likely to be close to $m = \sum_{r=0} r p_r = \sum_{r=0} r m_r/m$, the mean of the population.

Example 4.13. The r.v. Y/n in the above example is also called the *sample mean* associated with the r.v.'s $X_1, \ldots, X_n$ and is denoted by $\bar{X}$. In Example 4.12, if the coefficient of variation for each X_i is v, the coefficient of variation $v_{\bar{X}}$ of $\bar{X}$ is easily derived from Equations 4.38 and 4.41 to be

$$v_{\bar{X}} = v/\sqrt{n} \tag{4.45}$$

Equation 4.45 is the basis for the *law of $\sqrt{n}$* proposed by Schrödinger, which states that the laws of physics are accurate within a probable relative error of the order of $1/\sqrt{n}$, where n is the number of molecules that cooperate in a physical process. Basically, what Equation 4.45 suggests is that, if the action of each molecule exhibits a random variation measured by v, then a physical process resulting from additive actions of n molecules will possess a random variation measured by $v/\sqrt{n}$. It decreases as n increases. Since n is generally very large in the workings of physical processes, this result leads to the conjecture that the laws of physics can be exact laws despite local disorder.

IV.5 Characteristic Functions

The expectation $E\{e^{jtX}\}$ of a r.v. X is defined as the *characteristic function* of X. Denoted by $\phi_X(t)$, it is given by

$$\phi_X(t) = E\{e^{jtX}\} = \sum_i e^{jtx_i} p_X(x_i) \qquad (X \text{ discrete}) \tag{4.46}$$

$$\phi_X(t) = E\{e^{jtX}\} = \int_{-\infty}^{\infty} e^{jtx} f_X(x)\, dx \qquad (X \text{ continuous}) \tag{4.47}$$

where t is an arbitrary real-valued parameter and $j = \sqrt{-1}$. The characteristic function is thus the expectation of a complex function and is generally complex valued. Since

$$|e^{jtX}| = |\cos tX + j \sin tX| = 1$$

the sum and the integral in Equations 4.46 and 4.47 exist and therefore $\phi_X(t)$ always exists. Furthermore, we note

$$\phi_X(0) = 1$$
$$\phi_X(-t) = \phi_X^*(t) \qquad (4.48)$$
$$|\phi_X(t)| \leq 1$$

where the superscript * denotes complex conjugate. The first two properties are self-evident. The third relation follows from the observation that, since $f_X(x) \geq 0$,

$$|\phi_X(t)| = \left| \int_{-\infty}^{\infty} e^{jtx} f_X(x) dx \right| \leq \int_{-\infty}^{\infty} f_X(x) dx = 1$$

The proof is the same for discrete random variables.

We single this expectation out for discussion because it possesses a number of important properties that make it a powerful tool in random variable analysis and probabilistic modeling.

IV.5.1 GENERATION OF MOMENTS

One of the important uses of characteristic functions is in the determination of moments of a random variable. Expanding $\phi_X(t)$ in the MacLaurin series shows that (suppressing subscript X for convenience)

$$\phi(t) = \phi(0) + \phi'(0)t + \phi''(0)\frac{t^2}{2} + \cdots + \phi^{(n)}(0)\frac{t^n}{n!} + \cdots \qquad (4.49)$$

where the primes denote derivatives. The coefficients are, using Equation 4.47,

$$\phi(0) = \int_{-\infty}^{\infty} f_X(x) dx = 1$$

$$\phi'(0) = \left.\frac{d\phi(t)}{dt}\right|_{t=0} = \int_{-\infty}^{\infty} jx f_X(x) dx = j\alpha_1$$

$$\vdots \qquad (4.50)$$

$$\phi^{(n)}(0) = \left.\frac{d^n \phi(t)}{dt^n}\right|_{t=0} = \int_{-\infty}^{\infty} j^n x^n f_X(x) dx = j^n \alpha_n$$

Thus,

$$\phi(t) = 1 + \sum_{n=1}^{\infty} \frac{(jt)^n \alpha_n}{n!} \qquad (4.51)$$

Same results are obtained when X is discrete.

Equation 4.51 shows that moments of all orders, if they exist, are contained in the expansion for $\phi(t)$, and these moments can be found from $\phi(t)$ by simple differentiation. Specifically, Equations 4.50 give

$$\boxed{\alpha_n = j^{-n} \phi^{(n)}(0) \qquad n = 1, 2, \ldots} \qquad (4.52)$$

Example 4.14. Determine $\phi(t)$, mean, and variance of a r.v. X if it has the binomial distribution

$$p_X(k) = \binom{n}{k} p^k (1-p)^{n-k} \qquad k = 0, 1, \ldots, n$$

According to Equation 4.46,

$$\begin{aligned}
\phi_X(t) &= \sum_{k=0}^{n} e^{jtk} \binom{n}{k} p^k (1-p)^{n-k} \\
&= \sum_{k=0}^{n} \binom{n}{k} (pe^{jt})^k (1-p)^{n-k} \\
&= [pe^{jt} + (1-p)]^n
\end{aligned} \qquad (4.53)$$

Using Equation 4.52, we have

$$\begin{aligned}
\alpha_1 &= \frac{1}{j} \frac{d}{dt} [pe^{jt} + (1-p)]^n \bigg|_{t=0} = n[pe^{jt} + (1-p)]^{n-1} (pe^{jt}) \bigg|_{t=0} \\
&= np \\
\alpha_2 &= -\frac{d^2}{dt^2} [pe^{jt} + (1-p)]^n \bigg|_{t=0} = np[(n-1)p + 1]
\end{aligned}$$

and

$$\sigma_X^2 = \alpha_2 - \alpha_1^2 = np[(n-1)p + 1] - n^2 p^2 = np(1-p)$$

The results for the mean and the variance are the same as those obtained in Examples 4.1 and 4.5.

Example 4.15. Repeat the above when X is exponentially distributed with density function

$$f_X(x) = ae^{-ax} \quad x \geq 0$$
$$= 0 \quad \text{elsewhere}$$

The characteristic function $\phi_X(t)$ in this case is

$$\phi_X(t) = \int_0^\infty e^{jtx}(ae^{-ax})dx = a\int_0^\infty e^{-(a-jt)x}dx = \frac{a}{a-jt} \qquad (4.54)$$

The moments are

$$\alpha_1 = \frac{1}{j}\frac{d}{dt}\left(\frac{a}{a-jt}\right)\bigg|_{t=0} = \frac{1}{j}\left[\frac{ja}{(a-jt)^2}\right]\bigg|_{t=0} = \frac{1}{a}$$

$$\alpha_2 = -\frac{d^2}{dt^2}\left(\frac{a}{a-jt}\right)\bigg|_{t=0} = \frac{2}{a^2}$$

$$\sigma_X^2 = \alpha_2 - \alpha_1^2 = \frac{1}{a^2}$$

which agree with the moment calculations carried out in Examples 4.2 and 4.6.

Another useful expansion is the power series representation of the logarithm of the characteristic function, that is,

$$\log \phi_X(t) = \sum_{n=1}^\infty \frac{(jt)^n \lambda_n}{n!} \qquad (4.55)$$

where the coefficients λ_n are again obtained from

$$\lambda_n = j^{-n}\frac{d^n}{dt^n}\log \phi_X(t)\bigg|_{t=0} \qquad (4.56)$$

EXPECTATIONS AND MOMENTS

The relations between the coefficients λ_n and the moments α_n can be established by forming the exponential of log $\phi_X(t)$, expanding this in a power series of jt, and equating coefficients to those of corresponding powers in Equation 4.51. We obtain

$$\begin{aligned}\lambda_1 &= \alpha_1 \\ \lambda_2 &= \alpha_2 - \alpha_1^2 \\ \lambda_3 &= \alpha_3 - 3\alpha_1\alpha_2 + 2\alpha_1^3 \\ \lambda_4 &= \alpha_4 - 3\alpha_2^2 - 4\alpha_1\alpha_3 + 12\alpha_1^2\alpha_2 - 6\alpha_1^4 \\ &\ldots\end{aligned} \qquad (4.57)$$

It is seen that λ_1 is the mean, λ_2 is the variance, and λ_3 is the third central moment. The higher order λ_n's are related to the moments of the same order or lower, but in a more complex way. The coefficients λ_n are called the *cumulants* of X and, with a knowledge of these cumulants, we may obtain its moments and its central moments.

IV.5.2 INVERSION FORMULAS

Another important use of characteristic functions follows from the inversion formulas to be developed below.

Consider first a continuous r.v. X. We observe that Equation 4.47 also defines $\phi_X(t)$ as the inverse Fourier transform of $f_X(x)$. The other half of the Fourier transform pair is

$$f_X(x) = \frac{1}{2\pi}\int_{-\infty}^{\infty} e^{-jtx}\phi_X(t)dt \qquad (4.58)$$

This inversion formula shows that the knowledge of characteristic function specifies the distribution of X. Furthermore, it follows from the theory of Fourier transforms that $f_X(x)$ is uniquely determined from Equation 4.58, that is, no two distinct density functions can have the same characteristic function.

This property of the characteristic function provides us with an alternative way of arriving at the distribution of a random variable. In many physical problems, it is often more convenient to determine the density function of a random variable by first determining its characteristic function and then performing the Fourier transform as indicated by Equation 4.58. Furthermore, we shall see that the characteristic function has properties that render it particularly useful for determining the distribution of a sum of independent random variables.

While the inversion formula (4.58) follows immediately from the theory of Fourier transforms, it is of interest to give a derivation of this equation from a probabilistic point of view.

Proof of Equation 4.58.

An integration formula that can be found in any table of integrals is

$$\frac{1}{\pi} \int_{-\infty}^{\infty} \frac{\sin at}{t} \, dt = \begin{cases} -1 & a < 0 \\ 0 & a = 0 \\ 1 & a > 0 \end{cases} \quad (4.59)$$

This leads to

$$\frac{1}{\pi} \int_{-\infty}^{\infty} \frac{\sin at + j(1 - \cos at)}{t} \, dt = \begin{cases} -1 & a < 0 \\ 0 & a = 0 \\ 1 & a > 0 \end{cases} \quad (4.60)$$

because the function $(1 - \cos at)/t$ is an odd function of t so that its integral over a symmetric range vanishes. Upon replacing a by $X - x$ in Equation 4.60, we have

$$\frac{1}{2} - \frac{j}{2\pi} \int_{-\infty}^{\infty} \frac{1 - e^{j(X-x)t}}{t} \, dt = \begin{cases} 1 & X < x \\ \frac{1}{2} & X = x \\ 0 & X > x \end{cases} \quad (4.61)$$

For a fixed value of x, Equation 4.61 is a function of the r.v. X and it may be regarded as defining a new random variable Y. The r.v. Y is seen to be discrete, taking on values $1, \frac{1}{2}$, and 0 with probabilities $P(X < x)$, $P(X = x)$, and $P(X > x)$, respectively. The mean of Y is thus equal to

$$E\{Y\} = (1)P(X < x) + (\tfrac{1}{2})P(X = x) + (0)P(X > x)$$

However, notice that, since X is continuous, $P(X = x) = 0$ if x is a point of continuity in the distribution of X. Hence, using Equation 4.47,

$$E\{Y\} = P(X < x) = F_X(x)$$

$$= \frac{1}{2} - \frac{j}{2\pi} \int_{-\infty}^{\infty} \frac{1 - E\{e^{j(X-x)t}\}}{t} \, dt \quad (4.62)$$

$$= \frac{1}{2} - \frac{j}{2\pi} \int_{-\infty}^{\infty} \frac{1 - e^{-jtx}\phi_X(t)}{t} \, dt$$

EXPECTATIONS AND MOMENTS

The above defines the probability distribution function of X. Its derivative gives the inversion formula

$$f_X(x) = \frac{1}{2\pi} \int_{-\infty}^{\infty} e^{-jtx} \phi_X(t) dt \qquad (4.63)$$

and we have Equation 4.58.

The inversion formula when X is a discrete random variable is

$$p_X(x) = \lim_{u \to \infty} \int_{-u}^{u} e^{-jtx} \phi_X(t) dt \qquad (4.64)$$

A proof of this relation can be constructed along the same line as the one given above for the continuous case. For this case, we first note the standard integration formula

$$\frac{1}{2u} \int_{-u}^{u} e^{jat} dt = \begin{cases} \dfrac{\sin au}{au} & a \neq 0 \\ 1 & a = 0 \end{cases} \qquad (4.65)$$

Replacing a by $X - x$ and taking the limit as $u \to \infty$, we have a new r.v. Y defined by

$$Y = \lim_{u \to \infty} \frac{1}{2u} \int_{-u}^{u} e^{j(X-x)t} dt = \begin{cases} 0 & X \neq x \\ 1 & X = x \end{cases}$$

The mean of Y is

$$E\{Y\} = (1)P(X = x) + (0)P(X \neq x) = P(X = x) \qquad (4.66)$$

and therefore

$$\begin{aligned} P_X(x) &= \lim_{u \to \infty} \frac{1}{2u} \int_{-u}^{u} E\{e^{j(X-x)t}\} dt \\ &= \lim_{u \to \infty} \frac{1}{2u} \int_{-u}^{u} e^{-jtx} \phi_X(t) dt \end{aligned} \qquad (4.67)$$

which gives the desired inversion formula.

In summary, the transform pairs given by Equations 4.46, 4.47, 4.58, and 4.64 are collected and presented below for easy reference. For a continuous r.v. X,

$$\phi_X(t) = \int_{-\infty}^{\infty} e^{jtx} f_X(x) dx$$

$$f_X(x) = \frac{1}{2\pi} \int_{-\infty}^{\infty} e^{-jtx} \phi_X(t) dt$$

(4.68)

and for a discrete r.v. X,

$$\phi_X(t) = \sum_i e^{jtx_i} p_X(x_i)$$

$$p_X(x) = \lim_{u \to \infty} \frac{1}{2u} \int_{-u}^{u} e^{-jtx} \phi_X(t) dt$$

(4.69)

Of the two sets, Equations 4.68 for the continuous case are more important in terms of applicability. As we shall see in Chapter V, probability mass functions for discrete random variables can be found directly without resorting to their characteristic functions.

As we have mentioned before, the characteristic function is particularly useful for the study of a sum of independent random variables. In this connection, let us state the following important theorem.

Theorem.

The characteristic function of a sum of independent random variables is equal to the product of the characteristic functions of the individual random variables.

Proof.

Let

$$Y = X_1 + X_2 + \cdots + X_n \tag{4.70}$$

Then, by definition,

$$\phi_Y(t) = E\{e^{jtY}\} = E\{e^{jt(X_1 + X_2 + \cdots + X_n)}\}$$
$$= E\{e^{jtX_1} e^{jtX_2} \cdots e^{jtX_n}\}$$

EXPECTATIONS AND MOMENTS   101

Since $X_1, X_2, \ldots, X_n$ are mutually independent, Equation 4.36 leads to

$$E\{e^{jtX_1}e^{jtX_2}\cdots e^{jtX_n}\} = E\{e^{jtX_1}\}E\{e^{jtX_2}\}\cdots E\{e^{jtX_n}\}$$

We thus have

$$\phi_Y(t) = \phi_{X_1}(t)\phi_{X_2}(t)\cdots\phi_{X_n}(t) \qquad (4.71)$$

In Section IV.4 we have obtained moments of a sum of random variables; Equation 4.71, coupled with the inversion formula (4.58) or (4.64), enables us to determine the distribution of a sum of random variables from the knowledge of the distributions of $X_j, j = 1, 2, \ldots, n$, provided that they are mutually independent.

Example 4.16. Let X_1 and X_2 be two independent random variables, both having exponential distributions with parameter a, and let $Y = X_1 + X_2$. Determine the distribution of Y.

The characteristic function of an exponentially distributed random variable has been obtained in Example 4.15. From Equation 4.54, we have

$$\phi_{X_1}(t) = \phi_{X_2}(t) = \frac{a}{a - jt}$$

According to Equation 4.71, the characteristic function of Y is simply

$$\phi_Y(t) = \phi_{X_1}(t)\phi_{X_2}(t) = \frac{a^2}{(a - jt)^2}$$

Hence, the density function of Y is, as seen from the inversion formula (4.68),

$$\begin{aligned} f_Y(y) &= \frac{1}{2\pi}\int_{-\infty}^{\infty} e^{-jty}\phi_Y(t)dt \\ &= \frac{a^2}{2\pi}\int_{-\infty}^{\infty}\frac{e^{-jty}}{(a-jt)^2}dt \\ &= \begin{cases} a^2 y e^{-ay} & y \geq 0 \\ 0 & \text{elsewhere} \end{cases} \end{aligned} \qquad (4.72)$$

The distribution given by Equation 4.72 is called a gamma distribution, which will be discussed extensively in Chapter VII.

Example 4.17. In 1827, Robert Brown, an English botanist, noticed that small particles of matter from plants undergo erratic movements when suspended in fluids. It was soon discovered that the erratic motion was caused by impacts on the particles by the molecules of the fluid in which they were suspended. This phenomenon, which can also be observed in gases, is called *Brownian motion*. The explanation of Brownian motion was one of the major successes of statistical mechanics. In this example, we study Brownian motion in an elementary way using the one-dimensional random walk as an adequate mathematical model.

Consider a particle taking steps on a straight line. It moves either one step to the right with probability p or one step to the left with probability q ($p + q = 1$). The steps are always of unit length, positive to the right and negative to the left, and they are taken independently. We wish to determine the probability mass function of its position after n steps.

Let X_j be the random variable associated with the jth step and define

$$X_j = \begin{cases} 1 & \text{if it is to the right} \\ -1 & \text{if it is to the left} \end{cases} \qquad (4.73)$$

Then the r.v. Y defined by

$$Y = X_1 + X_2 + \cdots + X_n$$

gives the position of the particle after n steps. It is clear that Y takes integer values between $-n$ and n.

To determine $p_Y(k)$, $-n \le k \le n$, we first find its characteristic function. The characteristic function of each X_j is

$$\phi_{X_j}(t) = E\{e^{jtX_j}\} = pe^{jt} + qe^{-jt} \qquad (4.74)$$

It then follows from Equation 4.71 that, in view of independence,

$$\begin{aligned}\phi_Y(t) &= \phi_{X_1}(t)\phi_{X_2}(t) \cdots \phi_{X_n}(t) \\ &= (pe^{jt} + qe^{-jt})^n \end{aligned} \qquad (4.75)$$

Let us rewrite it as

$$\begin{aligned}\phi_Y(t) &= e^{-jnt}(pe^{2jt} + q)^n \\ &= \sum_{i=0}^{n} \binom{n}{i} p^i q^{n-i} e^{j(2i-n)t}\end{aligned}$$

EXPECTATIONS AND MOMENTS

Letting $k = 2i - n$, we get

$$\phi_Y(t) = \sum_{k=-n}^{n} \binom{n}{\frac{n+k}{2}} p^{(n+k)/2} q^{(n-k)/2} e^{jkt} \qquad (4.76)$$

Comparing Equation 4.76 with the definition (4.46) yields the mass function

$$p_Y(k) = \binom{n}{\frac{n+k}{2}} p^{(n+k)/2} q^{(n-k)/2} \qquad k = -n, -(n-2), \ldots, n \qquad (4.77)$$

We note that if n is even, k must also be even and, if n is odd, k must be odd.

Considerable importance is attached to the symmetric case in which $k \ll n$ and $p = q = \frac{1}{2}$. In order to consider this special case, we need to use Stirling's formula, which states that, for large n,

$$n! \cong \sqrt{2\pi} \, e^{-n} n^{n+\frac{1}{2}} \qquad (4.78)$$

Substituting this approximation into Equation 4.77 gives

$$p_Y(k) \cong \sqrt{\frac{2}{n\pi}} \, e^{-k^2/2n} \qquad k = -n, \ldots, n \qquad (4.79)$$

A further simplification results when the length of each step is small. Assuming that r steps occur in a unit time (i.e., $n = rt$) and letting a be the length of each step, then, as n becomes large, the r.v. Y approaches a continuous random variable and we can show that Equation 4.79 becomes

$$f_Y(y) = \frac{1}{(2\pi a^2 rt)^{1/2}} \exp\left(-\frac{y^2}{2a^2 rt}\right) \qquad -\infty < y < \infty \qquad (4.80)$$

where $y = ka$. On letting

$$D = \frac{a^2 r}{2}$$

we have

$$f_Y(y) = \frac{1}{(4\pi Dt)^{1/2}} \exp\left(-\frac{y^2}{4Dt}\right) \qquad -\infty < y < \infty \qquad (4.81)$$

The density function given above belongs to a *Gaussian* or *normal* random variable. This result is an illustration of the central limit theorem, discussed in Chapter VII.

Our derivation of Equation 4.81 has been purely analytical. In his theory of Brownian motion, Einstein also obtained this result with

$$D = \frac{2RT}{Nf} \qquad (4.82)$$

where R is the universal gas constant, T is the absolute temperature, N is the Avogadro's number, and f is the coefficient of friction which, for liquid or gas at ordinary pressure, can be expressed in terms of its viscosity and the particle size. Perrin, a French physicist, was awarded the Nobel Prize in 1926 for his success in determining from experiments the Avogadro's number.

IV.5.3 JOINT CHARACTERISTIC FUNCTIONS

The concept of characteristic functions finds usefulness as well in the case of two or more random variables. The development below is concerned with continuous random variables only but the principal results are equally valid in the case of discrete random variables. We also eliminate a bulk of the derivations involved since they follow closely those developed for the single-r.v. case.

The *joint characteristic function* of two r.v.'s X and Y, $\phi_{XY}(t, s)$, is defined by

$$\boxed{\phi_{XY}(t, s) = E\{e^{j(tX + sY)}\} = \int_{-\infty}^{\infty} \int_{-\infty}^{\infty} e^{j(tx + sy)} f_{XY}(x, y) \, dx \, dy} \qquad (4.83)$$

where t and s are two arbitrary real variables. This function always exists and some of its properties are noted below that are similar to those noted in Equations 4.48 corresponding to the single-r.v. case.

$$\phi_{XY}(0, 0) = 1$$
$$\phi_{XY}(-t, -s) = \phi_{XY}^*(t, s) \qquad (4.84)$$
$$|\phi_{XY}(t, s)| \leq 1$$

Furthermore, it is easy to verify that the joint characteristic function $\phi_{XY}(t, s)$ is related to the marginal characteristic functions $\phi_X(t)$ and $\phi_Y(s)$ by

$$\begin{aligned}\phi_X(t) &= \phi_{XY}(t, 0) \\ \phi_Y(s) &= \phi_{XY}(0, s)\end{aligned} \qquad (4.85)$$

EXPECTATIONS AND MOMENTS 105

If the r.v.'s X and Y are independent, then we also have

$$\phi_{XY}(t, s) = \phi_X(t)\phi_Y(s) \tag{4.86}$$

To show the above, we simply substitute $f_X(x)f_Y(y)$ for $f_{XY}(x, y)$ in Equation 4.83. The double integral on the right-hand side separates and we have

$$\phi_{XY}(t, s) = \int_{-\infty}^{\infty} e^{jtx} f_X(x) dx \int_{-\infty}^{\infty} e^{jsy} f_Y(y) dy$$

$$= \phi_X(t)\phi_Y(s)$$

and we have the desired result.

Analogous to the one-r.v. case, the joint characteristic function $\phi_{XY}(t, s)$ is often called on for the determination of the joint density function $f_{XY}(x, y)$ of X and Y and their joint moments. The density function $f_{XY}(x, y)$ is uniquely determined in terms of $\phi_{XY}(t, s)$ by the two-dimensional Fourier transform

$$\boxed{f_{XY}(x, y) = \frac{1}{4\pi^2} \int_{-\infty}^{\infty} \int_{-\infty}^{\infty} e^{-j(tx+sy)} \phi_{XY}(t, s) dt\, ds} \tag{4.87}$$

and the moments $E\{X^n Y^m\} = \alpha_{nm}$, if they exist, are related to $\phi_{XY}(t, s)$ by

$$\boxed{\begin{aligned}\frac{\partial^{n+m}}{\partial t^n \partial s^m} \phi_{XY}(t, s) \bigg|_{t,s=0} &= j^{n+m} \int_{-\infty}^{\infty} \int_{-\infty}^{\infty} x^n y^m f_{XY}(x, y) dx\, dy \\ &= j^{n+m} \alpha_{nm}\end{aligned}} \tag{4.88}$$

The MacLaurin series expansion of $\phi_{XY}(t, s)$ thus takes the form

$$\phi_{XY}(t, s) = \sum_{i=0}^{\infty} \sum_{k=0}^{\infty} \frac{\alpha_{ik}}{i!k!} (jt)^n (js)^m \tag{4.89}$$

The above development can be generalized to the case of more than two random variables in an obvious manner.

Example 4.18. Let us consider again the Brownian motion problem discussed in Example 4.17 and form two random variables X' and Y' as

$$\begin{aligned} X' &= X_1 + X_2 + \cdots + X_{2n} \\ Y' &= X_{n+1} + X_{n+2} + \cdots + X_{3n} \end{aligned} \tag{4.90}$$

They are, respectively, the position of the particle after $2n$ steps and its position after $3n$ steps relative to where it was after n steps. We wish to determine the jpdf $f_{XY}(x, y)$ of the r.v.'s $X = X'/\sqrt{n}$ and $Y = Y'/\sqrt{n}$ for large values of n.

For the simple case of $p = q = \frac{1}{2}$, the characteristic function of each X_k is (see Equation 4.74)

$$\phi(t) = E\{e^{jtX_k}\} = \tfrac{1}{2}(e^{jt} + e^{-jt}) = \cos t \tag{4.91}$$

and, following Equation 4.83, the joint characteristic function of X and Y is

$$\phi_{XY}(t, s) = E\{e^{j(tX + sY)}\} = E\left\{\exp\left[j\left(\frac{tX'}{\sqrt{n}} + \frac{sY'}{\sqrt{n}}\right)\right]\right\}$$

$$= E\left\{\exp\left\{\left(\frac{j}{\sqrt{n}}\right)\left[t\sum_{k=1}^{n} X_k + (t + s)\sum_{k=n+1}^{2n} X_k + s\sum_{k=2n+1}^{3n} X_k\right]\right\}\right\}$$

$$= \left\{\phi(t/\sqrt{n})\phi\left[\frac{s+t}{\sqrt{n}}\right]\phi(s/\sqrt{n})\right\}^n \tag{4.92}$$

where $\phi(t)$ is given by Equation 4.91. The last expression in Equation 4.92 is obtained based on the fact that X_k, $k = 1, 2, \ldots, 3n$, are mutually independent. It should be clear that X and Y are not independent, however.

We are now in the position to obtain $f_{XY}(x, y)$ from Equation 4.92 using the inverse formula given by Equation 4.87. First, however, some simplifications are in order. As n becomes large,

$$\left[\phi\left(\frac{t}{\sqrt{n}}\right)\right]^n = \cos^n\left(\frac{t}{\sqrt{n}}\right)$$

$$= \left(1 - \frac{t^2}{n2!} + \frac{t^4}{n^24!} - \cdots\right)^n$$

$$\cong e^{-t^2/2} \tag{4.93}$$

Hence, as $n \to \infty$,

$$\phi_{XY}(t, s) \cong e^{-(t^2 + ts + s^2)} \tag{4.94}$$

Now, substituting Equation 4.94 into Equation 4.87 gives

$$f_{XY}(x, y) = \frac{1}{4\pi^2}\int_{-\infty}^{\infty}\int_{-\infty}^{\infty} e^{-j(tx+sy)}e^{-(t^2+ts+s^2)}\,dt\,ds \tag{4.95}$$

which can be evaluated following a change of variables defined by

$$t = \frac{t' + s'}{\sqrt{2}} \qquad s = \frac{t' - s'}{\sqrt{2}} \tag{4.96}$$

The result is

$$f_{XY}(x, y) = \frac{1}{2\pi\sqrt{3}} \exp\left[-\frac{x^2 - xy + y^2}{3}\right] \tag{4.97}$$

The above is an example of a *bivariate normal distribution*, discussed in Chapter VII.

Incidentally, the joint moments of X and Y can be readily found by means of Equation 4.88. For large n, the means of X and Y, α_{10} and α_{01}, are

$$\alpha_{10} = -j \frac{\partial \phi_{XY}(t, s)}{\partial t}\bigg|_{t,s=0} = -j(-2t - s)e^{-(t^2 + ts + s^2)}\bigg|_{t,s=0} = 0$$

$$\alpha_{01} = -j \frac{\partial \phi_{XY}(t, s)}{\partial s}\bigg|_{t,s=0} = 0$$

Similarly, the second moments are

$$\alpha_{20} = E\{X^2\} = -\frac{\partial^2 \phi_{XY}(t, s)}{\partial t^2}\bigg|_{t,s=0} = 2$$

$$\alpha_{02} = E\{Y^2\} = -\frac{\partial^2 \phi_{XY}(t, s)}{\partial s^2}\bigg|_{t,s=0} = 2$$

$$\alpha_{11} = E\{XY\} = -\frac{\partial^2 \phi_{XY}(t, s)}{\partial t \, \partial s}\bigg|_{t,s=0} = 1$$

REFERENCES AND COMMENTS

As we mentioned in Section IV.2, Chebyshev inequality can be improved upon if some additional distribution features of a random variable are known beyond its first two moments. Some generalizations can be found in

1. C. L. Mallows, "Generalizations of Tchebycheff's Inequalities," *J. Royal Statistical Societies*, Series B, **18**, 139–176, 1956.

In many introductory texts, the discussion of characteristic functions of random variables is bypassed in favor of moment generating functions. The moment generating function $M_X(t)$ of a r.v. X is defined by

$$M_X(t) = E\{e^{tX}\}$$

In comparison with characteristic functions, the use of $M_X(t)$ is simpler since it avoids computations involving complex numbers and it generates moments of X in a similar fashion. However, there are two serious disadvantages in using $M_X(t)$. The first is that it may not exist for all values of t while $\phi_X(t)$ always exists. In addition, powerful inversion formulas associated with characteristic functions no longer exist for moment generating functions. For a discussion of the moment generating function, see, for example,

2. P. L. Meyer, *Introductory Probability and Statistical Applications*, 2nd ed., Addison-Wesley, Reading, Mass., 1970, pp. 210–217.

PROBLEMS

4.1. For each of the PDF's given in Problem 3.1, determine the mean and variance, if they exist, of its associated random variable.

4.2. For each of the pdf's given in Problem 3.4, determine the mean and variance, if they exist, of its associated random variable.

4.3. According to the PDF given in Problem 3.10, determine the average duration of a long-distance telephone call.

4.4. It is found that resistance of aircraft structural parts, R, in a nondimensionalized form, follows the distribution

$$f_R(r) = \frac{2\sigma_R^3}{0.9996\pi[\sigma_R^2 + (r-1)^2]^2} \qquad r \geq 0.33$$

$$= 0 \qquad \text{elsewhere}$$

where $\sigma_R = 0.0564$. Determine the mean of R.

EXPECTATIONS AND MOMENTS

4.5. A r.v. X has the exponential distribution

$$f_X(x) = ae^{-x/2} \qquad x \geq 0$$
$$= 0 \qquad \text{elsewhere}$$

Determine:
(a) The value of a.
(b) The mean and variance of X.
(c) The mean and variance of $Y = X/2 - 1$.

4.6. Let the mean and variance of X be m and σ^2. For what values of a and b does the r.v. $Y = aX + b$ have mean 0 and variance 1?

4.7. Suppose that a r.v. X is distributed (arbitrarily) over the interval $a \leq X \leq b$. Show that
(a) m_X is bounded by the same limits.
(b) $\sigma_X^2 \leq \dfrac{(b-a)^2}{4}$

4.8. Show that, given a r.v. X, $P(X = m_X) = 1$ if $\sigma_X^2 = 0$.

4.9. The waiting time T of a customer at an airline ticket counter can be characterized by a mixed distribution function (see figure)

$$F_T(t) = 0 \qquad\qquad\qquad\qquad t < 0$$
$$= p + (1-p)(1 - e^{-\lambda t}) \qquad t \geq 0$$

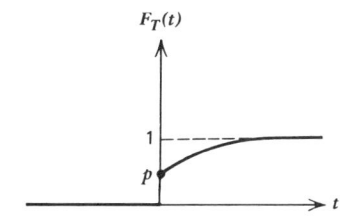

Determine:
(a) The average waiting time of an arrival ($E\{T\}$).
(b) The average waiting time of an arrival that waits ($E\{T|T > 0\}$).

4.10. For the commuter described in Problem 3.18, assuming that he or she makes one of the trains, what is the average arrival time at the destination?

4.11. A trapped miner has to choose one of two directions to find safety. If the miner goes to the right, then he will return to his original position after 3 minutes. If he goes to the left, he will with probability $\frac{1}{3}$ reach safety and with probability $\frac{2}{3}$ return to his original position after 5 minutes of traveling. Assuming that he is at all times equally likely to choose either direction, determine the expected number of minutes that the miner will be trapped.

4.12. Show that
(a) $E\{X|Y = y\} = E\{X\}$ if X and Y are independent.
(b) $E\{XY|Y = y\} = yE\{X|Y = y\}$.
(c) $E\{XY\} = E\{YE[X|Y]\}$.

4.13. Let r.v. X be uniformly distributed over the interval $0 \le x \le 2$. Determine a lower bound for $P(|X - 1| \le 0.75)$ using the Chebyshev inequality and compare it with the exact value of this probability.

4.14. For the r.v. X defined in Problem 4.13, plot $P(|X - m_X| \le h)$ as a function of h and compare it with its lower bound as determined by the Chebyshev inequality. Show that the lower bound becomes better approximation of $P(|X - m_X| \le h)$ as h becomes large.

4.15. Let a r.v. X take only nonnegative values, show that, for any $a > 0$,

$$P(X \ge a) \le \frac{m_X}{a}$$

This is known as *Markov's inequality*.

4.16. The yearly snowfall of a given region is a random variable with mean equal to 70 inches.
(a) What can be said about the probability that this year's snowfall will be between 55 and 85 inches?
(b) Can your answer be improved if, in addition, the standard deviation is known to be 10 inches?

4.17. The number X of airplanes arriving at an airport during a given period of time is distributed according to

$$p_X(k) = \frac{100^k e^{-100}}{k!} \quad k = 0, 1, 2, \ldots$$

Use the Chebyshev inequality to determine a lower bound for the probability $P(80 \le X \le 120)$ during this period of time.

4.18. For each joint distribution given in Problem 3.12, determine m_X, m_Y, σ_X^2, σ_Y^2, and ρ_{XY} of the r.v.'s X and Y.

4.19. Let the jpdf of X and Y be given by

$$f_{XY}(x, y) = 2xy \quad 0 < x < 1 \quad 0 < y < 2$$
$$= 0 \quad \text{elsewhere}$$

Determine the mean of $Z = \sqrt{X^2 + Y^2}$.

4.20. The product of two r.v.'s X and Y occurs frequently in applied problems. Let $Z = XY$ and assume that X and Y are independent. Determine the mean and variance of Z in terms of m_X, m_Y, σ_X^2, and σ_Y^2.

4.21. Let $X = X_1 + X_2$ and $Y = X_2 + X_3$. Determine the correlation coefficient ρ_{XY} of X and Y in terms of σ_{X_1}, σ_{X_2}, and σ_{X_3} when X_1, X_2, and X_3 are uncorrelated.

4.22. Let X and Y be discrete r.v.'s with jpmf given by the following table. Show that $\rho_{XY} = 0$ but X and Y are not independent.

Table for $p_{XY}(x, y)$
$(a + b = \frac{1}{4})$

y \ x	−1	0	1
−1	a	b	a
0	b	0	b
1	a	b	a

4.23. In a simple frame structure such as the one shown, the total horizontal displacement of the top story Y is the sum of the displacements of the individual stories X_1 and X_2. Assume that X_1 and X_2 are independent and

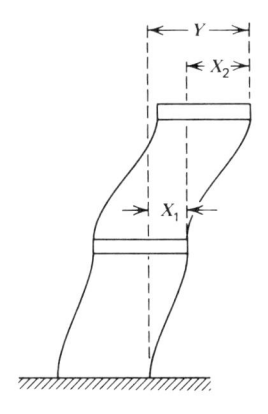

let m_{X_1}, m_{X_2}, $\sigma_{X_1}^2$, and $\sigma_{X_2}^2$ be their respective means and variances.
(a) Find the mean and variance of Y.
(b) Find the correlation coefficient between X_2 and Y. Discuss the result if $\sigma_{X_2}^2 \gg \sigma_{X_1}^2$.

4.24. Let $X_1, \ldots, X_n$ be a set of independent r.v.'s each of which having pdf of the form

$$f_{X_j}(x_j) = \frac{1}{\sqrt{2\pi}} e^{-x_j^2/2} \qquad j = 1, 2, \ldots, n \qquad -\infty < x_j < \infty$$

Determine the mean and variance of $Y = \sum_{j=1}^{n} X_j^2$.

4.25. Let $X_1, X_2, \ldots, X_n$ be independent r.v.'s and let σ_j^2 and μ_j be the respective variance and third central moment of X_j. Let σ^2 and μ denote the corresponding quantities for Y, where $Y = X_1 + X_2 + \cdots + X_n$.
(a) Show that $\sigma^2 = \sigma_1^2 + \sigma_2^2 + \cdots + \sigma_n^2$ and $\mu = \mu_1 + \mu_2 + \cdots + \mu_n$.
(b) Show that this additive property does not apply to the fourth- or higher-order central moments.

4.26. Determine the characteristic function corresponding to each of the PDF given in Problem 3.1 (a) through (e). Use it to generate the first two moments and compare them with results obtained in Problem 4.1. [Let $a = 2$ in part (e).]

4.27. We have shown that the characteristic function $\phi_X(t)$ of r.v. X facilitates the determination of the moments of X. Another function $M_X(t)$, defined by

$$M_X(t) = E\{e^{tX}\}$$

and called the moment generating function of X can also be used to obtain moments of X. Derive the relationships between $M_X(t)$ and the moments of X.

4.28. Let $Y = a_1 X_1 + a_2 X_2 + \cdots + a_n X_n$ where $X_1, X_2, \ldots, X_n$ are mutually independent. Show that

$$\phi_Y(t) = \phi_{X_1}(a_1 t) \phi_{X_2}(a_2 t) \cdots \phi_{X_n}(a_n t).$$

V
Functions of Random Variables

The basic topic to be discussed in this chapter is one of determining the relationship between probability distributions of two random variables X and Y when they are related by $Y = g(X)$. The functional form of $g(X)$ is given and deterministic. Generalizing to the case of many random variables, we are interested in the determination of the joint probability distribution of Y_j, $j = 1, 2, \ldots, m$, which are functionally dependent on X_k, $k = 1, 2, \ldots, n$, according to

$$Y_j = g_j(X_1, \ldots, X_n) \qquad j = 1, 2, \ldots, m \qquad m \leq n \qquad (5.1)$$

when the joint probabilistic behavior of X_k, $k = 1, 2, \ldots, n$, is known.

Some problems of this type (i.e., transformations of random variables) have been addressed in several places in Chapter IV. For example, Example 4.11 considers the transformation $Y = X_1 + \cdots + X_n$ and Example 4.18 deals with the transformation of $3n$ random variables $(X_1, X_2, \ldots, X_{3n})$ to two random variables (X', Y') defined by Equations 4.90. In science and engineering, most phenomena are based on functional relationships in which one or more dependent variables are expressed in terms of one or more independent variables. For example, force is a function of cross-sectional area and stress, distance traveled over a time interval is a function of the velocity, and so on. The techniques presented in this chapter thus permit us to determine probabilistic behavior of random variables that are functionally dependent on some others with known probabilistic properties.

In what follows, transformations of random variables are treated in a systematic manner. In Equation 5.1, we are basically interested in the joint distributions and joint moments of $Y_1, \ldots, Y_m$ given appropriate information on $X_1, \ldots, X_n$.

V.1 Functions of One Random Variable

Consider first a simple transformation involving only one random variable and let

$$Y = g(X) \qquad (5.2)$$

where $g(X)$ is assumed to be a continuous function of X. Given the probability distribution of X in terms of its PDF, pmf, or pdf, we are interested in the corresponding distribution for Y and its moment properties.

V.1.1 PROBABILITY DISTRIBUTIONS

Given the probability distribution of X, the quantity Y, being a function of X as defined by Equation 5.2, is thus also a random variable. Let R_X be the *range space* associated with r.v. X, defined as the set of all possible values assumed by X, and R_Y be the corresponding range space associated with Y. A basic procedure of determining the probability distribution of Y consists of the steps developed below.

For any outcome such as $X = x$, it follows from Equation 5.2 that $Y = y = g(x)$. As shown schematically in Figure 5.1, Equation 5.2 defines a mapping of

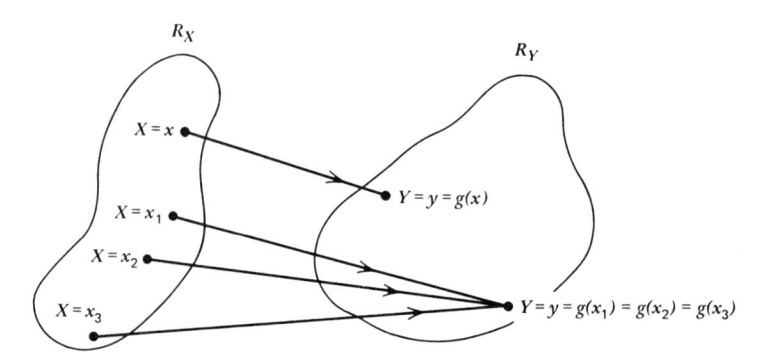

Figure 5.1 Transformation $y = g(x)$.

FUNCTIONS OF RANDOM VARIABLES

values in the range space R_X into corresponding values in the range space R_Y. Probabilities asociated with each point (in the case of discrete r.v. X) or with each region (in the case of continuous r.v. X) in R_X are carried over to its corresponding point or region in R_Y. The probability distribution of Y is determined on completing this transfer process for every point or every region of nonzero probability in R_X. Note that many-to-one transformations are possible as also shown in Figure 5.1. The procedure of determining the probability distribution of Y is thus critically dependent on the functional form of g in Equation 5.2.

Discrete Random Variables. Let us first dispose of the case when X is a discrete random variable since it requires only simple point-to-point mapping. Suppose that the possible values taken by X can be enumerated as $x_1, x_2, \ldots$. Equation 5.2 shows that the corresponding possible values of Y may be enumerated as $y_1 = g(x_1), y_2 = g(x_2), \ldots$. Let the pmf of X be given by

$$p_X(x_i) = p_i \qquad i = 1, 2, \ldots \tag{5.3}$$

The pmf of Y is simply determined as

$$p_Y[g(x_i)] = p_i \qquad i = 1, 2, \ldots \tag{5.4}$$

Example 5.1. The pmf of a r.v. X is given as

$$\begin{aligned}
p_X(x) &= \tfrac{1}{2} & x &= -1 \\
&= \tfrac{1}{4} & &= 0 \\
&= \tfrac{1}{8} & &= 1 \\
&= \tfrac{1}{8} & &= 2
\end{aligned}$$

Determine the pmf of Y if Y is related to X by $Y = 2X + 1$.

The corresponding values of Y are $g(-1) = 2(-1) + 1 = -1$, $g(0) = 1$, $g(1) = 3$, and $g(2) = 5$. Hence, the pmf of Y is given by

$$\begin{aligned}
p_Y(y) &= \tfrac{1}{2} & y &= -1 \\
&= \tfrac{1}{4} & &= 1 \\
&= \tfrac{1}{8} & &= 3 \\
&= \tfrac{1}{8} & &= 5
\end{aligned}$$

Example 5.2. For the same X as given in Example 5.1, determine the pmf of Y if $Y = 2X^2 + 1$.

In this case, the corresponding values of Y are $g(-1) = 2(-1)^2 + 1 = 3$, $g(0) = 1$, $g(1) = 3$ and $g(2) = 9$, resulting in

$$\begin{aligned} p_Y(y) &= \tfrac{1}{4} & y &= 1 \\ &= \tfrac{5}{8} \left(= \tfrac{1}{2} + \tfrac{1}{8}\right) & &= 3 \\ &= \tfrac{1}{8} & &= 9 \end{aligned}$$

Continuous Random Variables. A more frequently encountered case arises when X is continuous with known PDF, $F_X(x)$, or pdf, $f_X(x)$. To carry out the mapping steps as outlined at the beginning of this section, care must be exercised in choosing appropriate corresponding regions in the range spaces R_X and R_Y, this mapping being governed by the transformation $Y = g(X)$. Thus, the degree of complexity in determining the probability distribution of Y is a function of the complexity in the transformation $g(X)$.

Let us start by considering a simple relationship

$$Y = g(X) = 2X + 1 \tag{5.5}$$

The transformation $y = g(x)$ is presented graphically in Figure 5.2. Consider the PDF of Y, $F_Y(y)$, it is defined by

$$F_Y(y) = P(Y \le y) \tag{5.6}$$

The region defined by $Y \le y$ in the range space R_Y covers the heavier portion of the

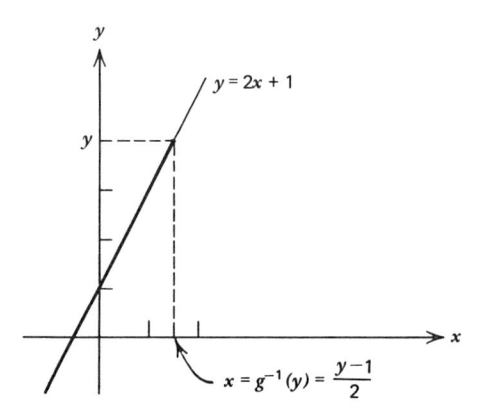

Figure 5.2 Transformation defined by Equation 5.5.

FUNCTIONS OF RANDOM VARIABLES

transformation curve as shown in Figure 5.2 which, in the range space R_X, corresponds to the region $g(X) \leq y$ or $X \leq g^{-1}(y)$ where

$$g^{-1}(y) = \frac{y-1}{2}$$

is the inverse function of $g(x)$ or the solution for x of Equation 5.5 in terms of y. Hence,

$$F_Y(y) = P(Y \leq y) = P(g(X) \leq y) = P[X \leq g^{-1}(y)] = F_X[g^{-1}(y)] \quad (5.7)$$

Equation 5.7 gives the relationship between the PDF of X and that of Y, our desired result.

The relationship between the pdf's of X and Y are obtained by differentiating both sides of Equation 5.7 with respect to y. We have

$$f_Y(y) = \frac{dF_Y(y)}{dy} = \frac{d}{dy}\{F_X[g^{-1}(y)]\} = f_X[g^{-1}(y)]\frac{dg^{-1}(y)}{dy} \quad (5.8)$$

It is clear that Equations 5.7 and 5.8 hold not only for the particular transformation given by Equation 5.5 but for all continuous $g(x)$ that are *strictly monotonic increasing functions* of x, that is, $g(x_2) > g(x_1)$ whenever $x_2 > x_1$.

Consider now a slightly different situation in which the transformation is given by

$$Y = g(X) = -2X + 1 \quad (5.9)$$

Starting again with $F_Y(y) = P(Y \leq y)$ and reasoning as before, the region $Y \leq y$ in the range space R_Y is now mapped into the region $X > g^{-1}(y)$ as indicated in Figure 5.3. Hence, we have in this case

$$\begin{aligned}F_Y(y) &= P(Y \leq y) = P[X > g^{-1}(y)] \\ &= 1 - P[X \leq g^{-1}(y)] = 1 - F_X[g^{-1}(y)]\end{aligned} \quad (5.10)$$

In comparison with Equation 5.7, Equation 5.10 yields a different relationship between the PDF's of X and Y due to a different $g(X)$.

The relationship between the pdf's of X and Y for this case is again obtained by differentiating both sides of Equation 5.10 with respect to y, giving

$$\begin{aligned}f_Y(y) &= \frac{dF_Y(y)}{dy} = \frac{d}{dy}\{1 - F_X[g^{-1}(y)]\} \\ &= -f_X[g^{-1}(y)]\frac{dg^{-1}(y)}{dy}\end{aligned} \quad (5.11)$$

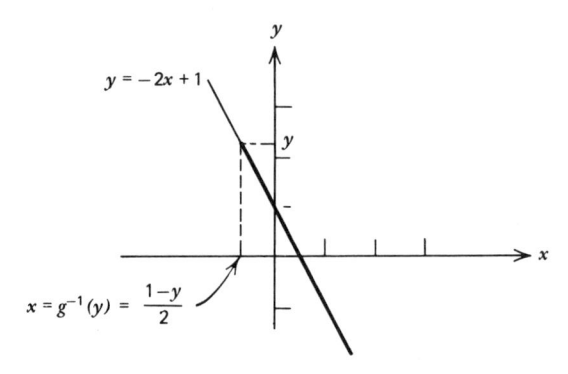

Figure 5.3 Transformation defined by Equation 5.9.

Again, we observe that Equations 5.10 and 5.11 hold for all continuous $g(x)$ that are *strictly monotonic decreasing functions* of x, that is, $g(x_2) < g(x_1)$ whenever $x_2 > x_1$.

Since the derivative $dg^{-1}(y)/dy$ in Equation 5.8 is always positive as $g(x)$ is strictly monotonic increasing and it is always negative in Equation 5.11 as $g(x)$ is strictly monotonic decreasing, the results expressed by these two equations can be combined to arrive at the following theorem.

Theorem

Let X be a continuous r.v. and $Y = g(X)$ where $g(X)$ is continuous in X and strictly monotone. Then

$$f_Y(y) = f_X[g^{-1}(y)] \left| \frac{dg^{-1}(y)}{dy} \right| \tag{5.12}$$

where $|u|$ denotes absolute value of u.

Example 5.3. The pdf of X is given by (Cauchy distribution)

$$f_X(x) = \frac{a}{\pi(x^2 + a^2)} \qquad -\infty < x < \infty \tag{5.13}$$

Determine the density function of Y where

$$Y = 2X + 1 \tag{5.14}$$

FUNCTIONS OF RANDOM VARIABLES

The transformation given by Equation 5.14 is strictly monotone. Equation 5.12 thus applies and we have

$$g^{-1}(y) = \frac{y-1}{2} \quad \text{and} \quad \frac{dg^{-1}(y)}{dy} = \frac{1}{2}$$

Following Equation 5.12, the result is

$$\begin{aligned} f_Y(y) &= f_X\left[\frac{y-1}{2}\right]\left(\frac{1}{2}\right) \\ &= \frac{a}{2\pi} \frac{1}{[(y-1)^2/4 + a^2]} \\ &= \frac{2a}{\pi} \frac{1}{(y-1)^2 + 4a^2} \quad -\infty < y < \infty \end{aligned} \quad (5.15)$$

It is valid over the entire range $-\infty < y < \infty$ as it is in correspondence with the range $-\infty < x < \infty$ defined in the range space R_X.

Example 5.4. The angle Φ of a pendulum as measured from the vertical is a r.v. uniformly distributed over the interval $(-\pi/2 < \Phi < \pi/2)$. Determine the pdf of Y, the horizontal distance, as shown in Figure 5.4.

The transformation equation in this case is

$$Y = \tan \Phi \quad (5.16)$$

where

$$\begin{aligned} f_\Phi(\phi) &= \frac{1}{\pi} \quad -\frac{\pi}{2} < \phi < \frac{\pi}{2} \\ &= 0 \quad \text{elsewhere} \end{aligned} \quad (5.17)$$

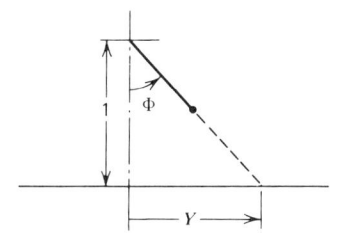

Figure 5.4 Example 5.4.

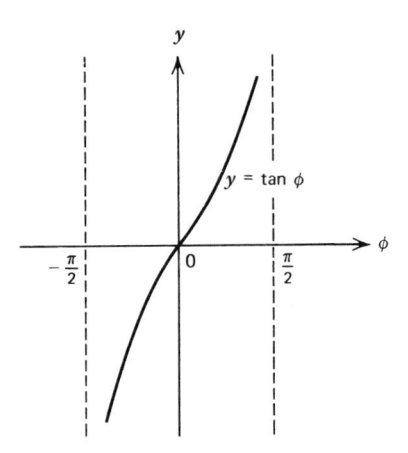

Figure 5.5 Transformation defined by Equation 5.16.

As shown in Figure 5.5, Equation 5.16 is monotone within the range $-\pi/2 < \phi < \pi/2$. Hence, Equation 5.12 again applies and we have

$$g^{-1}(y) = \tan^{-1} y \qquad \frac{dg^{-1}(y)}{dy} = \frac{1}{1+y^2}$$

The pdf of Y is thus given by

$$f_Y(y) = \frac{f_\Phi(\tan^{-1} y)}{1+y^2}$$
$$= \frac{1}{\pi(1+y^2)} \qquad -\infty < y < \infty \qquad (5.18)$$

The range in range space R_Y corresponding to $-\pi/2 < \phi < \pi/2$ is $-\infty < y < \infty$. The pdf given above is thus valid for the whole range of y. The r.v. Y has a Cauchy distribution and is plotted in Figure 5.6.

Example 5.5. The resistance R in the circuit shown in Figure 5.7 is random and has a triangular distribution as shown in Figure 5.8. With a constant current $i = 0.1$ amps and a constant resistance $r_0 = 100$ ohms, determine the pdf of voltage V.

FUNCTIONS OF RANDOM VARIABLES

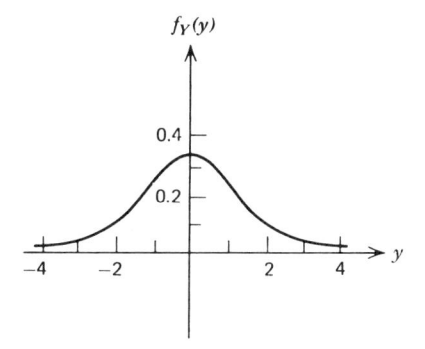

Figure 5.6 $f_Y(y)$ in Example 5.4.

The relationship between V and R is

$$V = i(R + r_0) = 0.1(R + 100) \tag{5.19}$$

and

$$\begin{aligned} f_R(r) &= 0.005(r - 90) & 90 \leq r \leq 110 \\ &= 0 & \text{elsewhere} \end{aligned} \tag{5.20}$$

The range $90 \leq r \leq 110$ corresponds to $19 \leq v \leq 21$ in the range space R_V. It is clear that $f_V(v)$ is zero outside the interval $19 \leq v \leq 21$. In this interval, since Equation 5.19 represents a strictly monotone function, we obtain by means of Equation 5.12

$$f_V(v) = f_R[g^{-1}(v)] \frac{dg^{-1}(v)}{dv} \qquad 19 \leq v \leq 21$$

where

$$g^{-1}(v) = -100 + 10v \qquad \text{and} \qquad \frac{dg^{-1}(v)}{dv} = 10$$

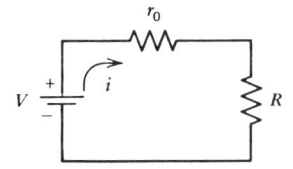

Figure 5.7 Circuit in Example 5.5.

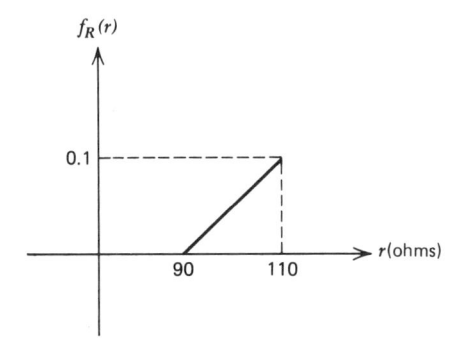

Figure 5.8 $f_R(r)$ in Example 5.5.

We thus have

$$f_V(v) = 0.005(-100 + 10v - 90)(10)$$
$$= 0.5(v - 19) \qquad 19 \leq v \leq 21$$

and

$$f_V(v) = 0 \qquad \text{elsewhere}$$

The pdf of V is plotted in Figure 5.9.

In the examples given above, it is easy to verify that all density functions obtained satisfy required properties.

Let us now turn our attention to the more general case where the function $Y = g(X)$ is not necessarily strictly monotone. Two examples are given in Figures 5.10 and 5.11. In Figure 5.10, monotonic property of the transformation holds for $y < y_1$ and $y > y_2$ and Equation 5.12 can be used to determine the pdf of Y in these

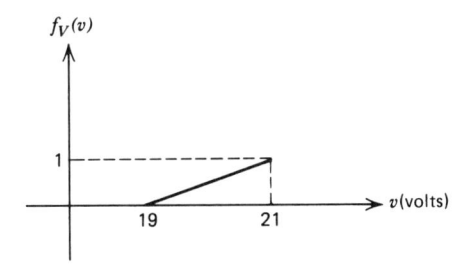

Figure 5.9 $f_V(v)$ in Example 5.5.

FUNCTIONS OF RANDOM VARIABLES

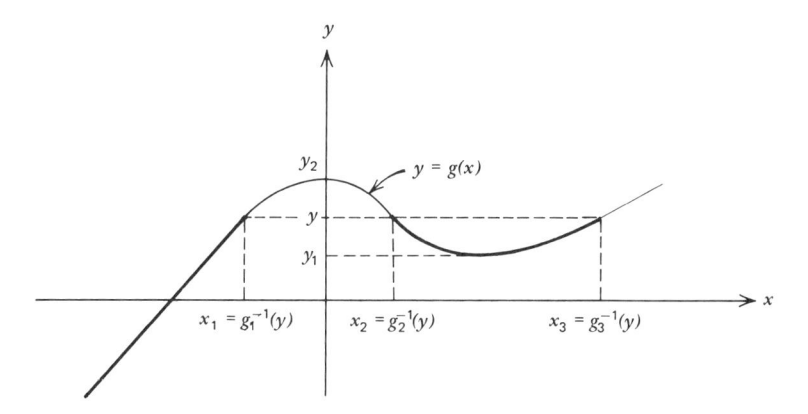

Figure 5.10 An example of nonmonotonic function $y = g(x)$.

intervals of y. For $y_1 \leq y \leq y_2$, however, we must start from the beginning and consider $F_Y(y) = P(Y \leq y)$. The region defined by $Y \leq y$ in the range space R_Y covers the heavier portions of the function $y = g(x)$ as shown in Figure 5.10. Thus,

$$\begin{aligned} F_Y(y) = P(Y \leq y) &= P[X \leq g_1^{-1}(y)] + P[g_2^{-1}(y) < X \leq g_3^{-1}(y)] \\ &= P[X \leq g_1^{-1}(y)] + P[X \leq g_3^{-1}(y)] - P[X \leq g_2^{-1}(y)] \\ &= F_X[g_1^{-1}(y)] + F_X[g_3^{-1}(y)] - F_X[g_2^{-1}(y)] \end{aligned} \quad (5.21)$$

$$y_1 \leq y \leq y_2$$

where $x_1 = g_1^{-1}(y)$, $x_2 = g_2^{-1}(y)$, and $x_3 = g_3^{-1}(y)$ are roots of the function $y = g(x)$ in terms of y.

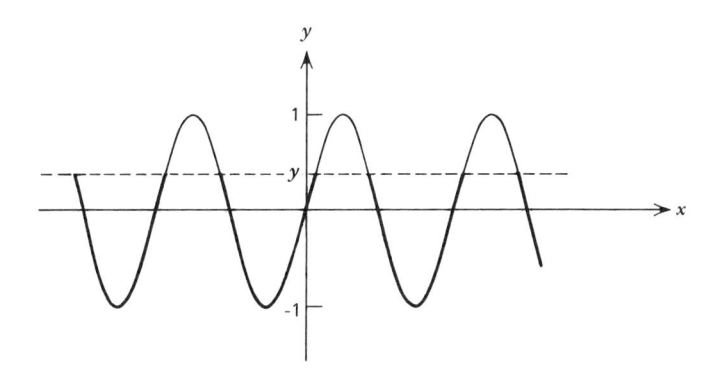

Figure 5.11 An example of nonmonotonic function $y = g(x)$.

As before, the relationship between the pdf's of X and Y is found by differentiating Equation 5.21 with respective to y. It is given by

$$f_Y(y) = f_X[g_1^{-1}(y)] \frac{dg_1^{-1}(y)}{dy} + f_X[g_3^{-1}(y)] \frac{dg_3^{-1}(y)}{dy}$$

$$-f_X[g_2^{-1}(y)] \frac{dg_2^{-1}(y)}{dy} \qquad y_1 \leq y \leq y_2 \qquad (5.22)$$

Since the derivative $dg_2^{-1}(y)/dy$ is negative while the others are positive, Equation 5.22 takes the convenient form

$$f_Y(y) = \sum_{j=1}^{3} f_X[g_j^{-1}(y)] \left| \frac{dg_j^{-1}(y)}{dy} \right| \qquad y_1 \leq y \leq y_2 \qquad (5.23)$$

Figure 5.11 represents the transformation $y = \sin x$, this equation has infinite (but countable) number of roots $x_1 = g_1^{-1}(y)$, $x_2 = g_2^{-1}(y)$, ... for any y in the interval $-1 \leq y \leq 1$. Following the procedure outlined above, an equation similar to Equation 5.21 (but with infinite number of terms) can be established for $F_Y(y)$ and, as seen from Equation 5.23, the pdf of Y now has the form

$$f_Y(y) = \sum_{j=1}^{\infty} f_X[g_j^{-1}(y)] \left| \frac{dg_j^{-1}(y)}{dy} \right| \qquad -1 \leq y \leq 1 \qquad (5.24)$$

It is clear from Figure 5.11 that $f_Y(y) = 0$ elsewhere.

A general pattern now emerges when the function $Y = g(X)$ is nonmonotonic. Equations 5.23 and 5.24 lead to the following important theorem.

Theorem.

Let X be a continuous r.v. and $Y = g(X)$, where $g(X)$ is continuous in X, and $y = g(x)$ admits at most a countable number of roots $x_1 = g_1^{-1}(y)$, $x_2 = g_2^{-1}(y)$, ... Then

$$\boxed{f_Y(y) = \sum_{j=1}^{r} f_X[g_j^{-1}(y)] \left| \frac{dg_j^{-1}(y)}{dy} \right|} \qquad (5.25)$$

where r is the number of roots for x of the equation $y = g(x)$. Clearly, Equation 5.12 is contained in this theorem as a special case ($r = 1$).

FUNCTIONS OF RANDOM VARIABLES

Example 5.6. In Example 5.4, let r.v. Φ now be uniformly distributed over the interval $-\pi < \Phi < \pi$. Determine the pdf of $Y = \tan \Phi$.

The pdf of X is now

$$f_\Phi(\phi) = \frac{1}{2\pi} \qquad -\pi < \phi < \pi$$

$$= 0 \qquad \text{elsewhere}$$

and the relevant portion of the transformation equation is plotted in Figure 5.12. For each y, the two roots ϕ_1 and ϕ_2 of $y = \tan \phi$ are (see Figure 5.12)

$$\left.\begin{aligned}\phi_1 = g_1^{-1}(y) = \tan^{-1} y & \qquad -\frac{\pi}{2} < \phi_1 \leq 0 \\ \phi_2 = g_2^{-1}(y) = \tan^{-1} y & \qquad \frac{\pi}{2} < \phi_2 \leq \pi\end{aligned}\right\} y \leq 0$$

$$\left.\begin{aligned}\phi_1 = \tan^{-1} y & \qquad -\pi < \phi_1 \leq -\frac{\pi}{2} \\ \phi_2 = \tan^{-1} y & \qquad 0 < \phi_2 \leq \frac{\pi}{2}\end{aligned}\right\} y > 0$$

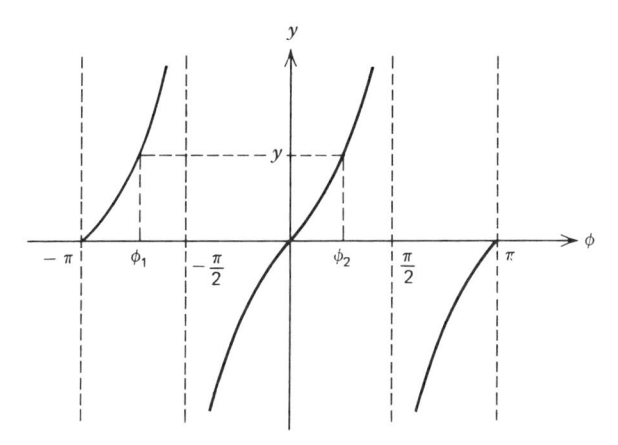

Figure 5.12 Transformation $y = \tan \phi$

For all y, Equation 5.25 yields

$$f_Y(y) = \sum_{j=1}^{2} f_\Phi[g_j^{-1}(y)] \left| \frac{dg_j^{-1}(y)}{dy} \right|$$

$$= \frac{1}{2\pi}\left(\frac{1}{1+y^2}\right) + \frac{1}{2\pi}\left(\frac{1}{1+y^2}\right)$$

$$= \frac{1}{\pi(1+y^2)} \qquad -\infty < y < \infty \qquad (5.26)$$

a result identical to the solution for Example 5.4.

Example 5.7. Determine the pdf of $Y = X^2$ where X is normally distributed according to

$$f_X(x) = \frac{1}{\sqrt{2\pi}} e^{-x^2/2} \qquad -\infty < x < \infty \qquad (5.27)$$

As shown in Figure 5.13, $f_Y(y) = 0$ for $y < 0$ since the transformation equation has no real roots in this range. For $y \geq 0$, the two roots of $y = x^2$ are

$$x_{1,2} = g_{1,2}^{-1}(y) = \pm\sqrt{y}$$

Hence, using Equation 5.25,

$$f_Y(y) = \sum_{j=1}^{2} f_X[g_j^{-1}(y)] \left| \frac{dg_j^{-1}(y)}{dy} \right|$$

$$= \frac{f_X(-\sqrt{y})}{2\sqrt{y}} + \frac{f_X(\sqrt{y})}{2\sqrt{y}}$$

$$= \frac{1}{\sqrt{2\pi y}} e^{-y/2}$$

or

$$f_Y(y) = \frac{1}{\sqrt{2\pi y}} e^{-y/2} \qquad y \geq 0$$

$$= 0 \qquad \text{elsewhere} \qquad (5.28)$$

This is a so-called χ^2 distribution, to be discussed in more detail in Chapter VII.

FUNCTIONS OF RANDOM VARIABLES   127

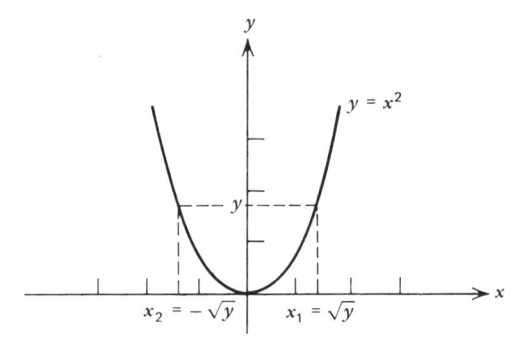

Figure 5.13 Transformation $y = x^2$.

Example 5.8. A random voltage V_1 having a uniform distribution over the interval $90 \leq V_1 \leq 110$ volts is put into a nonlinear device (a limiter) as shown in Figure 5.14. Determine the probability distribution of the output voltage V_2.

The relationship between V_1 and V_2 is

$$V_2 = g(V_1) \tag{5.29}$$

where

$$g(V_1) = 0 \qquad V_1 < 95$$

$$g(V_1) = \frac{V_1 - 95}{10} \qquad 95 \leq V_1 \leq 105$$

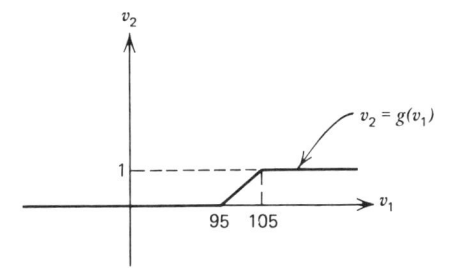

$$g(V_1) = 1 \qquad V_1 > 105$$

Figure 5.14 Transformation defined by Equation 5.29.

The theorems stated in this section do not apply in this case to the portions $v_1 < 95$ and $v_1 > 105$ because infinite and noncountable number of roots for v_1 exist in these regions. However, we deduce immediately from Figure 5.14 that

$$P(V_2 = 0) = P(V_1 \leq 95) = F_{V_1}(95)$$
$$= \int_{90}^{95} f_{V_1}(v_1) dv_1 = \tfrac{1}{4}$$
$$P(V_2 = 1) = P(V_1 > 105) = 1 - F_{V_1}(105)$$
$$= \tfrac{1}{4}$$

For the middle portion, Equation 5.7 leads to

$$F_{V_2}(v_2) = F_{V_1}[g^{-1}(v_2)]$$
$$= F_{V_1}(10v_2 + 95) \qquad 0 < v_2 < 1$$

Now,

$$F_{V_1}(v_1) = \frac{v_1 - 90}{20} \qquad 90 \leq v_1 \leq 110$$

We thus have

$$F_{V_2}(v_2) = \tfrac{1}{20}(10v_2 + 95 - 90) = \frac{2v_2 + 1}{4} \qquad 0 < v_2 < 1$$

The PDF $F_{V_2}(v_2)$ is shown in Figure 5.15, an example of a mixed distribution.

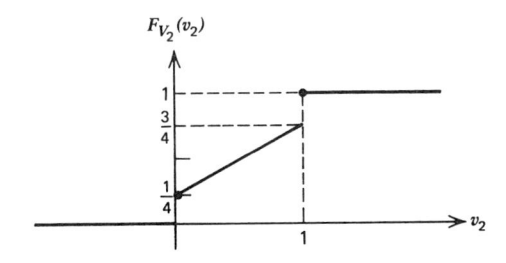

Figure 5.15 $F_{V_2}(v_2)$ in Example 5.8.

V.1.2 MOMENTS

Having developed methods of determining the probability distribution of $Y = g(X)$, it is a straightforward matter to calculate all the desired moments of Y if they exist. However, this procedure—the determination of moments of Y on finding the probability law of Y—is cumbersome and unneccessary if *only* the moments of Y are of interest.

A more expedient and direct way of finding the moments of $Y = g(X)$, given the probability law of X, is to express moments of Y as expectations of appropriate functions of X; they can then be evaluated directly within the probability domain of X. In fact, all machinery for proceeding along this line is contained in Equations 4.1 and 4.2.

Let $Y = g(X)$ and assume that all desired moments of Y exist. The nth moment of Y can be expressed as

$$E\{Y^n\} = E\{g^n(X)\} \tag{5.30}$$

It follows from Equations 4.1 and 4.2 that, in terms of pmf or pdf of X,

$$\boxed{\begin{aligned} E\{Y^n\} &= E\{g^n(X)\} = \sum_i g^n(x_i) p_X(x_i) & (X \text{ discrete}) \\ E\{Y^n\} &= E\{g^n(X)\} = \int_{-\infty}^{\infty} g^n(x) f_X(x) dx & (X \text{ continuous}) \end{aligned}} \tag{5.31}$$

An alternate approach is to determine the characteristic function of Y from which all moments of Y can be generated through differentiation. As we see from the definition (Equations 4.46 and 4.47), the characteristic function of Y can be expressed by

$$\phi_Y(t) = E\{e^{jtY}\} = E\{e^{jtg(X)}\} = \sum_i e^{jtg(x_i)} p_X(x_i) \quad (X \text{ discrete})$$

$$\phi_Y(t) = E\{e^{jtg(X)}\} = \int_{-\infty}^{\infty} e^{jtg(x)} f_X(x) dx \quad (X \text{ continuous}) \tag{5.32}$$

Upon evaluating $\phi_Y(t)$, the moments of Y are given by (Equation 4.52)

$$E\{Y^n\} = j^{-n} \phi_Y^{(n)}(0) \qquad n = 1, 2, \ldots \tag{5.33}$$

Example 5.9. A r.v. X is discrete and its pmf is given in Example 5.1. Determine the mean and variance of $Y = 2X + 1$

Using the first of Equations 5.31, we obtain

$$E\{Y\} = E\{2X + 1\} = \sum_i (2x_i + 1)p_X(x_i)$$

$$= (-1)(\tfrac{1}{2}) + (1)(\tfrac{1}{4}) + (3)(\tfrac{1}{8}) + (5)(\tfrac{1}{8})$$

$$= \tfrac{3}{4} \qquad (5.34)$$

$$E\{Y^2\} = E\{(2X + 1)^2\} = \sum_i (2x_i + 1)^2 p_X(x_i)$$

$$= (1)(\tfrac{1}{2}) + (1)(\tfrac{1}{4}) + (9)(\tfrac{1}{8}) + (25)(\tfrac{1}{8})$$

$$= 5 \qquad (5.35)$$

and

$$\sigma_Y^2 = E\{Y^2\} - E^2\{Y\} = 5 - (\tfrac{3}{4})^2 = \tfrac{71}{16} \qquad (5.36)$$

As a second approach, let us use the method of characteristic functions described by Equations 5.32 and 5.33. The characteristic function of Y is

$$\phi_Y(t) = \sum_i e^{jt(2x_i + 1)} p_X(x_i)$$

$$= e^{-jt}(\tfrac{1}{2}) + e^{jt}(\tfrac{1}{4}) + e^{3jt}(\tfrac{1}{8}) + e^{5jt}(\tfrac{1}{8})$$

$$= \tfrac{1}{8}(4e^{-jt} + 2e^{jt} + e^{3jt} + e^{5jt})$$

and we have

$$E\{Y\} = j^{-1}\phi_Y^{(1)}(0) = j^{-1}\left(\frac{j}{8}\right)(-4 + 2 + 3 + 5) = \tfrac{3}{4}$$

$$E\{Y^2\} = -\phi_Y^{(2)}(0) = \tfrac{1}{8}(4 + 2 + 9 + 25) = 5$$

As expected, these answers agree with the results obtained earlier (Equations 5.34 and 5.35).

Let us again remark that the procedures described above do not require the knowledge of $f_Y(y)$. One can determine $f_Y(y)$ before moment calculations but it is less expedient when only moments of Y are desired. Another remark to be made is

FUNCTIONS OF RANDOM VARIABLES 131

that, since the transformation is linear ($Y = 2X + 1$) in this case, only the first two moments of X are needed in finding the first two moments of Y, that is,

$$E\{Y\} = E\{2X + 1\} = 2E\{X\} + 1$$
$$E\{Y^2\} = E\{(2X + 1)^2\} = 4E\{X^2\} + 4E\{X\} + 1$$

as seen from Equations 5.34 and 5.35. When the transformation is nonlinear, on the other hand, moments of X of different orders will be needed.

Example 5.10. In Example 5.7, determine the mean and variance of $Y = X^2$. The mean of Y is, in terms of $f_X(x)$,

$$E\{Y\} = E\{X^2\} = \frac{1}{\sqrt{2\pi}} \int_{-\infty}^{\infty} x^2 e^{-x^2/2}\, dx = 1 \qquad (5.37)$$

and the second moment of Y is given by

$$E\{Y^2\} = E\{X^4\} = \frac{1}{\sqrt{2\pi}} \int_{-\infty}^{\infty} x^4 e^{-x^2/2}\, dx = 3 \qquad (5.38)$$

Thus,

$$\sigma_Y^2 = E\{Y^2\} - E^2\{Y\} = 3 - 1 = 2 \qquad (5.39)$$

In this case, the complete knowledge of $f_X(x)$ is not needed but we need to know the second and fourth moments of X.

V.2 Functions of Two or More Random Variables

In this section we extend earlier results to a more general case. The r.v. Y is now a function of n jointly distributed random variables $X_1, X_2, \ldots, X_n$. Formulas will be developed for the corresponding distribution for Y.

As in the single r.v. case, the case in which $X_1, X_2, \ldots,$ and X_n are discrete random variables presents no problem and we will demonstrate this by way of an example (Example 5.13). Our basic interest here lies in the determination of the distribution of Y when all $X_j, j = 1, 2, \ldots, n$, are continuous random variables.

Consider the transformation

$$Y = g(X_1, \ldots, X_n) \qquad (5.40)$$

where the joint distribution of $X_1, X_2, \ldots,$ and X_n is assumed to be specified in terms of their jpdf $f_{X_1 \cdots X_n}(x_1, \ldots, x_n)$ or their JPDF $F_{X_1 \cdots X_n}(x_1, \ldots, x_n)$. In a more compact notation, they can be written as $f_\mathbf{X}(\mathbf{x})$ and $F_\mathbf{X}(\mathbf{x})$, respectively, where $\mathbf{X}$ is an n-dimensional random vector with components $X_1, X_2, \ldots, X_n$.

The starting point of the derivation is the same as in the single r.v. case, that is, we consider $F_Y(y) = P(Y \leq y)$. In terms of $\mathbf{X}$, this probability is equal to $P[g(\mathbf{X}) \leq y]$. Thus

$$F_Y(y) = P(Y \leq y) = P[g(\mathbf{X}) \leq y]$$
$$= F_\mathbf{X}[\mathbf{x}: g(\mathbf{x}) \leq y] \qquad (5.41)$$

The last expression in the above represents the JPDF of $\mathbf{X}$ whose argument $\mathbf{x}$ satisfies $g(\mathbf{x}) \leq y$. In terms of $f_\mathbf{X}(\mathbf{x})$, it is given by

$$\boxed{F_\mathbf{X}[\mathbf{x}: g(\mathbf{x}) \leq y] = \int \cdots \int_{(R^n: g(\mathbf{x}) \leq y)} f_\mathbf{X}(\mathbf{x}) d\mathbf{x}} \qquad (5.42)$$

where the limits of integrals are determined by an n-dimensional region R^n within which $g(\mathbf{x}) \leq y$ is satisfied. In view of Equations 5.41 and 5.42, the PDF of Y, $F_Y(y)$, can be determined by evaluating the n-dimensional integral in Equation 5.42. The crucial step in this derivation is clearly the identification of R^n, which must be carried out on a problem-to-problem basis. As n becomes large, this can present a formidable obstacle.

The foregoing outlined procedure can be best demonstrated through examples.

Example 5.11. Let $Y = X_1 X_2$. Determine the pdf of Y in terms of $f_{X_1 X_2}(x_1, x_2)$.

From Equations 5.41 and 5.42, we have

$$F_Y(y) = \iint_{(R^2: x_1 x_2 \leq y)} f_{X_1 X_2}(x_1, x_2) dx_1 \, dx_2 \qquad (5.43)$$

The equation $x_1 x_2 = y$ is shown in Figure 5.16 in which the shaded area represents R^2 or $x_1 x_2 \leq y$. The limits of the double integral can thus be determined and Equation 5.43 becomes

$$F_Y(y) = \int_0^\infty \int_{-\infty}^{y/x_2} f_{X_1 X_2}(x_1, x_2) dx_1 \, dx_2$$
$$+ \int_{-\infty}^0 \int_{y/x_2}^\infty f_{X_1 X_2}(x_1, x_2) dx_1 \, dx_2 \qquad (5.44)$$

FUNCTIONS OF RANDOM VARIABLES

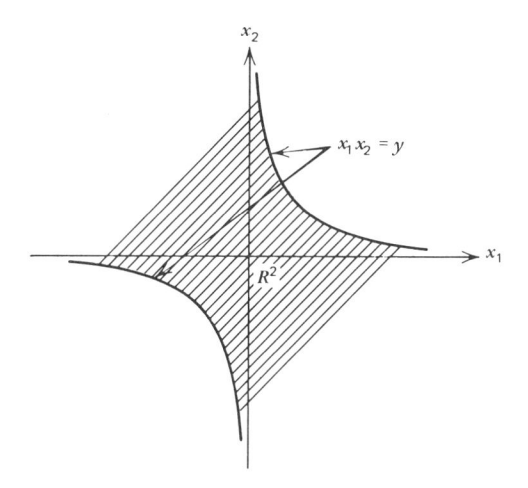

Figure 5.16 Region R^2 in Example 5.11.

Substituting $f_{X_1X_2}(x_1, x_2)$ into Equation 5.44 enables us to determine $F_Y(y)$ and, on differentiating with respect to y, gives $f_Y(y)$.

For the special case where X_1 and X_2 are independent, we have $f_{X_1X_2}(x_1, x_2) = f_{X_1}(x_1) f_{X_2}(x_2)$ and Equation 5.44 simplifies to

$$F_Y(y) = \int_0^\infty F_{X_1}\left(\frac{y}{x_2}\right) f_{X_2}(x_2) dx_2 + \int_{-\infty}^0 \left[1 - F_{X_1}\left(\frac{y}{x_2}\right)\right] f_{X_2}(x_2) dx_2$$

and

$$f_Y(y) = \frac{dF_Y(y)}{dy} = \int_{-\infty}^\infty f_{X_1}\left(\frac{y}{x_2}\right) f_{X_2}(x_2) \left|\frac{1}{x_2}\right| dx_2 \tag{5.45}$$

As a numerical example, suppose that X_1 and X_2 are independent and

$$f_{X_1}(x_1) = 2x_1 \quad 0 \leq x_1 \leq 1$$
$$= 0 \quad \text{elsewhere}$$

$$f_{X_2}(x_2) = \frac{2 - x_2}{2} \quad 0 \leq x_2 \leq 2$$
$$= 0 \quad \text{elsewhere}$$

The pdf of Y is, following Equation 5.45,

$$f_Y(y) = \int_{-\infty}^{\infty} f_{X_1}\left(\frac{y}{x_2}\right) f_{X_2}(x_2) \left|\frac{1}{x_2}\right| dx_2$$

$$= \int_y^2 2\left(\frac{y}{x_2}\right)\left(\frac{2-x_2}{2}\right)\left(\frac{1}{x_2}\right) dx_2 \qquad 0 \le y \le 2$$

$$= 0 \qquad \qquad \text{elsewhere} \qquad (5.46)$$

In the above, the integration limits are determined from the fact that $f_{X_1}(x_1)$ and $f_{X_2}(x_2)$ are nonzero in the intervals $0 \le x_1 \le 1$ and $0 \le x_2 \le 2$. With the argument of $f_{X_1}(x_1)$ replaced by y/x_2 in the integral, we have $0 \le y/x_2 \le 1$ and $0 \le x_2 \le 2$, which are equivalent to $y \le x_2 \le 2$. Also, the range $0 \le y \le 2$ for the nonzero portion of $f_Y(y)$ is determined from the fact that, since $y = x_1 x_2$, the intervals $0 \le x_1 \le 1$ and $0 \le x_2 \le 2$ directly give $0 \le y \le 2$.

Finally, Equation 5.46 gives

$$f_Y(y) = 2 + y(\ln y - 1 - \ln 2) \qquad 0 \le y \le 2$$
$$= 0 \qquad \qquad \text{elsewhere} \qquad (5.47)$$

This is shown graphically in Figure 5.17. It is an easy exercise to show that

$$\int_0^2 f_Y(y) dy = 1$$

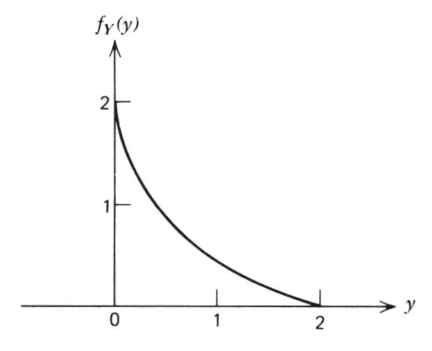

Figure 5.17 $f_Y(y)$ in Example 5.11.

FUNCTIONS OF RANDOM VARIABLES

Example 5.12. Let $Y = X_1/X_2$ where X_1 and X_2 are independent and identically distributed according to

$$f_{X_1}(x_1) = e^{-x_1} \quad x_1 > 0$$
$$= 0 \quad \text{elsewhere} \quad (5.48)$$

and similarly for X_2. Determine $f_Y(y)$.

It follows from Equations 5.41 and 5.42 that

$$F_Y(y) = \iint\limits_{(R^2: x_1/x_2 \leq y)} f_{X_1 X_2}(x_1, x_2) dx_1\, dx_2$$

The region R^2 for positive values of x_1 and x_2 is shown as the shaded area in Figure 5.18. Hence,

$$F_Y(y) = \int_0^\infty \int_0^{x_2 y} f_{X_1 X_2}(x_1, x_2) dx_1\, dx_2 \quad y > 0$$
$$= 0 \quad \text{elsewhere}$$

For independent X_1 and X_2,

$$F_Y(y) = \int_0^\infty \int_0^{x_2 y} f_{X_1}(x_1) f_{X_2}(x_2) dx_1\, dx_2$$
$$= \int_0^\infty F_{X_1}(x_2 y) f_{X_2}(x_2) dx_2 \quad y > 0$$
$$= 0 \quad \text{elsewhere} \quad (5.49)$$

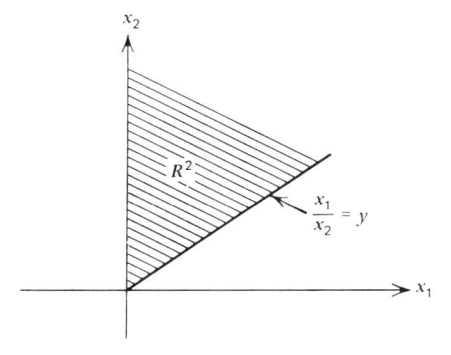

Figure 5.18 Region R^2 in Example 5.12.

The pdf of Y is thus given by, on differentiating Equation 5.49 with respect to y,

$$f_Y(y) = \int_0^\infty x_2 f_{X_1}(x_2 y) f_{X_2}(x_2) dx_2 \qquad y > 0$$
$$= 0 \qquad \text{elsewhere} \qquad (5.50)$$

and, on substituting Equation 5.48 into Equation 5.50, it takes the form

$$f_Y(y) = \int_0^\infty x_2 e^{-x_2 y} e^{-x_2} dx_2 = \frac{1}{(1+y)^2} \qquad y > 0 \qquad (5.51)$$
$$= 0 \qquad \text{elsewhere}$$

Again, it is easy to check that

$$\int_0^\infty f_Y(y) dy = 1$$

Example 5.13. To show that it is elementary to obtain solutions to problems discussed in this section when $X_1, X_2, \ldots,$ and X_n are discrete, consider again $Y = X_1/X_2$ given that X_1 and X_2 are discrete and their jpmf $p_{X_1 X_2}(x_1, x_2)$ is tabulated in Table 5.1. In this case, the pmf of Y is easily determined by assignment of probabilities $p_{X_1 X_2}(x_1, x_2)$ to the corresponding values of $y = x_1/x_2$. Thus, we obtain

$$\begin{aligned}
p_Y(y) &= 0.5 & y &= \tfrac{1}{2} \\
&= 0.24 + 0.04 = 0.28 & &= 1 \\
&= 0.04 & &= \tfrac{3}{2} \\
&= 0.06 & &= 2 \\
&= 0.12 & &= 3
\end{aligned}$$

Table 5.1. $p_{X_1 X_2}(x_1, x_2)$ in Example 5.13

x_2 \ x_1	1	2	3
1	0.04	0.06	0.12
2	0.5	0.24	0.04

FUNCTIONS OF RANDOM VARIABLES

Example 5.14. In structural reliability studies, the probability of failure p_f is defined by

$$p_f = P(R \leq S)$$

where R and S represent, respectively, structural resistance and applied force. Let R and S be independent random variables taking only positive values. Determine p_f in terms of the probability distributions associated with R and S.

Let $Y = R/S$. The probability p_f can be expressed by

$$p_f = P\left(\frac{R}{S} \leq 1\right) = P(Y \leq 1) = F_Y(1)$$

Identifying R and S with X_1 and X_2, respectively, in Example 5.12, it follows from Equation 5.49 that

$$p_f = F_Y(1) = \int_0^\infty F_R(s) f_S(s) ds$$

Example 5.15. Determine $F_Y(y)$ in terms of $f_{X_1 X_2}(x_1, x_2)$ when $Y = \min(X_1, X_2)$.

Now,

$$F_Y(y) = \iint\limits_{(R^2:\,\min(x_1, x_2)\,\leq\, y)} f_{X_1 X_2}(x_1, x_2) dx_1\, dx_2$$

where the region R^2 is shown in Figure 5.19. Thus

$$F_Y(y) = \int_{-\infty}^{y} \int_{-\infty}^{\infty} f_{X_1 X_2}(x_1, x_2) dx_1\, dx_2 + \int_{y}^{\infty} \int_{-\infty}^{y} f_{X_1 X_2}(x_1, x_2) dx_1\, dx_2$$

$$= \int_{-\infty}^{y} \int_{-\infty}^{\infty} f_{X_1 X_2}(x_1, x_2) dx_1\, dx_2 + \int_{-\infty}^{\infty} \int_{-\infty}^{y} f_{X_1 X_2}(x_1, x_2) dx_1\, dx_2$$

$$- \int_{-\infty}^{y} \int_{-\infty}^{y} f_{X_1 X_2}(x_1, x_2) dx_1\, dx_2$$

$$= F_{X_2}(y) + F_{X_1}(y) - F_{X_1 X_2}(y, y)$$

which is the desired solution. If r.v.'s X_1 and X_2 are independent, straightforward differentiation shows that

$$f_Y(y) = f_{X_1}(y)[1 - F_{X_2}(y)] + f_{X_2}(y)[1 - F_{X_1}(y)]$$

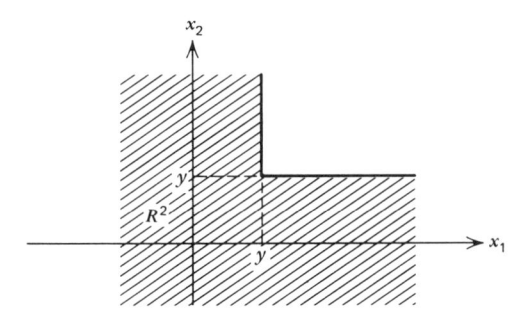

Figure 5.19 Region R^2 in Example 5.15.

Let us note here that the results given above can be obtained following a different, and more direct, procedure. Note that the event $[\min(X_1, X_2) \leq y]$ is equivalent to the event $(X_1 \leq y \cup X_2 \leq y)$. Hence,

$$F_Y(y) = P(Y \leq y) = P[\min(X_1, X_2) \leq y]$$
$$= P(X_1 \leq y \cup X_2 \leq y)$$

Since $P(A \cup B) = P(A) + P(B) - P(AB)$, we have

$$F_Y(y) = P(X_1 \leq y) + P(X_2 \leq y) - P(X_1 \leq y \cap X_2 \leq y)$$
$$= F_{X_1}(y) + F_{X_2}(y) - F_{X_1 X_2}(y, y)$$

If X_1 and X_2 are independent, we have

$$F_Y(y) = F_{X_1}(y) + F_{X_2}(y) - F_{X_1}(y)F_{X_2}(y)$$

and

$$f_Y(y) = \frac{dF_Y(y)}{dy} = f_{X_1}(y)[1 - F_{X_2}(y)] + f_{X_2}(y)[1 - F_{X_1}(y)]$$

We have not given examples in which more than two random variables are involved. While more complicated problems can be formulated in a similar fashion, it is in general more difficult to identify appropriate regions R^n required by Equation 5.42 and the integrals are, of course, more difficult to carry out. In principle, however, no intrinsic difficulties present themselves in cases of functions of more than two random variables.

FUNCTIONS OF RANDOM VARIABLES

V.2.1 SUMS OF RANDOM VARIABLES

One of the most important transformations we encounter is a sum of random variables. It has been discussed in Chapter IV in the context of characteristic functions. In fact, the technique of characteristic functions remains to be the most powerful technique for sums of *independent* random variables.

In this section, the procedure presented in the above is used to give an alternate method of attack.

Consider the sum

$$Y = g(X_1, \ldots, X_n) = X_1 + X_2 + \cdots + X_n \tag{5.52}$$

It suffices to determine $f_Y(y)$ for $n = 2$. The result for this case can then be applied successively to give the probability distribution of a sum of any number of random variables

For $Y = X_1 + X_2$, Equations 5.41 and 5.42 give

$$F_Y(y) = \iint\limits_{(R^2:\, x_1 + x_2 \leq y)} f_{X_1 X_2}(x_1, x_2) dx_1 \, dx_2$$

and, as seen from Figure 5.20,

$$F_Y(y) = \int_{-\infty}^{\infty} \int_{-\infty}^{y-x_2} f_{X_1 X_2}(x_1, x_2) dx_1 \, dx_2 \tag{5.53}$$

Upon differentiating with respect to y we obtain

$$f_Y(y) = \int_{-\infty}^{\infty} f_{X_1 X_2}(y - x_2, x_2) dx_2 \tag{5.54}$$

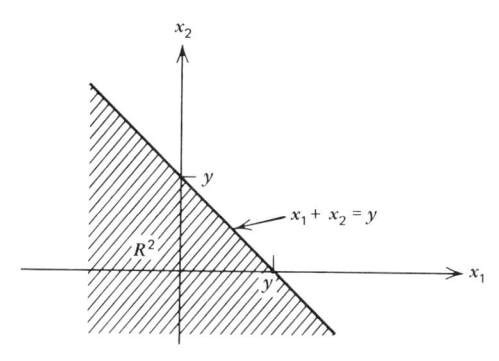

Figure 5.20 Region $R^2: x_1 + x_2 \leq y$.

When X_1 and X_2 are independent, the above result further reduces to

$$f_Y(y) = \int_{-\infty}^{\infty} f_{X_1}(y - x_2) f_{X_2}(x_2) dx_2 \tag{5.55}$$

Integrals of the form given above arise often in practice. It is called *convolution* of the functions $f_{X_1}(x_1)$ and $f_{X_2}(x_2)$.

Considerable importance is attached to the results expressed by Equations 5.54 and 5.55 because sums of random variables occur frequently in practical situations. By way of recognizing this fact, Equation 5.55 is repeated as follows as a theorem.

Theorem.

Let $Y = X_1 + X_2$ and let X_1 and X_2 be independent and continuous random variables. Then the pdf of Y is the convolution of the pdf's associated with X_1 and X_2, that is,

$$\boxed{f_Y(y) = \int_{-\infty}^{\infty} f_{X_1}(y - x_2) f_{X_2}(x_2) dx_2 = \int_{-\infty}^{\infty} f_{X_2}(y - x_1) f_{X_1}(x_1) dx_1} \tag{5.56}$$

Repeated applications of this formula determine $f_Y(y)$ when Y is a sum of any number of independent random variables.

Example 5.16. Determine $f_Y(y)$ of $Y = X_1 + X_2$ when X_1 and X_2 are independent and identically distributed according to

$$f_{X_1}(x_1) = ae^{-ax_1} \quad x_1 \geq 0$$
$$= 0 \quad \text{elsewhere} \tag{5.57}$$

and similarly for X_2.

Equation 5.56 in this case leads to

$$f_Y(y) = a^2 \int_0^y e^{-a(y - x_2)} e^{-ax_2} dx_2 \quad y \geq 0 \tag{5.58}$$

where the integration limits are determined from the requirements $y - x_2 > 0$ and $x_2 > 0$. The result is

$$f_Y(y) = a^2 y e^{-ay} \quad y \geq 0$$
$$= 0 \quad \text{elsewhere} \tag{5.59}$$

FUNCTIONS OF RANDOM VARIABLES

Let us note that this problem has also been solved in Example 4.16 by means of characteristic functions. It is to be stressed that the method of characteristic functions is another powerful technique for dealing with sums of *independent* random variables. In fact, when the number of random variables involved in a sum is large, the method of characteristic function is preferred since there is no need to consider only two at a time as required by Equation 5.56.

Example 5.17. The r.v.'s X_1 and X_2 are independent and uniformly distributed in the intervals $0 \leq x_1 \leq 1$ and $0 \leq x_2 \leq 2$. Determine the pdf of $Y = X_1 + X_2$.

The convolution of $f_{X_1}(x_1) = 1$, $0 \leq x_1 \leq 1$, and $f_{X_2}(x_2) = \frac{1}{2}$, $0 \leq x_2 \leq 2$, results in

$$f_Y(y) = \int_{-\infty}^{\infty} f_{X_1}(y - x_2) f_{X_2}(x_2) dx_2$$

$$= \int_0^y (1)(\tfrac{1}{2}) dx_2 = \frac{y}{2} \qquad 0 < y \leq 1$$

$$= \int_{y-1}^y (1)(\tfrac{1}{2}) dx_2 = \tfrac{1}{2} \qquad 1 < y \leq 2$$

$$= \int_{y-1}^2 (1)(\tfrac{1}{2}) dx_2 = \frac{3-y}{2} \qquad 2 < y \leq 3$$

$$= 0 \qquad \text{elsewhere}$$

In the above, the limits of the integrals are determined from the requirements $0 \leq y - x_2 \leq 1$ and $0 \leq x_2 \leq 2$. The shape of $f_Y(y)$ is that of a trapezoid as shown in Figure 5.21.

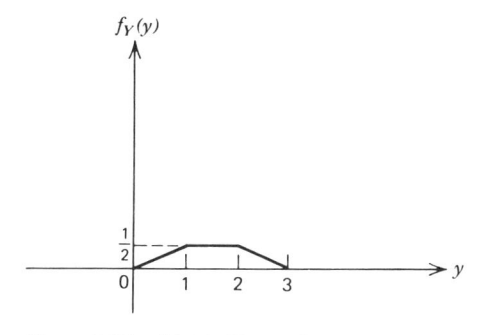

Figure 5.21 $f_Y(y)$ in Example 5.17.

V.3 n Functions of n Random Variables

We now consider the general transformation given by Equation 5.1, that is,

$$Y_j = g_j(X_1, \ldots, X_n) \qquad j = 1, 2, \ldots, m \qquad m \leq n \qquad (5.60)$$

The problem is to obtain the joint probability distribution of r.v.'s $Y_j, j = 1, 2, \ldots, m$, which arise as functions of n jointly distributed r.v.'s X_k, $k = 1, \ldots, n$. As before, we are primarily concerned with the case in which $X_1, \ldots, X_n$ are continuous random variables.

In order to develop pertinent formulas, the case of $m = n$ is first considered. We will see that the results obtained for this case encompass situations in which $m < n$.

Let **X** and **Y** be two n-dimensional random vectors with components $(X_1, \ldots, X_n)$ and $(Y_1, \ldots, Y_n)$, respectively. A vector equation representing Equation 5.60 is

$$\mathbf{Y} = \mathbf{g}(\mathbf{X}) \qquad (5.61)$$

where the vector $\mathbf{g}(\mathbf{X})$ has as components $g_1(\mathbf{X}), \ldots, g_n(\mathbf{X})$. We first consider the case in which the functions g_j in $\mathbf{g}$ are continuous with respect to each of their arguments, have continuous partial derivatives, and define one-to-one mappings. It then follows that the inverse functions g_j^{-1} of $\mathbf{g}^{-1}$ defined by

$$\mathbf{X} = \mathbf{g}^{-1}(\mathbf{Y}) \qquad (5.62)$$

exist and are unique. They also have continuous partial derivatives.

In order to determine $f_\mathbf{Y}(\mathbf{y})$ in terms of $f_\mathbf{X}(\mathbf{x})$, we observe that, if a closed region $R_\mathbf{X}^n$ in the range space of **X** is mapped into a closed region $R_\mathbf{Y}^n$ in the range space of **Y** under transformation $\mathbf{g}$, the conservation of probability gives

$$\int \cdots \int_{R_\mathbf{Y}^n} f_\mathbf{Y}(\mathbf{y}) d\mathbf{y} = \int \cdots \int_{R_\mathbf{X}^n} f_\mathbf{X}(\mathbf{x}) d\mathbf{x} \qquad (5.63)$$

where the integrals represent n-fold integrals with respect to the components of **x** and **y**, respectively. Following the usual rule of change of variables in multiple integrals, we can write [see, for example, Courant (1937)]

$$\int \cdots \int_{R_\mathbf{X}^n} f_\mathbf{X}(\mathbf{x}) d\mathbf{x} = \int \cdots \int_{R_\mathbf{Y}^n} f_\mathbf{X}[\mathbf{g}^{-1}(\mathbf{y})] |J| d\mathbf{y} \qquad (5.64)$$

FUNCTIONS OF RANDOM VARIABLES

where J is the Jacobian of transformation, defined as the determinant

$$J = \begin{vmatrix} \dfrac{\partial g_1^{-1}}{\partial y_1} & \dfrac{\partial g_1^{-1}}{\partial y_2} & \cdots & \dfrac{\partial g_1^{-1}}{\partial y_n} \\ \vdots & & & \vdots \\ \dfrac{\partial g_n^{-1}}{\partial y_1} & \dfrac{\partial g_n^{-1}}{\partial y_2} & \cdots & \dfrac{\partial g_n^{-1}}{\partial y_n} \end{vmatrix} \quad (5.65)$$

As a point of clarification, let us note that the vertical lines in Equation 5.65 denote determinant and those in Equation 5.64 represent absolute value.

Equations 5.63 and 5.64 then lead to the desired formula

$$f_\mathbf{Y}(\mathbf{y}) = f_\mathbf{X}[\mathbf{g}^{-1}(\mathbf{y})]|J| \quad (5.66)$$

This result is stated as the following theorem.

Theorem.

For the transformation given by Equation 5.61 where $\mathbf{X}$ is a continuous random vector and $\mathbf{g}$ is continuous with continuous partial derivatives and defines a one-to-one mapping, the jpdf of $\mathbf{Y}$, $f_\mathbf{Y}(\mathbf{y})$, is given by

$$\boxed{f_\mathbf{Y}(\mathbf{y}) = f_\mathbf{X}[\mathbf{g}^{-1}(\mathbf{y})]|J|} \quad (5.67)$$

where J is defined by Equation 5.65.

It is of interest to note that Equation 5.67 is an extension of Equation 5.12, which is for the special case of $n = 1$. Similarly, an extension is also possible of Equation 5.24 for the $n = 1$ case when the transformation admits more than one root. Reasoning as we have done in deriving Equation 5.24, we have the following result.

Theorem.

In the preceding theorem, suppose the transformation $\mathbf{y} = \mathbf{g}(\mathbf{x})$ admits at most a countable number of roots $\mathbf{x}_1 = \mathbf{g}_1^{-1}(\mathbf{y})$, $\mathbf{x}_2 = \mathbf{g}_2^{-1}(\mathbf{y})$, ... Then

$$\boxed{f_\mathbf{Y}(\mathbf{y}) = \sum_{j=1}^{r} f_\mathbf{X}[\mathbf{g}_j^{-1}(\mathbf{y})]|J_j|} \quad (5.68)$$

where r is the number of solutions for $\mathbf{x}$ of the equation $\mathbf{y} = \mathbf{g}(\mathbf{x})$ and J_j is defined by

$$J_j = \begin{vmatrix} \dfrac{\partial g_{j1}^{-1}}{\partial y_1} & \dfrac{\partial g_{j1}^{-1}}{\partial y_2} & \cdots & \dfrac{\partial g_{j1}^{-1}}{\partial y_n} \\ \vdots & & & \vdots \\ \dfrac{\partial g_{jn}^{-1}}{\partial y_1} & \dfrac{\partial g_{jn}^{-1}}{\partial y_2} & \cdots & \dfrac{\partial g_{jn}^{-1}}{\partial y_n} \end{vmatrix} \tag{5.69}$$

In the above, $g_{j1}, g_{j2}, \ldots,$ and g_{jn} are components of $\mathbf{g}_j$.

As we mentioned earlier, the results presented above can also be applied to the case in which the dimension of $\mathbf{Y}$ is smaller than that of $\mathbf{X}$. Consider the transformation (5.60) in which $m < n$. In order to use the formulas developed above, we first augment the m-dimensional random vector $\mathbf{Y}$ by another $(n - m)$-dimensional random vector $\mathbf{Z}$. The vector $\mathbf{Z}$ can be constructed as a simple function of $\mathbf{X}$ in the form

$$\mathbf{Z} = \mathbf{h}(\mathbf{X}) \tag{5.70}$$

where $\mathbf{h}$ satisfies conditions of continuity and continuity in partial derivatives. On combining Equations 5.60 and 5.70, we have now an n-r.v. to n-r.v. transformation and the jpdf of $\mathbf{Y}$ and $\mathbf{Z}$ can be obtained by means of Equation 5.67 or 5.68. The jpdf of $\mathbf{Y}$ alone is then found through integration with respect to the components of $\mathbf{Z}$.

Example 5.18. Let r.v.'s X_1 and X_2 be independent and identically and normally distributed according to

$$f_{X_1}(x_1) = \frac{1}{\sqrt{2\pi}} e^{-x_1^2/2} \qquad -\infty < x_1 < \infty$$

and similarly for X_2. Determine the jpdf of $Y_1 = X_1 + X_2$ and $Y_2 = X_1 - X_2$.

Equation 5.67 applies in this case. The solutions of x_1 and x_2 in terms of y_1 and y_2 are

$$x_1 = g_1^{-1}(\mathbf{y}) = \frac{y_1 + y_2}{2} \qquad x_2 = g_2^{-1}(\mathbf{y}) = \frac{y_1 - y_2}{2} \tag{5.71}$$

FUNCTIONS OF RANDOM VARIABLES

The Jacobian in this case takes the form

$$J = \begin{vmatrix} \dfrac{\partial g_1^{-1}}{\partial y_1} & \dfrac{\partial g_1^{-1}}{\partial y_2} \\ \dfrac{\partial g_2^{-1}}{\partial y_1} & \dfrac{\partial g_2^{-1}}{\partial y_2} \end{vmatrix} = \begin{vmatrix} \tfrac{1}{2} & \tfrac{1}{2} \\ \tfrac{1}{2} & -\tfrac{1}{2} \end{vmatrix} = -\tfrac{1}{2}$$

Hence, Equation 5.67 leads to

$$f_{Y_1 Y_2}(y_1, y_2) = f_{X_1}[g_1^{-1}(y)] f_{X_2}[g_2^{-1}(y)] |J|$$

$$= \frac{1}{4\pi} e^{-(y_1 + y_2)^2/8} e^{-(y_1 - y_2)^2/8}$$

$$= \frac{1}{4\pi} e^{-(y_1^2 + y_2^2)/4} \qquad (-\infty, -\infty) < (y_1, y_2) < (\infty, \infty) \quad (5.72)$$

It is of interest to note that the result given by Equation 5.72 can be written as

$$f_{Y_1 Y_2}(y_1, y_2) = f_{Y_1}(y_1) f_{Y_2}(y_2) \tag{5.73}$$

where

$$f_{Y_1}(y_1) = \frac{1}{\sqrt{4\pi}} e^{-y_1^2/4} \qquad -\infty < y_1 < \infty$$

$$f_{Y_2}(y_2) = \frac{1}{\sqrt{4\pi}} e^{-y_2^2/4} \qquad -\infty < y_2 < \infty$$

implying that, although Y_1 and Y_2 are both functions of X_1 and X_2, they are independent and identically and normally distributed

Example 5.19. For the same distributions assigned to X_1 and X_2 in Example 5.18, determine the jpdf of $Y_1 = \sqrt{X_1^2 + X_2^2}$ and $Y_2 = X_1/X_2$.

Let us first note that Y_1 takes values only in the positive range. Hence

$$f_{Y_1 Y_2}(y_1, y_2) = 0 \qquad y_1 < 0$$

For $y_1 \geq 0$, the transformation $\mathbf{y} = \mathbf{g}(\mathbf{x})$ admits two solutions. They are

$$x_{11} = g_{11}^{-1}(\mathbf{y}) = \frac{y_1 y_2}{\sqrt{1 + y_2^2}}$$

$$x_{12} = g_{12}^{-1}(\mathbf{y}) = \frac{y_1}{\sqrt{1 + y_2^2}}$$

and

$$x_{21} = g_{21}^{-1}(\mathbf{y}) = -x_{11}$$

$$x_{22} = g_{22}^{-1}(\mathbf{y}) = -x_{12}$$

Equation 5.68 now applies and we have

$$f_{Y_1 Y_2}(y_1, y_2) = f_{\mathbf{X}}[\mathbf{g}_1^{-1}(\mathbf{y})]|J_1| + f_{\mathbf{X}}[\mathbf{g}_2^{-1}(\mathbf{y})]|J_2| \qquad (5.74)$$

where

$$f_{\mathbf{X}}(\mathbf{x}) = f_{X_1}(x_1) f_{X_2}(x_2) = \frac{1}{2\pi} e^{-(x_1^2 + x_2^2)/2} \qquad (5.75)$$

$$J_1 = J_2 = \begin{vmatrix} \dfrac{\partial g_{11}^{-1}}{\partial y_1} & \dfrac{\partial g_{11}^{-1}}{\partial y_2} \\ \dfrac{\partial g_{12}^{-1}}{\partial y_1} & \dfrac{\partial g_{12}^{-1}}{\partial y_2} \end{vmatrix} = -\frac{y_1}{1 + y_2^2} \qquad (5.76)$$

On substituting Equations 5.75 and 5.76 into Equation 5.74, we have

$$f_{Y_1 Y_2}(y_1, y_2) = \left(\frac{y_1}{1 + y_2^2}\right)\left\{\frac{2}{2\pi} \exp\left[-\frac{y_1^2 y_2^2 + y_1^2}{2(1 + y_2^2)}\right]\right\}$$

$$= \frac{y_1}{(1 + y_2^2)\pi} e^{-y_1^2/2} \qquad y_1 \geq 0 \qquad -\infty < y_2 < \infty$$

$$= 0 \qquad \qquad \text{elsewhere} \qquad (5.77)$$

FUNCTIONS OF RANDOM VARIABLES

We note that the result can again be expressed as the product of $f_{Y_1}(y_1)$ and $f_{Y_2}(y_2)$ with

$$f_{Y_1}(y_1) = y_1 e^{-y_1^2/2} \qquad y_1 \geq 0$$
$$= 0 \qquad \text{elsewhere}$$

$$f_{Y_2}(y_2) = \frac{1}{\pi(1 + y_2^2)} \qquad -\infty < y_2 < \infty$$

Thus, r.v.'s Y_1 and Y_2 are again independent in this case where Y_1 has a so-called Raleigh distribution and Y_2 is Cauchy distributed. We remark that the factor $(1/\pi)$ is assigned to $f_{Y_2}(y_2)$ to make the area under each pdf equal to 1.

Example 5.20. Let us determine the pdf of Y considered in Example 5.11 using formulas developed in this section.

The transformation is

$$Y = X_1 X_2 \tag{5.78}$$

In order to conform with conditions stated in this section, we augment Equation 5.78 by some simple transformation such as

$$Z = X_2 \tag{5.79}$$

The random variables Y and Z now play the role of Y_1 and Y_2 in Equation 5.67 and we have

$$f_{YZ}(y, z) = f_{X_1 X_2}[g_1^{-1}(y, z), g_2^{-1}(y, z)]|J| \tag{5.80}$$

where

$$g_1^{-1}(y, z) = \frac{y}{z}$$

$$g_2^{-1}(y, z) = z$$

$$J = \begin{vmatrix} \frac{1}{z} & -\frac{y}{z^2} \\ 0 & 1 \end{vmatrix} = \frac{1}{z}$$

Using specific forms of $f_{X_1}(x_1)$ and $f_{X_2}(x_2)$ given in Example 5.11, Equation 5.80 becomes

$$f_{YZ}(y, z) = f_{X_1}\left(\frac{y}{z}\right) f_{X_2}(z)\left|\frac{1}{z}\right| = \left(\frac{2y}{z}\right)\left(\frac{2-z}{2z}\right)$$

$$= \frac{y(2-z)}{z^2} \qquad 0 \le y \le 2 \qquad y \le z \le 2$$

$$= 0 \qquad \text{elsewhere} \qquad (5.81)$$

Finally, the pdf $f_Y(y)$ is found by performing integration of Equation 5.81 with respect to z. It is

$$f_Y(y) = \int_{-\infty}^{\infty} f_{YZ}(y, z)\,dz = \int_{y}^{2}\left[\frac{y(2-z)}{z^2}\right]dz$$

$$= 2 + y(\ln y - 1 - \ln 2) \qquad 0 \le y \le 2$$

$$= 0 \qquad \text{elsewhere}$$

This result agrees with that given in Equation 5.47 in Example 5.11.

REFERENCE

1. R. Courant, *Differential and Integral Calculus*, Vol. II, Wiley-Interscience, New York, 1937.

PROBLEMS

5.1. Determine the PDF of $Y = 3X - 1$ if

(a) $F_X(x) = 0 \qquad x < 3$
$ = \frac{1}{3} \qquad 3 \le x < 6$
$ = 1 \qquad x \ge 6$

(b) $F_X(x) = 0 \qquad x < 3$
$ = \frac{x}{3} - 1 \qquad 3 \le x < 6$
$ = 1 \qquad x \ge 6$

5.2. Temperature C measured in degrees Celsius is related to temperature X in degrees Fahrenheit by $C = 5(X - 32)/9$. Determine the pdf of C if X is random and is distributed uniformly in the interval (86, 95).

5.3. The r.v. X has a triangular distribution as shown. Determine the pdf of $Y = 3X + 2$.

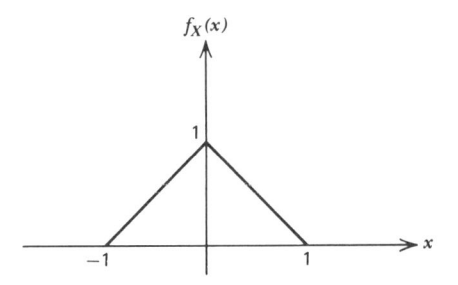

5.4. Determine $F_Y(y)$ in terms of $F_X(x)$ if $Y = \sqrt{X}$ where $F_X(x) = 0, x < 0$.

5.5. A r.v. Y has a "log-normal" distribution if it is related to X by $Y = e^X$ where X is normally distributed according to

$$f_X(x) = \frac{1}{\sigma\sqrt{2\pi}} e^{-(x-m)^2/2\sigma^2} \qquad -\infty < x < \infty$$

Determine the pdf of Y for $m = 0$ and $\sigma = 1$.

5.6. The risk R of an accident for a vehicle traveling at a "constant" speed V is given by

$$R = ae^{b(V-c)^2}$$

where a, b, and c are positive constants. Suppose that the speed V of a class of vehicles is random and is uniformly distributed between v_1 and v_2. Determine the pdf of R if (a) $(v_1, v_2) \geq c$ and (b) v_1 and v_2 are such that $c = (v_1 + v_2)/2$.

5.7. Let $Y = g(X)$ and X is uniformly distributed over the interval $a \leq x \leq b$. Suppose that the inverse function $X = g^{-1}(Y)$ is a single-valued function of Y in the interval $g(a) \leq y \leq g(b)$. Show that the pdf of Y is

$$f_Y(y) = \frac{1}{b-a}\left[\frac{1}{g'[g^{-1}(y)]}\right] \qquad g(a) \leq y \leq g(b)$$

$$= 0 \qquad \text{elsewhere}$$

where $g'(x) = dg(x)/dx$.

5.8. A rectangular plate of area a is situated in a flow stream at an angle Θ measured from the streamline as shown. Assuming that Θ is uniformly distributed from 0 to $\pi/2$, determine the pdf of the projected area perpendicular to the stream.

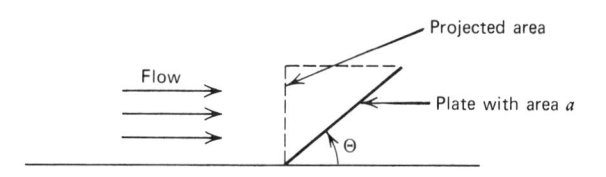

5.9. At a given location, the PDF of annual wind speed, V, in miles per hour is found to be

$$F_V(v) = \exp\left[-\left(\frac{v}{36.6}\right)^{-6.96}\right] \qquad v > 0$$

$$= 0 \qquad \qquad \text{elsewhere}$$

The wind force W exerted on structures is proportional to V^2. Let $W = aV^2$ and determine:
(a) The pdf of W and its mean and variance using $f_W(w)$.
(b) The mean and variance of W directly from the knowledge of $F_V(v)$.

5.10. An electrical device called a "full-wave" rectifier transforms input to the device X to output Y according to $Y = |X|$. If the input X has pdf in the form

$$f_X(x) = \frac{1}{\sqrt{2\pi}} e^{-(x-1)^2/2} \qquad -\infty < x < \infty$$

determine:
(a) The pdf of Y and its mean and variance using $f_Y(y)$.
(b) The mean and variance of Y directly from the knowledge of $f_X(x)$.

5.11. An electrical device gives as its output Y in terms of the input X by

$$Y = g(X) = \begin{cases} 1 & X > 0 \\ 0 & X \leq 0 \end{cases}$$

Is r.v. Y continuous or discrete? Determine its probability distribution in terms of the pdf of X.

FUNCTIONS OF RANDOM VARIABLES

5.12. The kinetic energy of a particle with mass m and velocity V is $X = mV^2/2$. Suppose that the velocity V is random with pdf given by

$$f_V(v) = av^2 e^{-bv^2} \quad v > 0$$
$$= 0 \quad \text{elsewhere}$$

Determine the pdf of X.

5.13. The radius R of a sphere is known to be distributed uniformly in the range $0.99r_0 \leq r \leq 1.01r_0$. Determine the pdf's of its surface area and its volume.

5.14. A resistor to be used as a component of a simple electrical circuit is randomly chosen from a stock whose resistance R has the pdf

$$f_R(r) = a^2 r e^{-ar} \quad r > 0$$
$$= 0 \quad \text{elsewhere}$$

Suppose that the voltage source v in the circuit is a deterministic constant.
(a) Find the pdf of current $I = v/R$ passing through the circuit.
(b) Find the pdf of power $W = I^2 R$ dissipated in the resistor.

5.15. The independent random variables X_1 and X_2 are uniformly and identically distributed with pdf's

$$f_{X_1}(x_1) = \tfrac{1}{2} \quad -1 \leq x_1 \leq 1$$
$$= 0 \quad \text{elsewhere}$$

and similarly for X_2. Let $Y = X_1 + X_2$.
(a) Determine the pdf of Y using Equation 5.56.
(b) Determine the pdf of Y using the method of characteristic functions developed in Section IV.5.

5.16. A discrete r.v. X has a binomial distribution with parameters (n, p) if its pmf has the form

$$p_X(k) = \binom{n}{k} p^k (1-p)^{n-k} \quad k = 0, 1, 2, \ldots, n$$

Show that, if X_1 and X_2 are independent and have binomial distributions with parameters (n_1, p) and (n_2, p), respectively, the sum $Y = X_1 + X_2$ has a binomial distribution with parameters $(n_1 + n_2, p)$.

5.17. Consider the sum of two independent random variables X_1 and X_2 where X_1 is discrete, taking values a and b with probabilities $P(X_1 = a) = p$ and $P(X_1 = b) = q$ ($p + q = 1$), and X_2 is continuous with pdf $f_{X_2}(x_2)$.

(a) Show that $Y = X_1 + X_2$ is a continuous random variable with pdf

$$f_Y(y) = p f_{Y_1}(y) + q f_{Y_2}(y)$$

where $f_{Y_1}(y)$ and $f_{Y_2}(y)$ are, respectively, the pdf's of $Y_1 = a + X_2$ and $Y_2 = b + X_2$.

(b) Plot $f_Y(y)$ by letting $a = 0$, $b = 1$, $p = \frac{1}{3}$, $q = \frac{2}{3}$, and

$$f_{X_2}(x_2) = \frac{1}{\sqrt{2\pi}} e^{-x_2^2/2} \qquad -\infty < x_2 < \infty$$

5.18. Consider a system with parallel arrangement as shown and let A be the primary component and B its redundant mate (backup component). The operating lives of A and B are denoted by T_1 and T_2, respectively, and they follow the exponential distributions

$$f_{T_1}(t_1) = a_1 e^{-a_1 t_1} \qquad t_1 > 0$$
$$= 0 \qquad \text{elsewhere}$$
$$f_{T_2}(t_2) = a_2 e^{-a_2 t_2} \qquad t_2 > 0$$
$$= 0 \qquad \text{elsewhere}$$

Let life of the system be denoted by T. Then $T = T_1 + T_2$ if the redundant part comes into operation only when the primary component fails (so-called "cold redundancy") and $T = \max(T_1, T_2)$ if the redundant part is

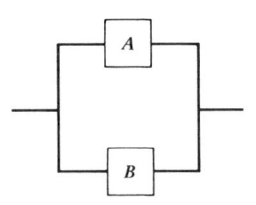

kept in a ready condition at all times so that delay is minimized in the event of change over from the primary component to its redundant mate (so-called "hot redundancy").

(a) Let $T_C = T_1 + T_2$ and $T_H = \max(T_1, T_2)$. Determine their respective pdf's.

(b) Suppose that we wish to maximize the probability $P(T \geq t)$ for some t. Which type of redundancy is preferred?

FUNCTIONS OF RANDOM VARIABLES

5.19. Consider a system with components arranged in series as shown and let T_1 and T_2 be independent random variables, representing operating lives of A and B, whose pdf's are given in Problem 5.18. Determine the pdf of system life $T = \min(T_1, T_2)$. Generalize to the case of n components in series.

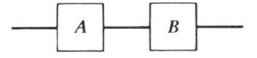

$$\boxed{A} \text{—} \boxed{B}$$

5.20. At a taxi stand, the number X_1 of taxis arriving during some time interval has a Poisson distribution with pmf given by

$$p_{X_1}(k) = \frac{\lambda^k e^{-\lambda}}{k!} \qquad k = 0, 1, 2, \ldots$$

where λ is a constant. Suppose that the demand X_2 at this location during the same time interval has the same distribution as X_1. Determine the pmf of $Y = X_2 - X_1$ where Y represents the excess of taxis in this time interval (positive and negative).

5.21. Determine the pdf of $Y = |X_1 - X_2|$ where X_1 and X_2 are independent random variables with respective pdf's $f_{X_1}(x_1)$ and $f_{X_2}(x_2)$.

5.22. The light intensity I at a given point X distance away from a light source is $I = C/X^2$ where C is the source candlepower. Determine the pdf of I if the pdf's of C and X are given by

$$f_C(c) = \tfrac{1}{36} \qquad 64 \leq c \leq 100$$
$$ = 0 \qquad \text{elsewhere}$$
$$f_X(x) = 1 \qquad 1 \leq x \leq 2$$
$$ = 0 \qquad \text{elsewhere}$$

5.23. Let X_1 and X_2 be independent and identically distributed according to

$$f_{X_1}(x_1) = \frac{1}{\sqrt{2\pi}} e^{-x_1^2/2} \qquad -\infty < x_1 < \infty$$

and similarly for X_2. By means of techniques developed in Section V.2, determine the pdf of $Y = \sqrt{X_1^2 + X_2^2}$. Check your answer with result obtained in Example 5.19. (Hint: Use polar coordinates to carry out integration.)

5.24. Extend the result in Problem 5.23 to the case of three independent and identically distributed random variables, that is, $Y = \sqrt{X_1^2 + X_2^2 + X_3^2}$. (Hint: Use spherical coordinates to carry out integration.)

5.25. The jpdf of r.v.'s X_1, X_2, and X_3 takes the form

$$f_{X_1X_2X_3}(x_1, x_2, x_3) = \frac{6}{(1 + x_1 + x_2 + x_3)^4} \qquad (x_1, x_2, x_3) > 0$$

$$= 0 \qquad \text{elsewhere}$$

Find the pdf of $Y = X_1 + X_2 + X_3$.

5.26. The pdf's of two independent random variables X_1 and X_2 are

$$f_{X_1}(x_1) = e^{-x_1} \qquad x_1 > 0$$
$$= 0 \qquad x_1 \leq 0$$
$$f_{X_2}(x_2) = e^{-x_2} \qquad x_2 > 0$$
$$= 0 \qquad x_2 \leq 0$$

Determine the jpdf of Y_1 and Y_2 defined by

$$Y_1 = X_1 + X_2 \qquad Y_2 = \frac{X_1}{Y_1}$$

and show that they are independent.

5.27. The jpdf of X and Y is given by

$$f_{XY}(x, y) = \frac{1}{2\pi\sigma^2} e^{-(x^2 + y^2)/2\sigma^2} \qquad -\infty < (x, y) < \infty$$

Determine the jpdf of R and Φ and their respective marginal pdf's where $R = \sqrt{X^2 + Y^2}$ is the vector length and $\Phi = \tan^{-1}(Y/X)$ is the phase angle. Are R and Φ independent?

5.28. Show that an alternate formula for Equation 5.67 is

$$f_\mathbf{Y}(\mathbf{y}) = f_\mathbf{X}[\mathbf{g}^{-1}(\mathbf{y})]|J'|^{-1}$$

FUNCTIONS OF RANDOM VARIABLES

where

$$J' = \begin{vmatrix} \dfrac{\partial g_1(\mathbf{x})}{\partial x_1} & \dfrac{\partial g_1(\mathbf{x})}{\partial x_2} & \cdots & \dfrac{\partial g_1(\mathbf{x})}{\partial x_n} \\ \vdots & & & \vdots \\ \dfrac{\partial g_n(\mathbf{x})}{\partial x_1} & \dfrac{\partial g_n(\mathbf{x})}{\partial x_2} & \cdots & \dfrac{\partial g_n(\mathbf{x})}{\partial x_n} \end{vmatrix}$$

evaluated at $\mathbf{x} = \mathbf{g}^{-1}(\mathbf{y})$. Similar alternate forms hold for Equations 5.12, 5.24, and 5.68.

VI
Some Important Discrete Distributions

This chapter deals with distributions of some discrete random variables that are important as models of scientific phenomena. The nature and applications of these distributions are discussed. An understanding of the situations in which these random variables arise enables us to choose an appropriate distribution for a scientific phenomenon. Thus, this chapter is also concerned with the induction step discussed in Chapter I, by which a model is chosen on the basis of factual understanding of the scientific system under study (step B to C in Figure 1.1).

Distributions of some important continuous random variables will be studied in Chapter VII.

VI.1 Bernoulli Trials

A large number of practical situations can be described by the repeated performance of a random experiment of the following basic nature.

A sequence of trials is performed so that (a) for each trial, there are only two possible outcomes, say, success and failure, (b) their probabilities remain the same throughout the trials, and (c) the trials are carried out independently. Trials performed under these conditions are called *Bernoulli trials*. In spite of simplicity of

SOME IMPORTANT DISCRETE DISTRIBUTIONS   **157**

the situation, mathematical models arising from this basic random experiment have wide applicabilities. In fact, we have encountered Bernoulli trials in the random walk problems described in Examples 3.4 and 4.17 and also in the traffic problem examined in Example 3.8. More examples will be given in the sections to follow.

Let us denote the event "success" by S and event "failure" by F. Also, let $P(S) = p$ and $P(F) = q$ $(p + q = 1)$. Possible outcomes resulting from performing a sequence of n Bernoulli trials can be symbolically represented by

$$SSFFSFSSS \cdots FF$$
$$FSFSSFFFS \cdots SF$$
$$\cdots$$
$$\cdots$$

and, due to independence, their probabilities are easily computed. Thus,

$$P(SSFFSF \cdots FF) = P(S)P(S)P(F)P(F)P(S)P(F) \cdots P(F)P(F)$$
$$= ppqqpq \cdots qq$$

A number of these possible outcomes with their associated probabilities are of practical interest. We introduce three important distributions in this connection.

VI.1.1 THE BINOMIAL DISTRIBUTION

The probability distribution of a r.v. X representing the number of successes in a sequence of n Bernoulli trials, regardless of the order in which they occur, is frequently of considerable interest. It is clear that X is a discrete random variable, assuming values $0, 1, 2, \ldots, n$. In order to determine its probability mass function, consider $p_X(k)$, the probability of having exactly k successes in n trials. This event can occur in as many ways as k letters S can be placed in n boxes. Now, we have n choices for the position of the first S, $n-1$ choices for the second S, ... and, finally, $n-k+1$ choices for the position of the kth S. The total number of possible arrangements is thus $n(n-1) \cdots (n-k+1)$. However, since no distinction is made of the S's that are in the occupied positions, we must divide the number obtained above by the number of ways in which k S's can be arranged in k boxes, that is, $k(k-1) \cdots 1 = k!$. Hence, the number of ways in which k successes can happen in n trials is

$$\frac{n(n-1) \cdots (n-k+1)}{k!} = \frac{n!}{k!(n-k)!} \tag{6.1}$$

and the probability associated with each is $p^k q^{n-k}$. Hence, we have

$$p_X(k) = \binom{n}{k} p^k q^{n-k} \qquad k = 0, 1, 2, \ldots, n \tag{6.2}$$

where

$$\binom{n}{k} = \frac{n!}{k!(n-k)!} \tag{6.3}$$

is the binomial coefficient in the binomial theorem

$$(a + b)^n = \sum_{k=0}^{n} \binom{n}{k} a^k b^{n-k} \tag{6.4}$$

In view of its similarity in appearance to the terms in the binomial theorem, the distribution defined by Equation 6.2 is called the *binomial distribution*.

The shape of a binomial distribution is determined by the values assigned to its two parameters, n and p. In general, n is given as a part of the problem statement and p must be estimated from observations.

A plot of the probability mass function $p_X(k)$ has been shown in Example 3.2 for $n = 10$ and $p = 0.2$. The peak of the distribution will shift to the right as p increases, reaching a symmetrical distribution when $p = 0.5$. More insight into the behavior of $p_X(k)$ can be gained by taking the ratio

$$\frac{p_X(k)}{p_X(k-1)} = \frac{(n-k+1)p}{kq} = 1 + \frac{(n+1)p - k}{kq} \tag{6.5}$$

We see from (6.5) that $P_X(k)$ is greater than the preceding one when $k < (n + 1)p$ and is smaller when $k > (n + 1)p$. Accordingly, if we define the integer k^* by

$$(n + 1)p - 1 < k^* \leq (n + 1)p \tag{6.6}$$

the value of $p_X(k)$ increases monotonically and attains its maximum value when $k = k^*$, then decreases monotonically. If $(n + 1)p$ happens to be an integer, the maximum value takes place at both $p_X(k^* - 1)$ and $p_X(k^*)$. The integer k^* is often called "the most probable number of successes."

Because of its wide usage, the pmf $p_X(k)$ is widely tabulated as a function of n and p. Table A.1 in Appendix A gives its values for $n = 2, \ldots, 10$ and $p = 0.01, \ldots, 0.50$. The calculation of $p_X(k)$ in Equation 6.2 is cumbersome as n becomes large. An approximate way of determining $p_X(k)$ for large n has been discussed in Example 4.17 by means of Stirling's formula. Poisson approximation to the binomial

SOME IMPORTANT DISCRETE DISTRIBUTIONS

distribution, to be discussed in Section VI.3.2, also facilitates probability calculations when n becomes large.

The PDF $F_X(x)$ for a binomial distribution is also widely tabulated. It is given by

$$F_X(x) = \sum_{k=0}^{m \leq x} \binom{n}{k} p^k q^{n-k} \qquad (6.7)$$

where m is the largest integer less than or equal to x.

Other important properties of a binomial distribution have been derived in Examples 4.1, 4.5, and 4.14. Without giving details, we have for the characteristic function, mean, and variance,

$$\begin{aligned} \phi_X(t) &= (pe^{jt} + q)^n \\ m_X &= np \\ \sigma_X^2 &= npq \end{aligned} \qquad (6.8)$$

The fact that the mean of X is np suggests that parameter p can be estimated based on the average value of observed data. This procedure is used in one of the following examples. We mention, however, that this parameter estimation problem needs to be examined much more rigorously and its systematic treatment will be taken up in Part B.

Let us remark here that another formulation leading to the binomial distribution is to define r.v. $X_j, j = 1, 2, \ldots, n$, to represent the outcome of the jth Bernoulli trial. If we let

$$\begin{aligned} X_j &= 0 \text{ if } j\text{th trial is a failure} \\ &= 1 \text{ if } j\text{th trial is a success} \end{aligned} \qquad (6.9)$$

then the sum

$$X = X_1 + X_2 + \cdots + X_n \qquad (6.10)$$

gives the number of successes in n trials. By definition, $X_1, \ldots,$ and X_n are independent random variables.

Moments and distribution of X can be easily found using Equation 6.10. Since

$$E\{X_j\} = 0(q) + 1(p) = p$$

it follows from Equation 4.38 that

$$E\{X\} = p + p + \cdots + p = np \qquad (6.11)$$

which is in agreement with that given in Equations 6.8. Similarly, its variance, characteristic function, and probability mass function are easily found following our discussion in Chapter IV concerning sums of independent random variables.

We have seen binomial distributions in Examples 3.4, 3.8, and 4.11. Its applications in other areas are further illustrated by the following additional examples.

Example 6.1. A homeowner has just installed 20 light bulbs in a new home. Suppose that each has a probability 0.2 of functioning more than three months. What is the probability that at least five of these function more than three months? What is the average number of bulbs the homeowner has to replace in three months?

It is safe to assume that the light bulbs perform independently. If X is the number of bulbs functioning more than three months (success), it has a binomial distribution with $n = 20$ and $p = 0.2$. The answer to the first question is thus

$$\sum_{k=5}^{20} p_X(k) = 1 - \sum_{k=0}^{4} p_X(k)$$

$$= 1 - \sum_{k=0}^{4} \binom{20}{k}(0.2)^k(0.8)^{20-k}$$

$$= 1 - (0.012 + 0.058 + 0.137 + 0.205 + 0.218) = 0.37$$

The average number of replacements is

$$20 - E\{X\} = 20 - np = 20 - 20(0.2) = 16$$

Example 6.2. Suppose that three telephone users share a party line, and we are interested in estimating the probability that more than one will use it at the same time. If independence of their telephone habit is assumed, the probability of exactly k persons requiring use of the telephone at the same time is given by the mass function $p_X(k)$ associated with the binomial distribution. Let it be given that, on the average, a telephone user is on the phone five minutes per hour, an estimate of p is $p = \frac{5}{60} = \frac{1}{12}$.

The solution to this problem is

$$p_X(2) + p_X(3) = \binom{3}{2}(\tfrac{1}{12})^2(\tfrac{11}{12}) + \binom{3}{3}(\tfrac{1}{12})^3$$

$$= \tfrac{17}{864} = 0.0197$$

Example 6.3. Let X_1 and X_2 be two independent random variables, both having binomial distributions with parameters (n_1, p) and (n_2, p), respectively, and let $Y = X_1 + X_2$. Determine the distribution of r.v. Y.

SOME IMPORTANT DISCRETE DISTRIBUTIONS

The characteristic functions of X_1 and X_2 are, according to the first of Equations 6.8,

$$\phi_{X_1}(t) = (pe^{jt} + q)^{n_1} \qquad \phi_{X_2}(t) = (pe^{jt} + q)^{n_2}$$

In view of Equation 4.71, the characteristic function of Y is simply the product of $\phi_{X_1}(t)$ and $\phi_{X_2}(t)$. Thus,

$$\phi_Y(t) = \phi_{X_1}(t)\phi_{X_2}(t) = (pe^{jt} + q)^{n_1 + n_2}$$

By inspection, it is the characteristic function corresponding to a binomial distribution with parameters $(n_1 + n_2, p)$. Hence, we have

$$p_Y(k) = \binom{n_1 + n_2}{k} p^k q^{n_1 + n_2 - k} \qquad k = 0, 1, \ldots, n_1 + n_2$$

Generalizing, we have the following important result:

Theorem

The binomial distribution generates itself under addition of independent random variables with the same p.

Example 6.4. If r.v.'s X and Y are independent binomial-distributed random variables with parameters (n_1, p) and (n_2, p), we wish to determine the conditional probability mass function of X given that $X + Y = m$ $(0 \leq m \leq n_1 + n_2)$.
For $k \leq \min(n_1, m)$, we have

$$P(X = k | X + Y = m) = \frac{P(X = k \cap X + Y = m)}{P(X + Y = m)}$$

$$= \frac{P(X = k \cap Y = m - k)}{P(X + Y = m)} = \frac{P(X = k)P(Y = m - k)}{P(X + Y = m)}$$

$$= \frac{\binom{n_1}{k} p^k (1 - p)^{n_1 - k} \binom{n_2}{m - k} p^{m-k} (1 - p)^{n_2 - m + k}}{\binom{n_1 + n_2}{m} p^m (1 - p)^{n_1 + n_2 - m}}$$

$$= \frac{\binom{n_1}{k} \binom{n_2}{m - k}}{\binom{n_1 + n_2}{m}}$$

$$k = 0, 1, \ldots, \min(n_1, m) \qquad (6.12)$$

where we have used the result given in Example 6.3 that $X + Y$ is binomially distributed with parameters $(n_1 + n_2, p)$.

The distribution given by Equation 6.12 is known as the *hypergeometric distribution*. It arises as distributions in such cases as the number of black balls that are chosen when a sample of m balls is randomly selected from a lot of n items having n_1 black balls and n_2 white balls $(n_1 + n_2 = n)$. Let the r.v. Z be this number. We have, from Equation 6.12 on replacing n_2 by $n - n_1$,

$$p_Z(k) = \frac{\binom{n_1}{k}\binom{n - n_1}{m - k}}{\binom{n}{m}} \qquad k = 0, 1, \ldots, \min(n_1, m) \qquad (6.13)$$

VI.1.2 THE GEOMETRIC DISTRIBUTION

Another event of interest arising from n Bernoulli trials is the number of trials to (and including) the first occurrence of success. If X is used to represent this number, it is a discrete random variable with possible integer values ranging from one to infinity. Its probability mass function is easily computed to be

$$p_X(k) = P(\underbrace{FF \cdots F}_{k-1}S) = \underbrace{P(F)P(F) \cdots P(F)}_{k-1}P(S)$$
$$= q^{k-1}p \qquad k = 1, 2, \ldots \qquad (6.14)$$

This distribution is known as the *geometric distribution* with parameter p, where the name stems from its similarity to the familiar term in geometric progression. A plot of $p_X(k)$ is given in Figure 6.1.

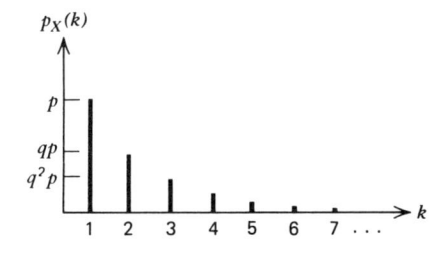

Figure 6.1 Geometric distribution.

SOME IMPORTANT DISCRETE DISTRIBUTIONS

The corresponding probability distribution function is

$$F_X(x) = \sum_{k=1}^{m \leq x} p_X(k) = p + qp + \cdots + q^{m-1}p$$
$$= (1 - q)(1 + q + q^2 + \cdots + q^{m-1}) = 1 - q^m \qquad (6.15)$$

where m is the largest integer less than or equal to x. The mean and variance of X can be found as follows:

$$E\{X\} = \sum_{k=1}^{\infty} kq^{k-1}p = p \sum_{k=1}^{\infty} \frac{d}{dq} q^k$$
$$= p \frac{d}{dq} \sum_{k=1}^{\infty} q^k = p \frac{d}{dq} \left(\frac{q}{1-q}\right) = \frac{1}{p} \qquad (6.16)$$

In the above, the interchange of summation and differentiation is allowed because $|q| < 1$. Following the same procedure, the variance has the form

$$\sigma_X^2 = \frac{1-p}{p^2} \qquad (6.17)$$

Example 6.5. A driver is eagerly eyeing a precious parking space some distance down the street. There are five cars in front of the driver, each of which having a probability 0.2 of taking it. What is the probability that the car immediately ahead will enter the parking space?

For this problem, we have a geometric distribution and need to evaluate $p_X(k)$ for $k = 5$ and $p = 0.2$. Thus,

$$p_X(5) = (0.8)^4(0.2) = 0.082$$

which may seem much smaller than what we experience in similar situations.

Example 6.6. Assume that the probability of a specimen failing during a given experiment is 0.1. What is the probability that it will take more than three specimens to have one surviving the experiment?

Let X denote the number of trials required for the first specimen to survive. It then has a geometric distribution with $p = 0.9$. The desired probability is

$$P(X > 3) = 1 - F_X(3) = 1 - (1 - q^3) = (0.1)^3 = 0.001$$

Example 6.7. Let the probability of occurrence of a flood of magnitude greater than a critical magnitude in any given year be 0.01. Assuming that floods occur independently, determine $E\{N\}$, the average return period. The *average return period* or simply *return period* is defined as the average number of years between floods whose magnitudes are greater than critical magnitude.

It is clear that N is a random variable with a geometric distribution and $p = 0.01$. The return period is then

$$E\{N\} = \frac{1}{p} = 100 \text{ years}$$

The critical magnitude which gives rise to $E\{N\} = 100$ years is often referred to as the "100-year flood."

VI.1.3 THE NEGATIVE BINOMIAL DISTRIBUTION

A natural generalization to the geometric distribution is the distribution of r.v. X representing the number of Bernoulli trials necessary for the rth success to occur, where r is a given positive integer.

In order to determine $p_X(k)$ for this case, let A be the event that the first $k - 1$ trials yield exactly $r - 1$ successes, regardless of their order, and B the event that a success turns up at the kth trial. Then, due to independence,

$$p_X(k) = P(A \cap B) = P(A)P(B) \tag{6.18}$$

Now, $P(A)$ obeys a binomial distribution with parameters $k - 1$ and $r - 1$, or

$$P(A) = \binom{k-1}{r-1} p^{r-1} q^{k-r} \quad k = r, r+1, \ldots \tag{6.19}$$

and $P(B)$ is simply

$$P(B) = p \tag{6.20}$$

Substituting Equations 6.19 and 6.20 into Equation 6.18 results in

$$\boxed{p_X(k) = \binom{k-1}{r-1} p^r q^{k-r} \quad k = r, r+1, \ldots} \tag{6.21}$$

SOME IMPORTANT DISCRETE DISTRIBUTIONS

We note that, as expected, it reduces to the geometric distribution when $r = 1$. The distribution defined by Equation 6.21 is known as the *negative binomial*, or *Pascal*, distribution with parameters r and p.

A useful variant of this distribution is obtained if we let $Y = X - r$. The r.v. Y is the number of Bernoulli trials *beyond* r needed for the realization of the rth success, or it can be interpreted as the number of failures before the rth success.

The probability mass function of Y, $p_Y(m)$, is obtained from Equation 6.21 upon replacing k by $m + r$. Thus,

$$p_Y(m) = \binom{m+r-1}{r-1} p^r q^m$$

$$= \binom{m+r-1}{m} p^r q^m \quad m = 0, 1, 2, \ldots \quad (6.22)$$

We see that r.v. Y has the convenient property that the range of m begins at zero rather than r for values associated with X.

Recalling a more general definition of the binomial coefficient

$$\binom{a}{j} = \frac{a(a-1)\cdots(a-j+1)}{j!} \quad (6.23)$$

for any real a and any positive integer j, direct evaluation shows that the binomial coefficient in Equation 6.22 can be written in the form

$$\binom{m+r-1}{m} = (-1)^m \binom{-r}{m} \quad (6.24)$$

Hence,

$$\boxed{P_Y(m) = \binom{-r}{m} p^r (-q)^m \quad m = 0, 1, 2, \ldots} \quad (6.25)$$

which is the reason for the name "negative binomial distribution."

The mean and variance of r.v. X can be determined either by following the standard procedure or by noting that X can be represented by

$$X = X_1 + X_2 + \cdots + X_r \quad (6.26)$$

where X_j is the number of trials between the $(j-1)$th and (including) the jth successes. These random variables are mutually independent, each having the geometric distribution with mean $1/p$ and variance $(1-p)/p^2$. Therefore, the mean and variance of the sum X are, according to Equations 4.38 and 4.41,

$$m_X = \frac{r}{p} \qquad \sigma_X^2 = \frac{r(1-p)}{p^2} \qquad (6.27)$$

Since $Y = X - r$, the corresponding moments of Y are

$$m_Y = \frac{r}{p} - r \qquad \sigma_Y^2 = \frac{r(1-p)}{p^2} \qquad (6.28)$$

Example 6.8. A curbside parking facility has a capacity for three cars. Determine the probability that it will be full within 10 minutes. It is estimated that six cars will pass this parking space within the time span and, on the average, 80% of all cars will want to park there.

The desired probability is simply the probability that the number of trials to the third success (taking the parking space) is less than or equal to 6. If X is this number, it has a negative binomial distribution with $r = 3$ and $p = 0.8$. Using Equation 6.21, we have

$$P(X \leq 6) = \sum_{k=3}^{6} p_X(k) = \sum_{k=3}^{6} \binom{k-1}{2}(0.8)^3 (0.2)^{k-3}$$
$$= (0.8)^3 [1 + (3)(0.2) + (6)(0.2)^2 + (10)(0.2)^3]$$
$$= 0.983$$

Let us note that an alternative way of arriving at this answer is to sum the probabilities of having three, four, five, and six successes in six Bernoulli trials using the binomial distribution. This observation leads to a general relationship between binomial and negative binomial distributions. Stated in general terms, if X_1 has a binomial distribution with parameters n and p and X_2 has a negative binomial distribution with parameters r and p, then

$$P(X_1 \geq r) = P(X_2 \leq n)$$
$$P(X_1 < r) = P(X_2 > n) \qquad (6.29)$$

Example 6.9. The negative binomial distribution is widely used in waiting time problems. Consider, for example, a car waiting on a ramp to merge into freeway traffic. Suppose that it is fifth in line to merge and the gaps between cars on freeway are such that there is a probability of 0.4 that they are large enough for merging. Then, if X is the waiting time before merging for this particular vehicle measured in terms of number of freeway gaps, it has a negative binomial distribution with $r = 5$ and $p = 0.4$. The mean waiting time is, as seen from Equations 6.27,

$$E\{X\} = \frac{5}{0.4} = 12.5 \text{ gaps}$$

VI.2 The Multinomial Distribution

Bernoulli trials can be generalized in several directions. A useful generalization is to relax the requirement that there be only two possible outcomes for each trial. Let there be r possible outcomes for each trial, denoted by $E_1, E_2, \ldots, E_r$, and let $P(E_i) = p_i, i = 1, \ldots, r$, and $p_1 + p_2 + \cdots + p_r = 1$. A typical result of n trials is a succession of symbols like

$$E_2 E_1 E_3 E_3 E_6 E_2 \cdots$$

If we let r.v. $X_i, i = 1, 2, \ldots, r$, represent the number of E_i's in a sequence of n trials, the joint probability mass function of $X_1, X_2, \ldots, X_r$ is given by

$$p_{X_1 X_2 \cdots X_r}(k_1, k_2, \ldots, k_r) = \frac{n!}{k_1! k_2! \cdots k_r!} p_1^{k_1} p_2^{k_2} \cdots p_r^{k_r} \qquad (6.30)$$

where $k_1, k_2, \ldots, k_r = 0, 1, 2, \ldots$, and $k_1 + k_2 + \cdots + k_r = n$.

Proof

We want to show that the coefficient in Equation 6.30 is equal to the number of ways of placing k_1 letters E_1, k_2 letters $E_2, \ldots$, and k_r letters E_r in n boxes. This can be easily verified by writing

$$\frac{n!}{k_1! k_2! \cdots k_r!} = \binom{n}{k_1} \binom{n - k_1}{k_2} \cdots \binom{n - k_1 - k_2 - \cdots - k_{r-1}}{k_r}$$

The first binomial coefficient is the number of ways of placing k_1 letters E_1 in n boxes, the second is the number of ways of placing k_2 letters E_2 in the remaining $n - k_1$ unoccupied boxes, and so on.

The formula given by Equation 6.30 is an important higher-dimension joint probability distribution. It is called the *multinomial distribution* because it has the form of the general term in the multinomial expansion of $(p_1 + p_2 + \cdots + p_r)^n$. We note that Equation 6.30 reduces to the binomial distribution when $r = 2$ and with $p_1 = p$, $p_2 = q$, $k_1 = k$, and $k_2 = n - k$.

Since each X_i defined above has a binomial distribution with parameters n and p_i, we have

$$m_{X_i} = np_i \qquad \sigma^2_{X_i} = np_i(1 - p_i) \qquad (6.31)$$

and it can be shown that

$$\text{cov}(X_i, X_j) = -np_ip_j \qquad i, j = 1, 2, \ldots, r \qquad i \neq j \qquad (6.32)$$

Example 6.10. Income levels are classified as low, medium, and high in a study of incomes of a given population. If, on the average, 10% of the population belongs to the low income group and 20% belongs to the high-income group, what is the probability that, of the 10 persons studied, three will be in the low-income group and the remaining in the medium-income group? What is the marginal distribution of the number of persons (out of 10) at the low-income level?

Let X_1 be the number of low-income persons in the group of 10 persons, X_2 be the number of medium-income persons, and X_3 be the number of high-income persons. Then X_1, X_2, and X_3 have a multinomial distribution with $p_1 = 0.1$, $p_2 = 0.7$, $p_3 = 0.2$, and $n = 10$.

Thus,

$$p_{X_1X_2X_3}(3, 7, 0) = \frac{10!}{3!7!0!} (0.1)^3(0.7)^7(0.2)^0$$

$$\cong 0.01$$

The marginal distribution of X_1 is binomial with $n = 10$ and $p = 0.1$.

We remark that the single-r.v. marginal distributions are binomial but, since $X_1, X_2, \ldots,$ and X_r are not independent, the multinomial distribution is *not* a product of binomial distributions.

VI.3 The Poisson Distribution

In this section we wish to consider a distribution that is used in a wide variety of physical situations. It is used as mathematical models for describing, in a specific interval of time, such events as emission of α-particles from a radioactive substance, passenger arrivals at an airline terminal, distribution of dust particles reaching a certain space, and car arrivals at an intersection.

SOME IMPORTANT DISCRETE DISTRIBUTIONS

To fix ideas in the following development, let us consider the problem of passenger arrivals at a terminal during a specified time interval. We shall use the notation $X(0, t)$ to represent the number of arrivals during time interval $[0, t)$, where the symbol $[\ )$ denotes a left-closed and right-open interval; it is a discrete random variable taking possible values $0, 1, 2, \ldots$, whose distribution clearly depends on t. For clarity, its probability mass function is written as

$$p_k(0, t) = P[X(0, t) = k] \qquad k = 0, 1, 2, \ldots \tag{6.33}$$

to show its explicit dependence on t. Note that this is different from our standard notation.

To study this problem we make the following basic assumptions:

1. The random variables $X(t_1, t_2), X(t_2, t_3), \ldots, X(t_{n-1}, t_n), t_1 < t_2 < \cdots < t_n$, are mutually independent, that is, the numbers of passenger arrivals in nonoverlapping time intervals are independent of each other.

2. For sufficiently small Δt,

$$p_1(t, t + \Delta t) = \lambda \Delta t + o(\Delta t) \tag{6.34}$$

where $o(\Delta t)$ stands for functions such that

$$\lim_{\Delta t \to 0} \frac{o(\Delta t)}{\Delta t} = 0 \tag{6.35}$$

This assumption says that, for a sufficiently small Δt, the probability of having exactly one arrival is directly proportional to the length of Δt. The parameter λ in Equation 6.34 is called *average density* or *mean rate of arrival*. It is assumed to be a constant for simplicity in our discussion; however, there is no difficulty in allowing it to vary with time.

3. For sufficiently small Δt,

$$\sum_{k=2}^{\infty} p_k(t, t + \Delta t) = o(\Delta t) \tag{6.36}$$

This condition implies that the probability of having two or more arrivals during a sufficiently small interval is negligible.

It follows from Equations 6.34 and 6.36 that

$$p_0(t, t + \Delta t) = 1 - \sum_{k=1}^{\infty} p_k(t, t + \Delta t)$$

$$= 1 - \lambda \Delta t + o(\Delta t) \tag{6.37}$$

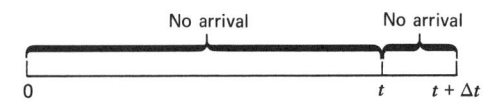

Figure 6.2 Interval $[0, t + \Delta t)$.

In order to determine the probability mass function $p_k(0, t)$ based on the assumptions stated above, let us first consider $p_0(0, t)$. Figure 6.2 shows two non-overlapping intervals $[0, t)$ and $[t, t + \Delta t)$. In order that there are no arrivals in the total interval $[0, t + \Delta t)$, we must have no arrivals in both subintervals. Due to independence of arrivals in nonoverlapping intervals, we thus can write

$$p_0(0, t + \Delta t) = p_0(0, t)p_0(t, t + \Delta t)$$
$$= p_0(0, t)[1 - \lambda \Delta t + o(\Delta t)] \qquad (6.38)$$

Rearranging Equation 6.38 and dividing both sides by Δt gives

$$\frac{p_0(0, t + \Delta t) - p_0(0, t)}{\Delta t} = -p_0(0, t)\left[\lambda - \frac{o(\Delta t)}{\Delta t}\right]$$

Upon letting $\Delta t \to 0$ we obtain the differential equation

$$\frac{dp_0(0, t)}{dt} = -\lambda p_0(0, t) \qquad (6.39)$$

Its solution satisfying initial condition $p_0(0, 0) = 1$ is

$$p_0(0, t) = e^{-\lambda t} \qquad (6.40)$$

The determination of $p_1(0, t)$ is similar. We first observe that one arrival in $[0, t + \Delta t)$ can only be accomplished by having no arrival in the subinterval $[0, t)$ and one arrival in $[t, t + \Delta t)$, or one arrival in $[0, t)$ and no arrival in $[t, t + \Delta t)$. Hence we have

$$p_1(0, t + \Delta t) = p_0(0, t)p_1(t, t + \Delta t) + p_1(0, t)p_0(t, t + \Delta t) \qquad (6.41)$$

Substituting Equations 6.34, 6.37, and 6.40 into Equation 6.41 and letting $\Delta t \to 0$ gives

$$\frac{dp_1(0, t)}{dt} = -\lambda p_1(0, t) + \lambda e^{-\lambda t} \qquad p_1(0, 0) = 0 \qquad (6.42)$$

SOME IMPORTANT DISCRETE DISTRIBUTIONS

which yields

$$p_1(0, t) = \lambda t e^{-\lambda t} \tag{6.43}$$

Continuing in this way we find, for the general term,

$$\boxed{p_k(0, t) = \frac{(\lambda t)^k e^{-\lambda t}}{k!} \qquad k = 0, 1, 2, \ldots} \tag{6.44}$$

Equation 6.44 gives the probability mass function of $X(0, t)$, the number of arrivals during time interval $[0, t)$ subject to the assumptions stated above. It is called the *Poisson distribution* with parameters λ and t. However, since λ and t appear in Equation 6.44 as a product λt, it can be replaced by a single parameter $v = \lambda t$ and we can also write

$$\boxed{p_k(0, t) = \frac{v^k e^{-v}}{k!} \qquad k = 0, 1, 2, \ldots} \tag{6.45}$$

The mean of $X(0, t)$ is given by

$$E\{X(0, t)\} = \sum_{k=0}^{\infty} k p_k(0, t) = e^{-\lambda t} \sum_{k=0}^{\infty} \frac{k(\lambda t)^k}{k!}$$

$$= \lambda t e^{-\lambda t} \sum_{k=1}^{\infty} \frac{(\lambda t)^{k-1}}{(k-1)!} = \lambda t e^{-\lambda t} e^{\lambda t} = \lambda t \tag{6.46}$$

Similarly, we can show that

$$\sigma_{X(0, t)}^2 = \lambda t \tag{6.47}$$

It is seen from Equation 6.46 that parameter λ is equal to the average number of arrivals per unit interval of time; the name "mean rate of arrival" for λ is thus justified. In determining the value of this parameter in a given problem, it can be estimated from observations by m/n, where m is the observed number of arrivals in n unit time intervals. Similarly, since $v = \lambda t$, v represents the average number of arrivals in time interval $[0, t)$.

It is also seen from Equation 6.47 that, as expected, the variance, as well as the mean, increases as the mean rate increases. The Poisson distribution for several values of λt is shown in Figure 6.3. In general, if we examine the ratio of $p_k(0, t)$ and $p_{k-1}(0, t)$ as we did for binomial distribution, it shows that $p_k(0, t)$ increases monotonically and then decreases monotonically as k increases, reaching its maximum when k is the largest integer not exceeding λt.

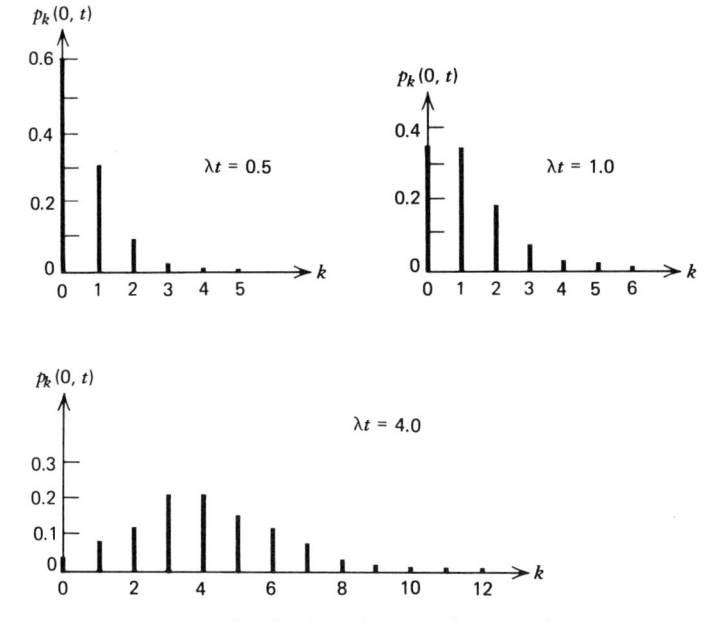

Figure 6.3 Poisson distributions for several values of λt.

Example 6.11. Traffic load in design of a pavement system is an important consideration. Vehicles arrive at some point on the pavement in a random manner both in space (amplitude and velocity) and in time (arrival rate). Considering the time aspect alone, the following observations are made at 30-second intervals.

No. of Vehicles/30 sec	Frequency (No. of Observations)
0	18
1	32
2	28
3	20
4	13
5	7
6	0
7	1
8	1
≥ 9	0
	Total = 120

SOME IMPORTANT DISCRETE DISTRIBUTIONS

Suppose that the rate of 10 vehicles per minute is the level of critical traffic load. We wish to determine the probability that this critical level is reached or exceeded.

Let $X(0, t)$ be the number of vehicles per minute crossing some point on the pavement. It can be assumed that all conditions for a Poisson distribution are satisfied in this case. The probability mass functions of $X(0, t)$ is thus given by Equation 6.44. From the data, the average number of vehicles per 30 seconds is

$$\frac{0(18) + 1(32) + 2(28) + \cdots + 9(0)}{120} \cong 2.08$$

Hence, an estimate of λt is $2.08(2) = 4.16$. The desired probability is then

$$P(X(0, t) \geq 10) = \sum_{k=10}^{\infty} p_k(0, t) = 1 - \sum_{k=0}^{9} p_k(0, t)$$

$$= 1 - \sum_{k=0}^{9} \frac{(4.16)^k e^{-4.16}}{k!}$$

$$\cong 1 - 0.992 = 0.008$$

The calculations involved in the computation above are tedious. Because of its wide applicability, the Poisson distribution for different values of λt is tabulated in the literature. Table A.2 in Appendix A gives its mass function for values of λt ranging from 0.1 to 10. Figure 6.4 is also convenient for determining the probability distribution function associated with a Poisson-distributed random variable. The answer for Example 6.11, for example, can be easily read off from Figure 6.4.

Example 6.12. Let X_1 and X_2 be two independent random variables, both having Poisson distribution with parameters v_1 and v_2, respectively, and let $Y = X_1 + X_2$. Determine the distribution of Y.

We proceed by determining first the characteristic functions of X_1 and X_2. They are

$$\phi_{X_1}(t) = E\{e^{jtX_1}\} = e^{-v_1} \sum_{k=0}^{\infty} \frac{e^{jtk} v_1^k}{k!}$$

$$= \exp[v_1(e^{jt} - 1)]$$

and

$$\phi_{X_2}(t) = \exp[v_2(e^{jt} - 1)]$$

Due to independence, the characteristic function of Y, $\phi_Y(t)$, is simply the product of $\phi_{X_1}(t)$ and $\phi_{X_2}(t)$ (see Equation 4.71). Hence,

$$\phi_Y(t) = \phi_{X_1}(t)\phi_{X_2}(t) = \exp[(v_1 + v_2)(e^{jt} - 1)]$$

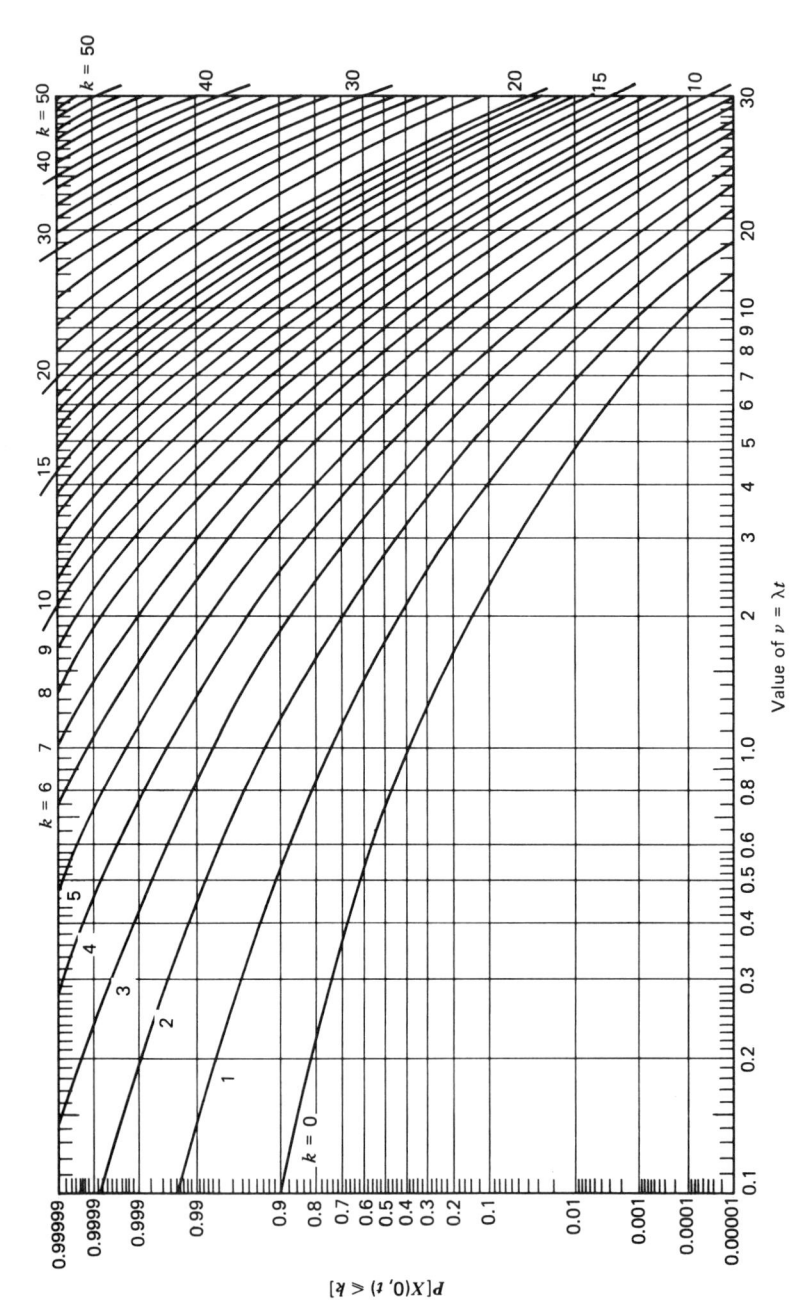

Figure 6.4 Probability distribution functions of Poisson distribution. [Taken from H. F. Dodge and H. G. Romig (1944). Copyright (1944) Bell Telephone Laboratories. Reprinted by permission.]

SOME IMPORTANT DISCRETE DISTRIBUTIONS

By inspection, it is the characteristic function corresponding to a Poisson distribution with parameter $v_1 + v_2$. Its probability mass function is thus

$$p_Y(k) = \frac{(v_1 + v_2)^k e^{-(v_1+v_2)}}{k!} \qquad k = 0, 1, 2, \ldots \qquad (6.48)$$

As in the case of binomial distributions, this result leads to the following important theorem:

Theorem

The Poisson distribution generates itself under addition of independent random variables.

Example 6.13. Suppose that the probability of an insect laying r eggs is $v^r e^{-v}/r!$, $r = 0, 1, \ldots$, and that the probability of an egg developing is p. Assuming mutual independence of individual developing processes, show that the probability of a total of k survivors is $(pv)^k e^{-pv}/k!$.

Let X be the number of eggs laid by the insect and Y be the number of eggs developed. Then, given $X = r$, the distribution of Y is binomial. Thus,

$$P(Y = k | X = r) = \binom{r}{k} p^k (1-p)^{r-k} \qquad k \leq r$$

Now, using the total probability theorem (Equation 2.27),

$$P(Y = k) = \sum_{r=k}^{\infty} P(Y = k | X = r) P(X = r)$$

$$= \sum_{r=k}^{\infty} \binom{r}{k} \frac{p^k (1-p)^{r-k} v^r e^{-v}}{r!} \qquad (6.49)$$

If we let $r = k + n$, Equation 6.49 becomes

$$P(Y = k) = \sum_{n=0}^{\infty} \binom{n+k}{k} \frac{p^k (1-p)^n v^{n+k} e^{-v}}{(n+k)!}$$

$$= \frac{(pv)^k e^{-v}}{k!} \sum_{n=0}^{\infty} \frac{(1-p)^n v^n}{n!}$$

$$= \frac{(pv)^k e^{-v} e^{(1-p)v}}{k!} = \frac{(pv)^k e^{-pv}}{k!} \qquad k = 0, 1, 2, \ldots \qquad (6.50)$$

An important observation can be made based on this result. It implies that, if a r.v. X is Poisson distributed with parameter v, then a r.v. Y, which is derived from X by selecting only with probability p each of the items counted by X, is also Poisson distributed with parameter pv. Other examples of application of this result include situations in which Y is the number of disaster-level hurricanes when X is the total number of hurricanes occurred in a given year, or Y is the number of passengers, due to overbooking, not being able to board a given flight when X is the number of passenger arrivals.

VI.3.1 SPATIAL DISTRIBUTIONS

The Poisson distribution has been derived based on arrivals developing in time, but the same argument applies to distribution of points in space. Consider the distribution of flaws in a material. The number of flaws in a given volume has a Poisson distribution if the aforementioned assumptions (1) to (3) are valid with time intervals replaced by volumes, *and* if it is reasonable to assume that the probability of finding k flaws in any region depends only on its volume and not on its shape.

Other physical situations in which the Poisson distribution has been used include bacteria counts on a Petri plate, distribution of airplane-spread fertilizers in a field, and distribution of industrial pollutants in a given region.

Example 6.14. A good example of this application is the study carried out by Clarke concerning the distribution of flying-bomb hits in a part of London during World War II. The area is divided into 576 small areas of 0.25 km² each. In Table 6.1 the number n_k of areas with exactly k hits is recorded and is compared with predicted number based on a Poisson distribution with $\lambda t =$ number of total hits/number of areas $= \frac{537}{576} = 0.932$. We see an excellent agreement between predicted and observed results.

VI.3.2 POISSON APPROXIMATION TO THE BINOMIAL DISTRIBUTION

Let X be a random variable having the binomial distribution

$$p_X(k) = \binom{n}{k} p^k (1-p)^{n-k} \qquad k = 0, 1, \ldots, n \tag{6.51}$$

Table 6.1. *Comparison of Observed and Theoretical Distributions—Example 6.14. Taken from R. D. Clarke (1946)*

k	0	1	2	3	4	5 or More
n_k (observed)	229	211	93	35	7	1
n_k (predicted)	226.7	211.4	98.5	30.6	7.1	1.6

SOME IMPORTANT DISCRETE DISTRIBUTIONS

Consider the case when $n \to \infty$ and $p \to 0$ in such a way that $np = v$ remains fixed. We note that v is the mean of X, which is assumed to remain constant. Then,

$$p_X(k) = \binom{n}{k} \left(\frac{v}{n}\right)^k \left(1 - \frac{v}{n}\right)^{n-k} \tag{6.52}$$

As $n \to \infty$, the factorials $n!$ and $(n - k)!$ appearing in the binomial coefficient can be approximated using Stirling's formula (4.78). We also note that

$$\lim_{n \to \infty} \left(1 + \frac{c}{n}\right)^n = e^c \tag{6.53}$$

Using these relationships in Equation 6.52 then gives, after some manipulation,

$$p_X(k) = \frac{v^k e^{-v}}{k!} \quad k = 0, 1, \ldots \tag{6.54}$$

This Poisson approximation to the binomial distribution can be used to advantage from the point of view of computational labor. It also establishes the fact that a close relationship exists between these two important distributions.

Example 6.15. Suppose that the probability of a resistor manufactured by a certain firm being defective is 0.015. What is the probability that there is no defective resistor in a batch of 100 resistors?

The desired probability is

$$P = \binom{100}{0}(0.015)^0 (0.985)^{100-0} = (0.985)^{100} = 0.2206$$

Since n is large and p is small in this case, the Poisson approximation is appropriate and we obtain

$$P = \frac{(1.5)^0 e^{-1.5}}{0!} = e^{-1.5} = 0.223$$

which is very close to the exact answer. In practice, the Poisson approximation is frequently used when $n > 10$ and $p < 0.1$.

Example 6.16. In oil exploration, the probability of an oil strike in the North Sea is one in 500 drillings. What is the probability of having exactly three oil-producing wells in 1000 explorations?

Table 6.2. Summary of Discrete Distributions

Distribution	Probability Mass Function	Parameters	Mean and Variance
Binomial	$\binom{n}{k} p^k (1-p)^{n-k} \quad k = 0, 1, \ldots, n$	n, p	$np, \, np(1-p)$
Hypergeometric	$\dfrac{\binom{n_1}{k}\binom{n-n_1}{m-k}}{\binom{n}{m}} \quad k = 0, 1, \ldots, \min(n_1, m)$	n, n_1, m	$\dfrac{mn_1}{n}, \, \dfrac{mn_1(n-n_1)(n-m)}{n^2(n-1)}$
Geometric	$(1-p)^{k-1} p \quad k = 1, 2, \ldots$	p	$\dfrac{1}{p}, \, \dfrac{1-p}{p^2}$
Negative binomial (Pascal)	$\binom{k-1}{r-1} p^r (1-p)^{k-r} \quad k = r, r+1, \ldots$	r, p	$\dfrac{r}{p}, \, \dfrac{r(1-p)}{p^2}$
Multinomial	$\dfrac{n!}{k_1! k_2! \cdots k_r!} p_1^{k_1} p_2^{k_2} \cdots p_r^{k_r},$ $k_1, \ldots, k_r = 0, 1, 2, \ldots \quad \sum_{i=1}^{r} k_i = n$	$n, p_i \quad i = 1, \ldots, r$ $\sum_{i=1}^{r} p_i = 1$	$np_i, \, np_i(1-p_i)$ $i = 1, \ldots, r$
Poisson	$\dfrac{(\lambda t)^k e^{-\lambda t}}{k!} \quad k = 0, 1, 2, \ldots$	λt	$\lambda t, \, \lambda t$

In this case, $n = 1000$ and $p = \frac{1}{500} = 0.002$ and the Poisson approximation is appropriate. Using Equation 6.54, we have $v = np = 2$ and the desired probability is

$$p_X(3) = \frac{2^3 e^{-2}}{3!} = 0.18$$

The examples above demonstrate that the Poisson distribution finds applications in problems where the probability of an event occurring is small. For this reason, it is often referred to as the *distribution of rare events*.

VI.4 Summary

We have introduced in this chapter several discrete distributions that are used extensively in science and engineering. Table 6.2 summarizes some of the important properties associated with these distributions for easy reference.

REFERENCES AND COMMENTS

1. R. D. Clarke, "An Application of the Poisson Distribution," *J. Inst. Actuaries*, **72**, 48–52, 1946.
2. H. F. Dodge and H. G. Romig, *Sampling Inspection Tables*, Wiley, New York, 1944.
3. C. B. Solloway, "A Simplified Statistical Model for Missile Launching—III," TM 312-287, Jet Propulsion Laboratory, Pasadena, California, 1963.

Binomial and Poisson distributions are widely tabulated in the literature. Additional references in which these tables can be found are:

4. H. Arkin and R. Colton, *Tables for Statisticians*, 2nd Ed., Barnes and Noble, New York, 1963.
5. W. H. Beyer, *Handbook of Tables for Probability and Statistics*, The Chemical Rubber Co., Cleveland, 1966.
6. E. C. Grant, *Statistical Quality Control*, 3rd Ed., McGraw-Hill, New York, 1964.
7. F. A. Haight, *Handbook of the Poisson Distribution*, Wiley, New York, 1967.

8. A. Hald, *Statistical Tables and Formulas*, Wiley, New York, 1952.
9. E. C. Molina, *Poisson's Exponential Binomial Limit*, Van Nostrand, New York, 1949.
10. National Bureau of Standards, *Tables of the Binomial Probability Distributions* (Applied Mathematics Series 6), Washington, D.C., GPO, 1949.
11. D. Owen, *Handbook of Statistical Tables*, Addison-Wesley, Reading, Mass., 1962.
12. E. S. Pearson and H. O. Hartley (Eds.), *Biometrika Tables for Statisticians*, Vol. 1, Cambridge University Press, Cambridge, England, 1954.

PROBLEMS

6.1. The r.v. X has a binomial distribution with parameters (n, p). Using the formulation given by Equation 6.10, derive its pmf, mean, and variance and compare them with results given in Equations 6.2 and 6.8. (Hint: See Example 4.17.)

6.2. Let X be the number of defective parts produced on a certain production line. It is known that, for a given lot, X is binomial with mean equal to 240 and variance 48. Determine the pmf of X and the probability that none of the parts is defective in this lot.

6.3. An experiment is repeated five times. Assuming that the probability of an experiment being successful is 0.75 and assuming independence of experimental outcomes:
   (a) What is the probability that all five experiments will be successful?
   (b) How many experiments are expected to succeed?

6.4. Suppose that the probability is 0.2 that air pollution level in a given region will be in the unsafe range. What is the probability that the level will be unsafe 7 days in a 30-day month? What is the average number of "unsafe" days in a 30-day month?

6.5. An airline estimates that 5 % of the people making reservations on a certain flight will not show up. Consequently, their policy is to sell 84 tickets for a flight that can only hold 80 passengers. What is the probability that there will be a seat available for every passenger that shows up? What is the average number of no-shows?

SOME IMPORTANT DISCRETE DISTRIBUTIONS

6.6. Assuming that each child has probability 0.51 of being a boy:
  (a) Find the probability that a family of four children will have (i) exactly one boy (ii) exactly one girl, (iii) at least one boy, and (iv) at least one girl.
  (b) Find the number of children a couple should have in order that the probability of their having at least two boys will be greater than 0.75.

6.7. A parking-by-permit-only facility has m parking spaces. A total of n ($n \geq m$) parking permits are issued while each permit holder has a probability p of using the facility in a given period. Determine:
  (a) The probability that a permit holder will be denied a parking space in the given time period.
  (b) The expected number of people turned away in the given time period.

6.8. For the hypergeometric distribution given by Equation 6.13, show that as $n \to \infty$ it approaches the binomial distribution with parameters m and n_1/n, that is,

$$p_Z(k) = \binom{m}{k}\left(\frac{n_1}{n}\right)^k \left(1 - \frac{n_1}{n}\right)^{m-k} \qquad k = 0, 1, \ldots, m$$

Thus, hypergeometric distribution can be approximated by a binomial distribution as $n \to \infty$.

6.9. A manufacturing firm receives a lot of 100 parts of which 5 are defective. Suppose that the firm accepts all 100 parts if and only if no defective ones are found in a sample of 10 parts randomly selected for inspection. Determine the probability that this lot will be accepted.

6.10. A shipment of 10 boxes of meat coming into this country contains two boxes of contaminated goods. An inspector randomly selects four boxes and let Z be the number of boxes of contaminated meat among the selected four.
  (a) What is the pmf of Z?
  (b) What is the probability that at least one of the four boxes is contaminated?
  (c) How many boxes must be selected so that the probability of having at least one contaminated box is larger than 0.75?

6.11. In a sequence of Bernoulli trials with probability p of success, determine the probability that r number of successes will occur before s number of failures.

6.12. Cars arrive independently at an intersection. Assuming that on the average 25% of the cars turn left and that the left-turn lane has a capacity for 5 cars, what is the probability that capacity is reached in the left-turn lane when 10 cars are delayed by a red signal?

6.13. Suppose that n independent steps must be taken in the sterilization procedure for a biological experiment, each of which having a probability p for success. If a failure in any of the n steps would cause contamination, what is the probability of contamination if $n = 10$ and $p = 0.99$?

6.14. The definition of the 100-year flood is given in Example 6.7. Determine:
   (a) The probability that exactly one flood equaling to or exceeding the 100-year flood will occur in a 100-year period.
   (b) The probability that one or more floods equaling to or exceeding the 100-year flood will occur in a 100-year period.

6.15. A shipment of electronic parts are sampled by testing items sequentially until the first defective part is found. If 10 or more parts are tested before the first defective part is found, the shipment is accepted as meeting specifications.
   (a) Determine the probability that the shipment will be accepted if it contains 10% defective parts.
   (b) How many need to be sampled if it is desired that a shipment with 25% defective parts be rejected with probability of at least 0.75?

6.16. In Example 6.10, determine the jpmf of X_1 and X_2. Determine the probability that, of the 10 persons studied, less than two persons will be in the low-income group and less than three persons will be in the middle-income group.

6.17. The following describes a simplified countdown procedure for launching three space vehicles from two pads.
   (a) Two vehicles are erected simultaneously on two pads and the countdown proceeds on one vehicle.
   (b) When the countdown has been successfully completed on the first vehicle, the countdown is initiated on the second vehicle, the following day.
   (c) Simultaneously, the vacated pad is immediately cleaned and prepared for the third vehicle. There is a (fixed) period of r days delay after a launching before the same pad may be utilized for a second launch attempt (the turnaround time).
   (d) After the third vehicle is erected on the vacated pad, the countdown procedure is *not* initiated until the day after the second vehicle is launched.
   (e) Each vehicle is independent of, and identical to, the others. On any single countdown attempt there is a probability p of a successful completion and a probability $q = 1 - p$ of failure. Any failure results in the termination of that countdown attempt and a new attempt is made the *following day*. That is, any failure leads to a one-day delay. It is assumed that a successful countdown attempt can be completed in one day.

(f) The failure to complete a countdown does not affect subsequent attempts in any way, that is, the trials are independent from day to day as well as from vehicle to vehicle.

Let X be the number of days until the third successful countdown. Show that the pmf of X is

$$p_X(k) = (k-r-1)p^2q^{k-r-2}(1-q^{r-1}) + \frac{(k-r)!}{2(k-r-2)!}p^3q^{k-3}$$

$$k = r+2, r+3, \ldots$$

6.18. Derive the variance of a Poisson distributed r.v. X as given by Equation 6.47.

6.19. Show that, for the Poisson distribution, $p_k(0, t)$ increases monotonically and then decreases monotonically as k increases, reaching its maximum when k is the largest integer not exceeding λt.

6.20. Assume that the number of traffic accidents in New York State during a four-day memorial day weekend is Poisson distributed with parameter $\lambda = 3.25$ per day. Determine the probability that the number of accidents is less than 10 in this four-day period.

6.21. Each air traffic controller at an airport is given the responsibility of monitoring at most 20 takeoffs and landings per hour. During a given period, the average rate of takeoffs and landings is one every two minutes. Assuming Poisson arrivals and departures, determine the probability that two controllers will be needed in this time period.

6.22. The number of vehicles crossing a certain point on a highway during a unit time period has a Poisson distribution with parameter λ. A traffic counter is used to record this number but, due to limited capacity, it registers the maximum number of 30 whenever the count equals to or exceeds 30. Determine the pmf of Y if Y is the number of vehicles recorded by the counter.

6.23. As an application of Poisson approximation to the binomial distribution, estimate the probability that in a class of 200 students exactly 20 will have birthdays on any given day.

6.24. A book of 500 pages contains on the average one misprint per page. Estimate the probability that:
(a) A given page contains at least one misprint.
(b) At least 3 pages will contain at least one misprint.

6.25. Earthquakes are registered at an average frequency of 250 per year in a given region. Suppose that the probability is 0.09 that any earthquake will have a magnitude greater than 5 on the Richter scale. Assuming independent occurrences of earthquakes, determine the pmf of X, the number of earthquakes greater than 5 on the Richter scale per year.

6.26. Let X be the number of accidents in which a driver is involved in t years. In proposing a distribution for X, the "accident likelihood" Λ varies from driver to driver and is considered as a random variable. Suppose that the conditional probability mass function $p_{X\Lambda}(x|\lambda)$ is given by the Poisson distribution

$$p_{X\Lambda}(k|\lambda) = \frac{(\lambda t)^k e^{-\lambda t}}{k!} \qquad k = 0, 1, 2, \ldots$$

and suppose that the pdf of Λ is of the form $(a, b > 0)$

$$f_\Lambda(\lambda) = \frac{a}{b\Gamma(a)} \left(\frac{a\lambda}{b}\right)^{a-1} e^{-a\lambda/b} \qquad \lambda \geq 0$$

$$= 0 \qquad \text{elsewhere}$$

where $\Gamma(a)$ is the gamma function defined by

$$\Gamma(a) = \int_0^\infty x^{a-1} e^{-x} \, dx$$

Show that the pmf of X has a negative binomial distribution in the form

$$p_X(k) = \frac{\Gamma(a+k)}{k!\,\Gamma(a)} \left(\frac{a}{a+bt}\right)^a \left(\frac{bt}{a+bt}\right)^k \qquad k = 0, 1, 2, \ldots$$

6.27. Suppose that λ, the mean rate of arrival, in the Poisson distribution is time dependent and is given by

$$\lambda = \frac{vt^{v-1}}{w}$$

Determine the pmf $p_k(0, t)$, the probability of exactly k arrivals in the time interval $[0, t)$. (Note that differential equations such as Equations 6.39 and 6.42 now have time-dependent coefficients.)

6.28. Derive the joint probability mass function of two Poisson r.v.'s $X_1 = X(0, t_1)$ and $X_2 = X(0, t_2)$, $t_2 > t_1$, with the same mean rate of arrival λ. Determine the probability $P(X_1 \leq \lambda t_1 \cap X_2 \leq \lambda t_2)$. This is the probability that the numbers of arrivals in intervals $[0, t_1)$ and $[0, t_2)$ are both equal to or less than the average arrivals in their respective intervals.

VII
Some Important Continuous Distributions

Let us now turn our attention to some important continuous probability distributions. Physical quantities such as time, length, area, temperature, pressure, load, intensity, etc., when they need to be described probabilistically, are modeled by continuous random variables. A number of important continuous distributions are introduced in this chapter and, as in Chapter VI, we are also concerned with the nature and applications of these distributions in science and engineering.

VII.1 The Uniform Distribution

A continuous r.v. X has a *uniform distribution* over an interval a to b ($b > a$) if it is equally likely to take on any value in this interval. The probability density function of X is constant over the interval (a, b) and has the form

$$\boxed{\begin{aligned} f_X(x) &= \frac{1}{b-a} \quad a \leq x \leq b \\ &= 0 \quad\quad\;\; \text{elsewhere} \end{aligned}} \tag{7.1}$$

As we see from Figure 7.1a, it is constant over (a, b) whose height must be $1/(b-a)$ in order that area under the density function be unity.

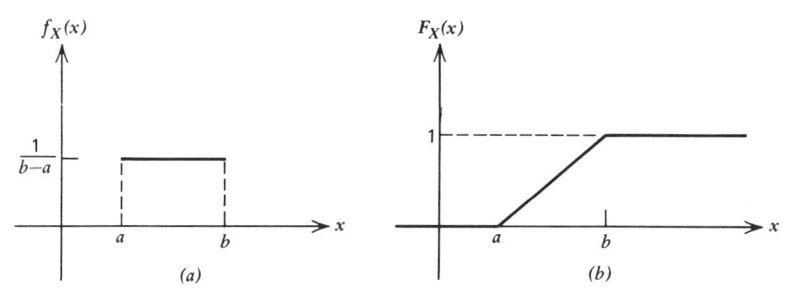

Figure 7.1 pdf and PDF of X.

The probability distribution function is, on integrating Equation 7.1,

$$\boxed{\begin{aligned} F_X(x) &= 0 & x &< a \\ &= \frac{x-a}{b-a} & a &\leq x \leq b \\ &= 1 & x &> b \end{aligned}} \quad (7.2)$$

which is graphically presented in Figure 7.1b.

The mean and variance of X are easily found to be

$$\begin{aligned} m_X &= \int_a^b x f_X(x) dx = \frac{1}{b-a} \int_a^b x\, dx = \frac{a+b}{2} \\ \sigma_X^2 &= \frac{1}{b-a} \int_a^b \left(x - \frac{a+b}{2} \right)^2 dx = \frac{(b-a)^2}{12} \end{aligned} \quad (7.3)$$

The uniform distribution is one of the simplest distributions and is commonly used in situations where there is no reason to give unequal likelihoods to possible ranges assumed by the random variable over a given interval. For example, the arrival time of a flight might be considered uniformly distributed over a certain time interval, or the distribution of the distance from location of live loads on a bridge to an end support might be adequately represented by a uniform distribution over the bridge span. Let us also comment that one often assigns a uniform distribution to a specific random variable simply because of a lack of information beyond knowing the range of values it spans.

Example 7.1. Due to unpredictable traffic situations, the time required by a certain student to travel from her home to her morning class is uniformly distri-

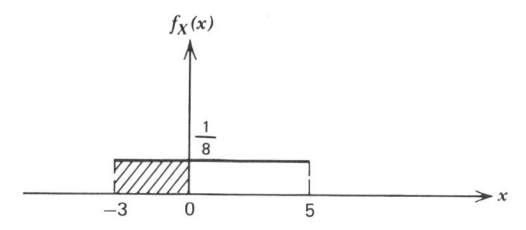

Figure 7.2 pdf of X in Example 7.1.

buted between 22 and 30 minutes. If she leaves home at precisely 7:35 A.M., what is the probability that she will not be late for class which begins promptly at 8:00 A.M.?

Let X be the class arrival time of the student in minutes after 8:00 A.M. It then has a uniform distribution given by

$$f_X(x) = \tfrac{1}{8} \quad -3 \leq x \leq 5$$
$$= 0 \quad \text{elsewhere}$$

We are interested in the probability $P(-3 \leq X \leq 0)$. As seen from Figure 7.2, it is clear that this probability is equal to the ratio of the shaded area and the unit total area. Hence,

$$P(-3 \leq X \leq 0) = 3(\tfrac{1}{8}) = \tfrac{3}{8}$$

It is also clear that, due to uniformity in the distribution, the solution can be found simply by taking the ratio of the length from -3 to 0 to the total length of the distribution interval. Stated in general terms, if a r.v. X is uniformly distributed over an interval A, then the probability of X taking values in a subinterval B is given by

$$P(X \text{ in } B) = \frac{\text{length of } B}{\text{length of } A} \tag{7.4}$$

VII.1.1 A BIVARIATE UNIFORM DISTRIBUTION

Let r.v. X be uniformly distributed over an interval (a_1, b_1) and r.v. Y uniformly distributed over an interval (a_2, b_2). Furthermore, let us assume that they are independent. Then the joint probability density function of X and Y is simply

$$f_{XY}(x, y) = f_X(x) f_Y(y) = \frac{1}{(b_1 - a_1)(b_2 - a_2)} \quad a_1 \leq x \leq b_1 \quad a_2 \leq y \leq b_2$$
$$= 0 \quad \text{elsewhere}$$

(7.5)

It takes the shape of a flat surface bounded by (a_1, b_1) along the x-axis and (a_2, b_2) along the y-axis. We have seen an application of this bivariate uniform distribution in Example 3.6. Indeed, Example 3.6 gives a typical situation in which the distribution given by Equation 7.5 is conveniently applied. Let us give one more example.

Example 7.2. A warehouse receives merchandise and fills a specific order for the same merchandise in any given day. Suppose that it receives merchandise with equal likelihood during equal intervals of time over the eight-hour working day and likewise for the order to be filled. (a) What is the probability that the order will arrive after the merchandise is received? and (b) What is the probability that the order will arrive within two hours after the receipt of merchandise?

Let X be the time of receipt of merchandise expressed in fraction of an eight-hour working day and Y be the time of receipt of the order similarly expressed. Then

$$f_X(x) = 1 \quad 0 \le x \le 1$$
$$= 0 \quad \text{elsewhere} \tag{7.6}$$

and similarly for $f_Y(y)$. The joint probability density function is, assuming independence,

$$f_{XY}(x, y) = 1 \quad 0 \le x \le 1 \quad 0 \le y \le 1$$
$$= 0 \quad \text{elsewhere}$$

and is shown in Figure 7.3.

To answer the first question, we integrate $f_{XY}(x, y)$ over an appropriate region in the (x, y)-plane satisfying $y \ge x$. Since $f_{XY}(x, y)$ is a constant over $0 \le (x, y) \le 1$,

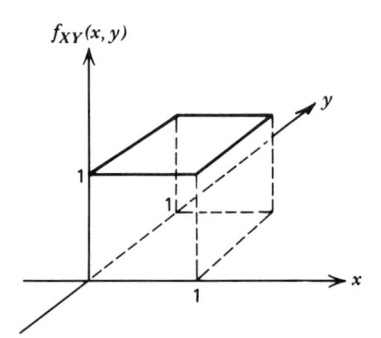

Figure 7.3 jpdf of X and Y in Example 7.2.

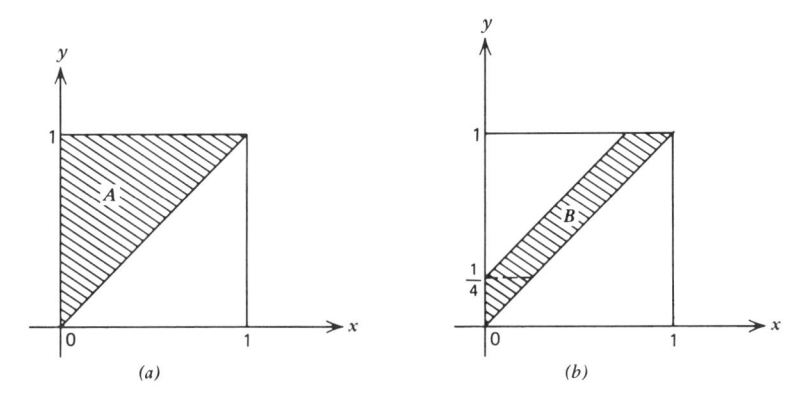

Figure 7.4 Regions A and B in (x, y)-plane in Example 7.2.

this is the same as taking the ratio of the area satisfying $y \geq x$ to the total area bounded by $0 \leq (x, y) \leq 1$, which is unity. As seen from Figure 7.4a, we have

$$P(Y \geq X) = \text{shaded area } A = \tfrac{1}{2}$$

We proceed the same way in answering the second question. It is easy to see that the appropriate region for this part is the shaded area B as shown in Figure 7.4b. The desired probability is, after dividing area B into two subregions as shown,

$$P(X \leq Y \leq X + \tfrac{1}{4}) = \text{shaded area } B$$
$$= \tfrac{1}{4}(\tfrac{3}{4}) + \tfrac{1}{2}(\tfrac{1}{4})(\tfrac{1}{4}) = \tfrac{7}{32}$$

We see from this example that calculations of various probabilities of interest in this situation involve taking ratios of appropriate areas. If r.v.'s X and Y are independent and uniformly distributed over a region A, then the probability of X and Y taking values in a subregion B is given by

$$P[(X, Y) \text{ in } B] = \frac{\text{area of } B}{\text{area of } A} \tag{7.7}$$

It is noteworthy that, if the independence assumption is removed, the joint probability density function of two uniformly distributed random variables will not take the simple form as given by Equation 7.5. In the extreme case when X and Y are perfectly correlated, the jpdf of X and Y degenerates from a surface into a line over the (x, y)-plane. For example, let X and Y be uniformly and identically

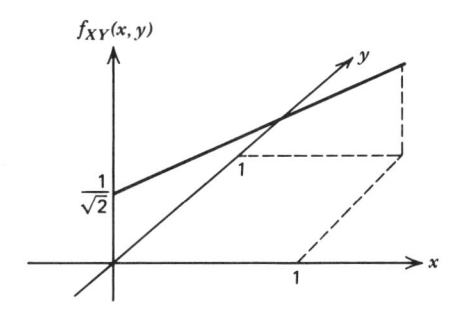

Figure 7.5 jpdf of X and Y given by Equation 7.8.

distributed over the interval $(0, 1)$ and let $X = Y$. Then the jpdf of X and Y has the form

$$f_{XY}(x, y) = \frac{1}{\sqrt{2}} \qquad x = y \quad \text{and} \quad 0 \leq (x, y) \leq 1 \tag{7.8}$$

which is graphically presented in Figure 7.5. More detailed discussions on correlated and uniformly distributed random variables can be found in Kramer (1940).

VII.2 The Gaussian or Normal Distribution

The most important probability distribution in theory as well as in application is the *Gaussian* or *normal* distribution. A r.v. X is *Gaussian* or *normal* if its probability density function $f_X(x)$ is of the form

$$f_X(x) = \frac{1}{\sqrt{2\pi}\sigma} \exp\left[-\frac{(x - m)^2}{2\sigma^2}\right] \qquad -\infty < x < \infty \tag{7.9}$$

where m and σ are two parameters with $\sigma > 0$. Our choice of these particular symbols for the parameters will become clear presently.

Its corresponding probability distribution function is

$$F_X(x) = \frac{1}{\sqrt{2\pi}\sigma} \int_{-\infty}^{x} \exp\left[-\frac{(u - m)^2}{2\sigma^2}\right] du \qquad -\infty < x < \infty \tag{7.10}$$

SOME IMPORTANT CONTINUOUS DISTRIBUTIONS

which cannot be expressed in closed form analytically but can be numerically evaluated for any x.

The pdf and PDF expressed by Equations 7.9 and 7.10 are graphed in Figure 7.6 with $m = 0$ and $\sigma = 1$. The graph of $f_X(x)$ in this particular case is the well-known bell-shaped curve, symmetrical about the origin.

Let us determine the mean and variance of X. By definition,

$$E\{X\} = \int_{-\infty}^{\infty} x f_X(x) dx = \frac{1}{\sqrt{2\pi}\sigma} \int_{-\infty}^{\infty} x \exp\left[-\frac{(x-m)^2}{2\sigma^2}\right] dx$$

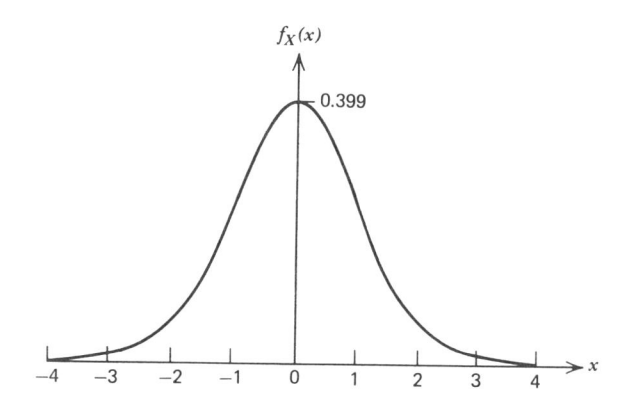

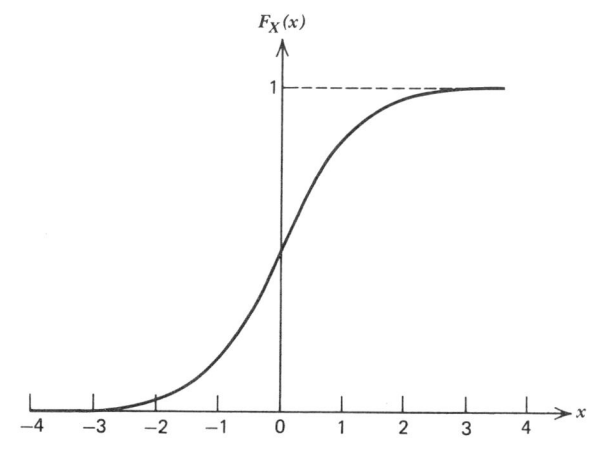

Figure 7.6 pdf and PDF of X with $m = 0$ and $\sigma = 1$.

which yields

$$E\{X\} = m$$

Similarly, we can show that (7.11)

$$\text{var}(X) = \sigma^2$$

We thus see that the two parameters m and σ in the probability distribution are, respectively, the mean and standard deviation of X. This observation justifies our choice of these special symbols for them and it also points out an important property of the normal distribution, that is, the knowledge of its mean and variance completely characterizes a normal distribution. Since the normal distribution shall be referred to frequently in our discussion, it is sometimes represented by the simple notation $N(m, \sigma^2)$. Thus, for example, X being $N(0, 9)$ implies that X has the probability density function given by Equation 7.9 with $m = 0$ and $\sigma = 3$.

Higher-order moments of X also take simple forms and can be derived in a straightforward fashion. Let us first state that, following the definition of characteristic functions discussed in Section IV.5, the characteristic function of a normal r.v. X is

$$\phi_X(t) = E\{e^{jtX}\} = \frac{1}{\sqrt{2\pi}\sigma} \int_{-\infty}^{\infty} \exp\left[jtx - \frac{(x-m)^2}{2\sigma^2}\right] dx$$

$$= \exp\left(jmt - \frac{\sigma^2 t^2}{2}\right) \quad (7.12)$$

The moments of X of any order can now be found from the above through differentiation. Expressed in terms of central moments, the use of Equation 4.52 gives us

$$\begin{aligned}\mu_n &= 0 & n \text{ odd} \\ &= 1(3)\cdots(n-1)\sigma^n & n \text{ even}\end{aligned} \quad (7.13)$$

Let us note in passing that γ_2, the coefficient of excess defined by Equation 4.12, for a normal distribution is zero. Hence, it is used as the reference distribution for γ_2.

VII.2.1 THE CENTRAL LIMIT THEOREM

The great practical importance associated with the normal distribution stems from the powerful central limit theorem stated below. Instead of giving the theorem in its entire generality, it serves our purposes quite well by stating a more restricted version due to Lindeberg (1922).

SOME IMPORTANT CONTINUOUS DISTRIBUTIONS

Central Limit Theorem

Let $\{X_n\}$ be a sequence of mutually independent and identically distributed r.v.'s with means m and variances σ^2. Let

$$Y = \sum_{j=1}^{n} X_j \qquad (7.14)$$

and let the normalized r.v. Z be defined as

$$Z = \frac{(Y - nm)}{\sqrt{n}\,\sigma} \qquad (7.15)$$

Then the probability distribution function of Z, $F_Z(z)$, converges to $N(0, 1)$ as $n \to \infty$ for every fixed z.

Proof

We first remark that, following our discussion in Section IV.4 on moments of sums of random variables, r.v. Y defined by Equation 7.14 has mean nm and standard deviation $\sqrt{n}\,\sigma$. Hence, Z is simply the "standardized" r.v. Y with zero mean and unit standard deviation. In terms of the characteristic functions $\phi_X(t)$ of r.v.'s X_j, the characteristic function of Y is simply

$$\phi_Y(t) = [\phi_X(t)]^n \qquad (7.16)$$

Consequently, Z possesses the characteristic function

$$\phi_Z(t) = \left[\exp\left(-\frac{jmt}{\sqrt{n}\,\sigma}\right)\phi_X\left(\frac{t}{\sqrt{n}\,\sigma}\right)\right]^n \qquad (7.17)$$

Expanding $\phi_X(t)$ in MacLaurin series as indicated by Equation 4.41, we can write

$$\phi_Z(t) = \left\{\exp\left(-\frac{jmt}{\sqrt{n}\,\sigma}\right)\left[1 + \frac{mjt}{\sqrt{n}\,\sigma} + \frac{(\sigma^2 + m^2)}{2}\left(\frac{jt}{\sqrt{n}\,\sigma}\right)^2 + \cdots\right]\right\}^n$$

$$= \left[1 - \frac{t^2}{2n} + o\left(\frac{t^2}{n}\right)\right]^n$$

$$\to e^{-t^2/2} \qquad (7.18)$$

as $n \to \infty$. In the last step we have used the elementary identity

$$\lim_{n \to \infty} \left(1 + \frac{c}{n}\right)^n = e^c \qquad (7.19)$$

for any real c.

Comparing the result given by Equation 7.18 with the form of characteristic function of a normal random variable given by Equation 7.12, we see that $\phi_Z(t)$ approaches the characteristic function of a zero-mean, unit-variance normal distribution. The proof is thus complete.

As we mentioned earlier, this result is a somewhat restrictive version of the central limit theorem. It can be extended in several directions, including cases in which Y is a sum of dependent as well as nonidentically distributed random variables.

The central limit theorem describes a very general class of random phenomena whose distributions can be approximated by the normal distribution. In words, when the randomness in a physical phenomenon is the cumulation of many small additive random effects, it tends to a normal distribution irrespective of the distributions of individual effects. For example, gasoline consumptions of all 1980 automobiles of a particular brand, supposedly manufactured under identical processes, differ from one automobile to another. This randomness stems from a wide variety of sources: inherent inaccuracies in manufacturing processes, nonuniformities in materials used, differences in weight and other specifications, difference in gasoline quality, different driver behavior, and many others. If one accepts the fact that all of these differences contribute to the randomness in gasoline consumption, the central limit theorem tells us that it tends to a normal distribution. By the same reasoning, temperature variations in a room, readout errors associated with an instrument, target errors of a certain weapon, etc., can also be reasonably approximated by normal distributions.

Let us also mention that, in view of the central limit theorem, our result in Example 4.17 concerning a one-dimensional random walk should be of no surprise. As the number of steps increases, it is expected that position of the particle becomes normally distributed in the limit.

VII.2.2 PROBABILITY TABULATIONS

Due to its importance, we are often called upon to evaluate probabilities associated with a normal r.v. $X: N(m, \sigma^2)$, such as

$$P(a < X \leq b) = \frac{1}{\sqrt{2\pi}\,\sigma} \int_a^b \exp\left[-\frac{(x-m)^2}{2\sigma^2}\right] dx \qquad (7.20)$$

SOME IMPORTANT CONTINUOUS DISTRIBUTIONS 195

However, as we commented earlier, the integral above cannot be evaluated by analytical means, and is generally performed numerically. For convenience, tables are provided which enable us to determine probabilities such as the one expressed by Equation 7.20.

The tabulation of PDF for the normal distribution with $m = 0$ and $\sigma = 1$ is given in Table A.3 in Appendix A. A r.v. with distribution $N(0, 1)$ is called the *standardized* normal r.v. and we shall denote it by U. Table A.3 gives $F_U(u)$ for points in the right half of the distribution only (i.e., for $u \geq 0$). The corresponding values for $u < 0$ are obtained from symmetry property of the normal distribution by the relationship (see Figure 7.6)

$$F_U(-u) = 1 - F_U(u) \qquad (7.21)$$

First of all, Table A.3 together with Equation 7.21 can be used to determine $P(a < U \leq b)$ for any a and b. Consider, for example, $P(-1.5 < U \leq 2.5)$. It is given by

$$P(-1.5 < U \leq 2.5) = F_U(2.5) - F_U(-1.5)$$

The value of $F_U(2.5)$ is found from Table A.3 to be 0.9938 and $F_U(-1.5) = 1 - F_U(1.5)$ with $F_U(1.5) = 0.9332$ as seen from Table A.3. Thus

$$P(-1.5 < U \leq 2.5) = F_U(2.5) - [1 - F_U(1.5)]$$
$$= 0.994 - 1 + 0.933 = 0.927$$

More important, Table A.3 and Equation 7.21 are also sufficient for determining probabilities associated with normal random variables with arbitrary means and variances. To do this, let us first state the following theorem.

Theorem

Let X be a normal r.v. with distribution $N(m, \sigma^2)$. Then $(X - m)/\sigma$ is the standardized normal r.v. with distribution $N(0, 1)$ or

$$U = \frac{X - m}{\sigma} \qquad (7.22)$$

Proof

The characteristic function of the r.v. $(X - m)/\sigma$ is

$$E\left\{\exp\left[\frac{jt(X - m)}{\sigma}\right]\right\} = e^{-jtm/\sigma} E\{e^{j(t/\sigma)X}\} = e^{-jtm/\sigma} \phi_X(t/\sigma)$$

From Equation 7.12, we have

$$\phi_X(t) = \exp\left(jmt - \frac{\sigma^2 t^2}{2}\right) \qquad (7.23)$$

Hence,

$$E\left\{\exp\left[\frac{jt(X-m)}{\sigma}\right]\right\} = \exp\left[-\frac{jmt}{\sigma} + \frac{jmt}{\sigma} - \frac{\sigma^2(t/\sigma)^2}{2}\right]$$
$$= e^{-t^2/2} \qquad (7.24)$$

The result given above takes the form of $\phi_X(t)$ with $m = 0$ and $\sigma = 1$ and the proof is complete.

The theorem above implies that

$$P(a < X \le b) = P[a < (U\sigma + m) \le b] = P\left(\frac{a-m}{\sigma} < U \le \frac{b-m}{\sigma}\right) \qquad (7.25)$$

The value of the right-hand side can now be found from Table A.3 with the aid of Equation 7.21 if necessary.

Example 7.3. Due to many independent error sources, the length of a manufactured machine part is normally distributed with $m = 11$ centimeters and $\sigma = 0.2$ centimeters. If specifications require that the length be between 10.6 centimeters and 11.2 centimeters, what proportion of the manufactured parts will be rejected on the average?

If X is used to denote the part length in centimeters, it is reasonable to assume that it is distributed according to $N(11, 0.04)$. Thus, on the average, the proportion of acceptable parts is $P(10.6 < X \le 11.2)$. Following Equation 7.25 and using Table A.3, we have

$$P(10.6 < X \le 11.2) = P\left(\frac{10.6 - 11}{0.2} < U \le \frac{11.2 - 11}{0.2}\right)$$
$$= P(-2 < U \le 1) = F_U(1) - [1 - F_U(2)]$$
$$= 0.8413 - (1 - 0.9772) = 0.8185$$

The desired answer is then $1 - 0.8185 = 0.1815$.

SOME IMPORTANT CONTINUOUS DISTRIBUTIONS

The use of normal distribution in this example raises an immediate concern. Normal r.v.'s assume values in both positive and negative ranges, while the length of a machine part as well as many other physical quantities cannot take negative values. However, from modeling point of view, it is a commonly accepted practice that normal r.v.'s are valid representations for nonnegative quantities in as much as the probability $P(X < 0)$ is sufficiently small. In Example 7.3, for example, this probability is

$$P(X < 0) = P\left(U < -\frac{11}{0.2}\right) = P(U < -55) \cong 0$$

Example 7.4. Let us compute $P(m - k\sigma < X \leq m + k\sigma)$ where X is distributed $N(m, \sigma^2)$. It follows from Equations 7.21 and 7.25 that

$$P(m - k\sigma < X \leq m + k\sigma) = P(-k < U \leq k)$$
$$= F_U(k) - F_U(-k) = 2F_U(k) - 1 \quad (7.26)$$

We note that this result is independent of m and σ and is only a function of k. Thus, the probability that X takes values within k standard deviations about its expected value depends only on k and is given by Equation 7.26. It is seen from Table A.3 that 68.3, 95.5, and 99.7% of the area under a normal density function are located, respectively, in the ranges $m \pm \sigma$, $m \pm 2\sigma$, and $m \pm 3\sigma$. This is illustrated in Figure 7.7. For example, the chances are about 99.7% that a randomly selected sample from a normal distribution is within the range of $m \pm 3\sigma$.

VII.2.3 THE MULTIVARIATE NORMAL DISTRIBUTION

Consider two random variables X and Y. They are said to be *jointly normal* if their joint density function takes the form

$$f_{XY}(x, y) = \frac{1}{2\pi\sigma_X\sigma_Y\sqrt{1-\rho^2}} \exp\left\{-\frac{1}{2(1-\rho^2)}\left[\left(\frac{x-m_X}{\sigma_X}\right)^2 - 2\rho\frac{(x-m_X)(y-m_Y)}{\sigma_X\sigma_Y} + \left(\frac{y-m_Y}{\sigma_Y}\right)^2\right]\right\} \quad -\infty < (x, y) < \infty$$

(7.27)

Equation 7.27 describes the *bivariate normal distribution*. There are five parameters associated with it: m_X, m_Y, $\sigma_X(>0)$, $\sigma_Y(>0)$, and $\rho(|\rho| \leq 1)$. A typical plot of this joint density function is given in Figure 7.8.

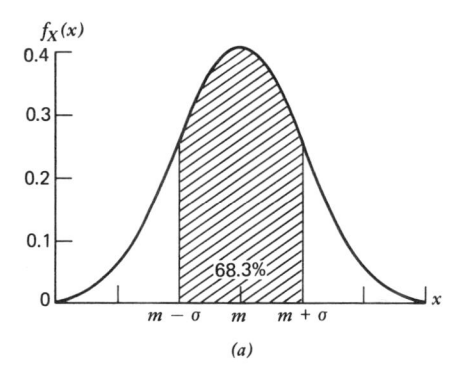

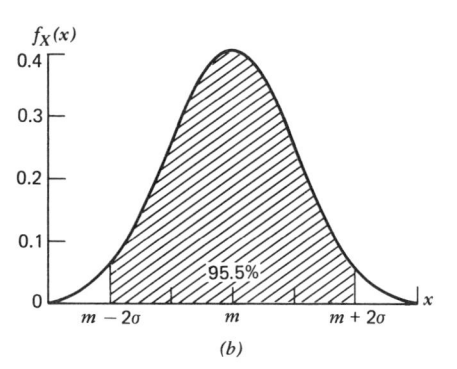

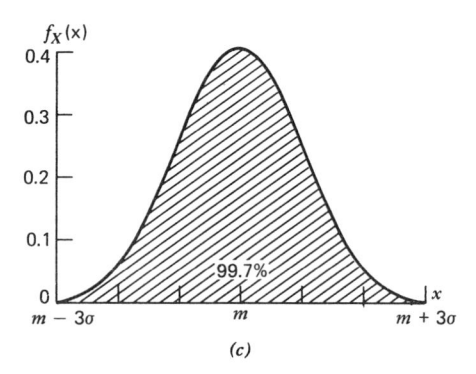

Figure 7.7 Areas under normal density function within ranges $m \pm \sigma$, $m \pm 2\sigma$, and $m \pm 3\sigma$.

SOME IMPORTANT CONTINUOUS DISTRIBUTIONS

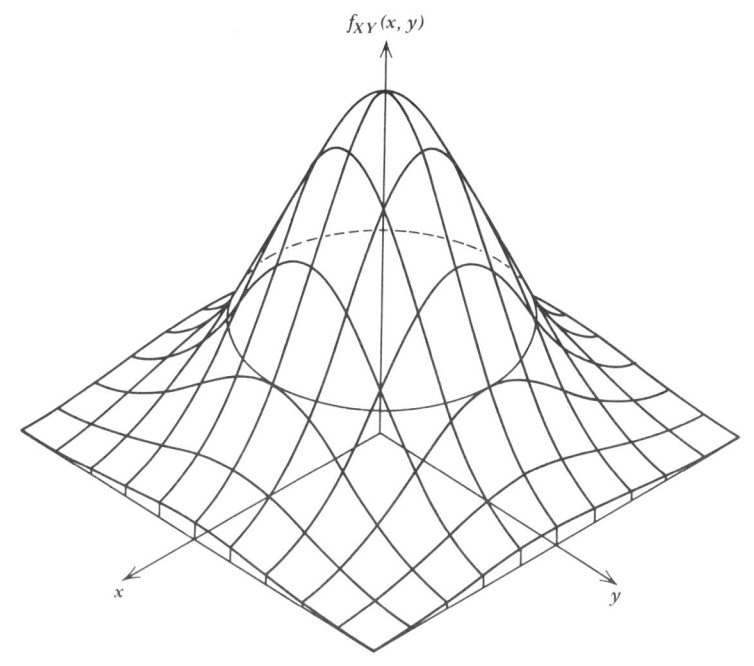

Figure 7.8 Bivariate normal distribution with $m_X = m_Y = 0$ and $\sigma_X = \sigma_Y$.

Let us determine the marginal density function of r.v. X. It is given by, following straightforward calculations,

$$f_X(x) = \int_{-\infty}^{\infty} f_{XY}(x, y)dy = \frac{1}{\sqrt{2\pi}\sigma_X} \exp\left[-\frac{(x - m_X)^2}{2\sigma_X^2}\right] \quad -\infty < x < \infty \quad (7.28)$$

Thus, r.v. X by itself has a normal distribution $N(m_X, \sigma_X^2)$. Similar calculations show that Y is also normal with distribution $N(m_Y, \sigma_Y^2)$ and $\rho = \mu_{XY}/\sigma_X\sigma_Y$ is the correlation coefficient of X and Y. We thus see that the five parameters contained in the bivariate density function $f_{XY}(x, y)$ represent five important moments associated with the random variables. This also leads us to observe that the bivariate normal distribution is completely characterized by the first- and second-order joint moments.

Another interesting and important property associated with jointly normally distributed random variables is noted below.

Theorem

Zero correlation implies independence when the random variables are jointly normal.

Proof

Let $\rho = 0$ in Equation 7.27. We easily get

$$f_{XY}(x, y) = \left\{\frac{1}{\sqrt{2\pi}\sigma_X}\exp\left[-\frac{(x-m_X)^2}{2\sigma_X^2}\right]\right\}\left\{\frac{1}{\sqrt{2\pi}\sigma_Y}\exp\left[-\frac{(y-m_Y)^2}{2\sigma_Y^2}\right]\right\}$$
$$= f_X(x)f_Y(y) \qquad (7.29)$$

which is the desired result. It is stressed again, as we have done in Section IV.3.1, that this property is not shared by random variables in general.

We have the *multivariate normal distribution* when the case of two random variables is extended to that involving n random variables. For compactness, the vector-matrix notation is used in the following development.

Consider a sequence of n random variables, $X_1, X_2, \ldots, X_n$. They are said to be *jointly normal* if its associated joint density function has the form

$$f_{X_1 X_2 \cdots X_n}(x_1, x_2, \ldots, x_n) = f_\mathbf{X}(\mathbf{x})$$
$$= (2\pi)^{-n/2}|\Lambda|^{-1/2}\exp[-\tfrac{1}{2}(\mathbf{x}-\mathbf{m})^T\Lambda^{-1}(\mathbf{x}-\mathbf{m})]$$
$$-\infty < \mathbf{x} < \infty \qquad (7.30)$$

where $\mathbf{m}^T = [m_1 \; m_2 \cdots \; m_n] = [E\{X_1\} \; E\{X_2\} \cdots \; E\{X_n\}]$ and $\Lambda = [\mu_{ij}]$ is the $n \times n$ covariance matrix of $\mathbf{X}$ with (see Equations 4.34 and 4.35)

$$\mu_{ij} = E\{(X_i - m_i)(X_j - m_j)\} \qquad (7.31)$$

The superscripts T and -1 denote, respectively, matrix transpose and matrix inverse. Again, we see that a joint normal distribution is completely specified by the first- and second-order joint moments.

It is instructive to derive the joint characteristic function associated with $\mathbf{X}$. As seen from Section IV.5.3, it is defined by

$$\phi_{X_1 X_2 \cdots X_n}(t_1, t_2, \ldots, t_n) = \phi_\mathbf{X}(\mathbf{t})$$
$$= E\{e^{j(t_1 X_1 + \cdots + t_n X_n)}\}$$
$$= \int_{-\infty}^{\infty} \cdots \int_{-\infty}^{\infty} e^{j\mathbf{t}^T\mathbf{x}} f_\mathbf{X}(\mathbf{x})d\mathbf{x} \qquad (7.32)$$

which gives, on substituting Equation 7.30 into Equation 7.32,

$$\phi_\mathbf{X}(\mathbf{t}) = \exp(j\mathbf{m}^T\mathbf{t} - \tfrac{1}{2}\mathbf{t}^T\Lambda\mathbf{t}) \qquad (7.33)$$

where $\mathbf{t}^T = [t_1 \; t_2 \cdots \; t_n]$.

SOME IMPORTANT CONTINUOUS DISTRIBUTIONS

Joint moments of $\mathbf{X}$ can be obtained by differentiating the joint characteristic function $\phi_\mathbf{X}(\mathbf{t})$ and setting $\mathbf{t} = 0$. The expectation $E\{X_1^{m_1} X_2^{m_2} \cdots X_n^{m_n}\}$, for example, is given by

$$E\{X_1^{m_1} X_2^{m_2} \cdots X_n^{m_n}\} = j^{-(m_1 + m_2 + \cdots + m_n)} \left[\frac{\partial^{m_1 + m_2 + \cdots + m_n}}{\partial t_1^{m_1} \partial t_2^{m_2} \cdots \partial t_n^{m_n}} \phi_\mathbf{X}(\mathbf{t}) \right]_{\mathbf{t}=0} \quad (7.34)$$

It is clear that, since joint moments of the first- and second-order completely specify the joint normal distribution, these moments also determine joint moments of orders higher than 2. We can show that, in the case when r.v.'s $X_1, X_2, \ldots, X_n$ have zero means, all odd-order moments of these random variables vanish and, for n even,

$$E\{X_1 X_2 \cdots X_n\} = \sum_{m_1, \ldots, m_n} E\{X_{m_1} X_{m_2}\} E\{X_{m_2} X_{m_3}\} \cdots E\{X_{m_{n-1}} X_{m_n}\} \quad (7.35)$$

The sum above is taken over all possible combinations of $n/2$ pairs of n r.v.'s. The number of terms in the summation is $(1)(3)(5) \cdots (n-3)(n-1)$.

VII.2.4 SUMS OF NORMAL RANDOM VARIABLES

We have seen through discussions and examples that sums of random variables arise in a number of problem formulations. In the case of normal random variables, we have the following important result.

Theorem

Let $X_1, X_2, \ldots,$ and X_n be n normally distributed random variables (not necessarily independent). Then the r.v. Y where

$$Y = c_1 X_1 + c_2 X_2 + \cdots + c_n X_n \quad (7.36)$$

is normally distributed.

Proof

For convenience, the proof will be given by assuming that all $X_j, j = 1, 2, \ldots, n$, have zero means. For this case, the mean of Y is clearly zero and its variance is, as seen from Equation 4.43,

$$\sigma_Y^2 = E\{Y^2\} = \sum_{i,j=1}^{n} c_i c_j \mu_{ij} \quad (7.37)$$

where $\mu_{ij} = \text{cov}(X_i, X_j)$.

Since the X_j's are normally distributed, their joint characteristic function is given by Equation 7.33, which is

$$\phi_{\mathbf{X}}(\mathbf{t}) = \exp\left(-\tfrac{1}{2}\sum_{i,j=1}^{n} \mu_{ij} t_i t_j\right) \qquad (7.38)$$

The characteristic function of Y is

$$\phi_Y(t) = E\{e^{jtY}\} = E\left\{\exp\left(jt \sum_{k=1}^{n} c_k X_k\right)\right\}$$

$$= \exp\left(-\tfrac{1}{2}t^2 \sum_{i,j=1}^{n} \mu_{ij} c_i c_j\right)$$

$$= \exp(-\tfrac{1}{2}\sigma_Y^2 t^2) \qquad (7.39)$$

which is the characteristic function associated with a normal r.v. Hence Y is also a normal r.v.

A further generalization of the above result is given below which we shall state without proof.

Theorem

Let $X_1, X_2, \ldots,$ and X_n be n normally distributed random variables (not necessarily independent). Then the r.v.'s $Y_1, Y_2, \ldots,$ and Y_m where

$$Y_j = \sum_{k=1}^{n} c_{jk} X_k \qquad j = 1, 2, \ldots, m \qquad (7.40)$$

are themselves jointly normally distributed.

VII.3 The Lognormal Distribution

We have seen that normal distributions arise from sums of many independent random actions. Consider now another common phenomenon which is the resultant of many *multiplicative* independent random effects. An example of multiplicative phenomena is in fatigue studies of materials where internal material damage at a given stage of loading is a random proportion of damage at the previous stage. In biology, the distribution of size of an organism is another example whose growth is subject to many small impulses, each of which is proportional to its momentary size. Other examples include size distribution of particles under impaction or impulsive forces, life distribution of mechanical components, distribution of personal incomes due to annual adjustments, and other similar phenomena.

Let us consider

$$Y = X_1 X_2 \cdots X_n \qquad (7.41)$$

and assume that $X_1, \ldots,$ and X_n are independent and identically distributed random variables. We are interested in the distribution of Y as n becomes large, when the r.v.'s $X_j, j = 1, 2, \ldots, n$, can take only positive values.

Taking logarithms of both sides, Equation 7.41 becomes

$$\ln Y = \ln X_1 + \ln X_2 + \cdots + \ln X_n \qquad (7.42)$$

The r.v. $\ln Y$ is seen as a sum of independent and identically distributed r.v.'s $\ln X_1, \ln X_2, \ldots,$ and $\ln X_n$. It thus follows from the central limit theorem that $\ln Y$ tends to a normal distribution as $n \to \infty$. The probability distribution of Y is thus determined from

$$Y = e^X \qquad (7.43)$$

where X is a normal r.v.

Definition. Let X be $N(m_X, \sigma_X^2)$. The r.v. Y as determined from Equation 7.43 is said to have a *lognormal* distribution.

The pdf of Y is easy to determine. Since Equation 7.43 gives Y as a monotonic function of X, Equation 5.12 in Chapter V immediately gives

$$f_Y(y) = \frac{1}{y\sigma_X\sqrt{2\pi}} \exp\left[-\frac{1}{2\sigma_X^2}(\ln y - m_X)^2\right] \quad y \geq 0$$
$$= 0 \qquad \text{elsewhere} \qquad (7.44)$$

Equation 7.44 shows that Y has a one-sided distribution (i.e., it takes values only in the positive range of y). This property makes it attractive for physical quantities that are restricted to having only positive values. In addition, $f_Y(y)$ takes many different shapes for different values of m_X and $\sigma_X(>0)$. As seen from Figure 7.9, the pdf of Y is skewed to the right, becoming more pronounced as σ_X increases.

It is noted that the parameters m_X and σ_X appearing in the pdf of Y are the mean and standard deviation of X, or $\ln Y$, but not of Y. To obtain a more natural pair

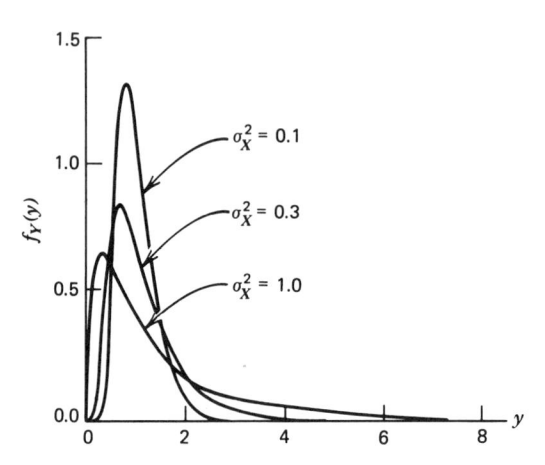

Figure 7.9 Lognormal distribution with $m_X = 0$ and several values of σ_X^2.

of parameters for $f_Y(y)$, we observe that, if medians of X and Y are denoted by θ_X and θ_Y, respectively, the definition of median of a random variable gives

$$0.5 = P(Y \leq \theta_Y) = P(X \leq \ln \theta_Y) = P(X \leq \theta_X)$$

or

$$\ln \theta_Y = \theta_X \qquad (7.45)$$

Since, due to symmetry of the normal distribution,

$$\theta_X = m_X$$

we can write

$$m_X = \ln \theta_Y \qquad (7.46)$$

Now, writing $\sigma_X = \sigma_{\ln Y}$, the pdf of Y can be written in the form

$$f_Y(y) = \frac{1}{y\sigma_{\ln Y}\sqrt{2\pi}} \exp\left[-\frac{1}{2\sigma_{\ln Y}^2} \ln^2\left(\frac{y}{\theta_Y}\right)\right] \qquad y \geq 0$$
$$= 0 \qquad \text{elsewhere} \qquad (7.47)$$

SOME IMPORTANT CONTINUOUS DISTRIBUTIONS

The mean and standard deviation of Y can be found either through direct integration using $f_Y(y)$ or using the relationship given by Equation 7.43 together with $f_X(x)$. In terms of θ_Y and $\sigma_{\ln Y}$, they take the forms

$$m_Y = \theta_Y \exp\left(\frac{\sigma_{\ln Y}^2}{2}\right) \tag{7.48}$$

$$\sigma_Y^2 = m_Y^2[\exp(\sigma_{\ln Y}^2) - 1]$$

VII.3.1 PROBABILITY TABULATIONS

Because of close ties that exist between the normal distribution and the lognormal distribution through Equation 7.43, probability calculations involving a lognormal distributed random variable can be carried out with the aid of probability tables provided for normal random variables.

The relationship between $f_Y(y)$ and $f_X(x)$ is, following Equation 5.12,

$$f_Y(y) = f_X[g^{-1}(y)]\left|\frac{dg^{-1}(y)}{dy}\right|$$

where $g^{-1}(y) = \ln y$. Hence

$$f_Y(y) = f_X(\ln y)\left(\frac{1}{y}\right) \quad y \geq 0$$

or, in terms of the standardized normal r.v. U,

$$f_Y(y) = \frac{1}{y\sigma_{\ln Y}} f_U\left[\frac{\ln(y/\theta_Y)}{\sigma_{\ln Y}}\right] \tag{7.49}$$

For probability distribution functions, we have

$$F_Y(y) = P(Y \leq y) = P(X \leq \ln y) = F_X(\ln y) \quad y \geq 0$$

and, in terms of U,

$$F_Y(y) = F_U\left[\frac{\ln(y/\theta_Y)}{\sigma_{\ln Y}}\right] \quad y \geq 0 \tag{7.50}$$

Since $f_U(u)$ and $F_U(u)$ are tabulated, Equations 7.49 and 7.50 can be used for probability calculations with the aid of normal probability tables.

Example 7.5. The annual maximum runoff Y of a certain river can be modeled by a lognormal distribution. Suppose that the observed mean and standard deviation of Y are $m_Y = 300$ cfs and $\sigma_Y = 200$ cfs. Determine the probability $P(Y > 400 \text{ cfs})$.

Using Equations 7.48, the parameters θ_Y and $\sigma_{\ln Y}$ are solutions of the equations

$$\theta_Y \exp\left(\frac{\sigma_{\ln Y}^2}{2}\right) = 300$$

$$\exp(\sigma_{\ln Y}^2) = \frac{4 \times 10^4}{9 \times 10^4} + 1$$

resulting in

$$\theta_Y = 250$$
$$\sigma_{\ln Y} = 0.61 \qquad (7.51)$$

The desired answer is, using Equation 7.50 and Table A.3,

$$P(Y > 400) = 1 - P(Y \le 400) = 1 - F_Y(400)$$

where

$$F_Y(400) = F_U\left[\frac{\ln(400/250)}{0.61}\right]$$
$$= F_U(0.77) = 0.7794$$

Hence,

$$P(Y > 400) = 1 - 0.7794 = 0.2206$$

VII.4 Gamma and Related Distributions

The gamma distribution describes another class of useful one-sided distributions. The density function associated with the gamma distribution is

$$f_X(x) = \frac{\lambda^\eta}{\Gamma(\eta)} x^{\eta-1} e^{-\lambda x} \qquad x \ge 0$$
$$= 0 \qquad \text{elsewhere} \qquad (7.52)$$

where $\Gamma(\eta)$ is the well-known gamma function

$$\Gamma(\eta) = \int_0^\infty u^{\eta-1} e^{-u} \, du \tag{7.53}$$

which is widely tabulated, and

$$\Gamma(\eta) = (\eta - 1)! \tag{7.54}$$

when η is a positive integer.

The parameters associated with the gamma distribution are η and λ, both are taken to be positive. Since the gamma distribution is one-sided, physical quantities that can take values only in, say, the positive range are frequently modeled by it. Furthermore, it serves as a useful model because of its versatility in the sense that a wide variety of shapes of the gamma density function can be obtained by varying the values of η and λ. This is illustrated in Figures 7.10 and 7.11 which show plots of Equation 7.52 for several values of η and λ. We notice from these figures that η determines the shape of the distribution and is thus a shape parameter whereas λ is a scale parameter of the distribution. In general, the gamma density function is unimodal with its peak at $x = 0$ for $\eta \leq 1$ and at $x = (\eta - 1)/\lambda$ for $\eta > 1$.

As we will verify in Section VII.4.1, it can also be shown that the gamma distribution is an appropriate model for time required for a total of exactly η

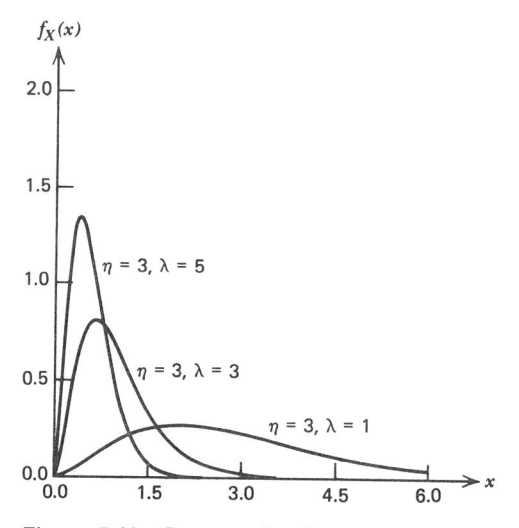

Figure 7.10 Gamma distribution with $\eta = 3$ and several values of λ.

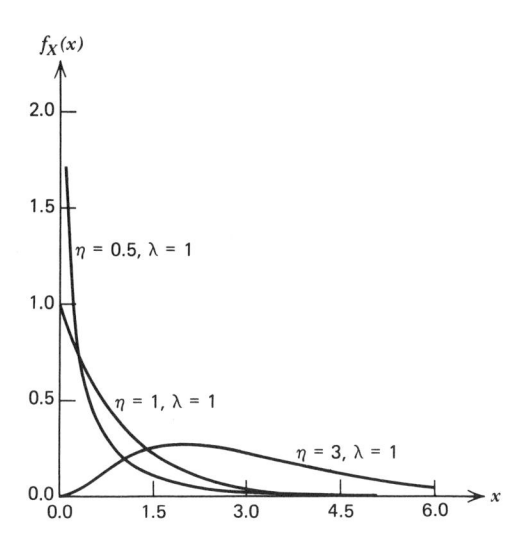

Figure 7.11 Gamma distribution with $\lambda = 1$ and several values of η.

Poisson arrivals. Because of wide applicabilities of Poisson arrivals, the gamma distribution also finds numerous applications.

The distribution function of a r.v. X having a gamma distribution is

$$F_X(x) = \int_0^x f_X(u)du = \frac{\lambda^\eta}{\Gamma(\eta)} \int_0^x u^{\eta-1} e^{-\lambda u} \, du$$

$$= \frac{\Gamma(\eta, \lambda x)}{\Gamma(\eta)} \qquad x \geq 0$$

$$= 0 \qquad \text{elsewhere} \qquad (7.55)$$

In the above, $\Gamma(\eta, u)$ is the incomplete gamma function

$$\Gamma(\eta, u) = \int_0^u x^{\eta-1} e^{-x} \, dx \qquad (7.56)$$

which is also widely tabulated.

The mean and variance of a gamma distributed r.v. X take quite simple forms. Without carrying out the necessary integrations, they are given by

$$m_X = \frac{\eta}{\lambda} \qquad \sigma_X^2 = \frac{\eta}{\lambda^2} \qquad (7.57)$$

A number of important distributions are special cases of the gamma distribution. Two of these are discussed below in more detail.

VII.4.1 THE EXPONENTIAL DISTRIBUTION

When $\eta = 1$, the gamma density function given by Equation 7.52 reduces to the exponential form

$$\boxed{\begin{aligned} f_X(x) &= \lambda e^{-\lambda x} & x \geq 0 \\ &= 0 & \text{elsewhere} \end{aligned}} \quad (7.58)$$

where $\lambda(>0)$ is the parameter of the distribution. Its associated probability distribution function, mean, and variance are obtained from Equations 7.55 and 7.57 by setting $\eta = 1$. They are

$$\begin{aligned} F_X(x) &= 1 - e^{-\lambda x} & x \geq 0 \\ &= 0 & \text{elsewhere} \end{aligned} \quad (7.59)$$

$$m_X = \frac{1}{\lambda} \qquad \sigma_X^2 = \frac{1}{\lambda^2} \quad (7.60)$$

Among many of its applications, two broad classes stand out. First, we will show that the exponential distribution describes interarrival time when arrivals obey the Poisson distribution. It also plays a central role in reliability where exponential distribution is one of the most important failure laws.

Interarrival Time. There is a very close tie between the Poisson and the exponential distributions. Let r.v. $X(0, t)$ be the number of arrivals in the time interval $[0, t)$ and assume that it is Poisson distributed. Our interest now is in the time between two successive arrivals, which is, of course, also a random variable. Let this interarrival time be denoted by T. Its probability distribution function, $F_T(t)$, is, by definition,

$$F_T(t) = P(T \leq t) = 1 - P(T > t) \qquad t \geq 0 \quad (7.61)$$

In terms of $X(0, t)$, the event $T > t$ is equivalent to the event that there are no arrivals during time interval $[0, t)$, or $X(0, t) = 0$. Hence, since $P[X(0, t) = 0] = e^{-\lambda t}$ as derived in Section VI.3, we have

$$\begin{aligned} F_T(t) &= 1 - e^{-\lambda t} & t \geq 0 \\ &= 0 & \text{elsewhere} \end{aligned} \quad (7.62)$$

Comparing this expression with Equation 7.59, it establishes the result that the interarrival time between Poisson arrivals has an exponential distribution; the parameter λ in the distribution of T is the mean arrival rate associated with Poisson arrivals.

Example 7.6. Referring to Example 6.11, determine the probability that the headway (spacing measured in time) between vehicles arriving on the pavement is at least two minutes. Also compute the mean headway.

In Example 6.11, the parameter λ was estimated to be 4.16 vehicles per minute. Hence, if T is the headway in minutes, we have

$$P(T \geq 2) = \int_2^\infty f_T(t)dt = 1 - F_T(2) = e^{-2(4.16)} = 0.00024$$

The mean headway is

$$m_T = \frac{1}{\lambda} = \frac{1}{4.16} = 0.24 \text{ minutes}$$

Since interarrival times for Poisson arrivals are independent, the time required for a total of n Poisson arrivals is a sum of n independent and exponentially distributed random variables. Let T_j, $j = 1, 2, \ldots, n$, be the interarrival time between the $(j - 1)$th and jth arrivals. The time required for a total of n arrivals, denoted by X_n, is

$$X_n = T_1 + T_2 + \cdots + T_n \tag{7.63}$$

where T_j, $j = 1, 2, \ldots, n$, are independent and exponentially distributed with the same parameter λ. In Example 4.16, we have shown that X_n has a gamma distribution with $\eta = 2$ when $n = 2$. The same procedure immediately shows that, for general n, X_n is gamma distributed with $\eta = n$. Thus, as stated in Section VII.4, the gamma distribution is appropriate for describing the time required for a total of η Poisson arrivals.

Example 7.7. Ferry boats depart for trips across a river as soon as nine vehicles are driven aboard. It is observed that vehicles arrive independently at an average rate of six per hour. Determine the probability that the time between trips will be less than one hour.

From our earlier discussion, the time between trips follows a gamma distribution with $\eta = 9$ and $\lambda = 6$. Hence, let X be the time between trips in hours, its

SOME IMPORTANT CONTINUOUS DISTRIBUTIONS

density function and distribution function are given by Equations 7.52 and 7.55. The desired result is, using Equation 7.55,

$$P(X \leq 1) = F_X(1) = \frac{\Gamma(\eta, \lambda)}{\Gamma(\eta)} = \frac{\Gamma(9, 6)}{\Gamma(9)}$$

Now, $\Gamma(9) = 8!$ and the incomplete gamma function $\Gamma(9, 6)$ can be obtained by table lookup. We obtain

$$P(X \leq 1) = 0.153$$

An alternate computation procedure for determining $P(X \leq 1)$ in Example 7.7 can be found by noting from Equation 7.63 that r.v. X can be represented by a sum of η independent random variables. Hence, according to the central limit theorem, its distribution approaches that of a normal r.v. when η is large. Thus, provided that η is large, computations such as that required in Example 7.7 can be carried out using Table A.3 for normal random variables. Let us again consider Example 7.7. Approximating X by a normal r.v., the desired probability is (see Equation 7.25)

$$P(X \leq 1) \simeq P\left(U \leq \frac{1 - m_X}{\sigma_X}\right)$$

where U is the standardized normal r.v. The mean and standard deviation of X are, using Equations 7.57,

$$m_X = \frac{\eta}{\lambda} = \frac{9}{6} \qquad \sigma_X = \frac{\sqrt{\eta}}{\lambda} = \frac{3}{6}$$

Hence, with the aid of Table A.3,

$$P(X \leq 1) \simeq P(U \leq -1) = F_U(-1) = 1 - F_U(1)$$
$$= 1 - 0.8413 = 0.159$$

which is quite close to the answer obtained in Example 7.7.

Reliability and Exponential Failure Law. One can infer from our discussion on interarrival time that many analogous situations can be treated by applying the exponential distribution. In reliability studies, the time to failure for a unit or a system is expected to be exponentially distributed if the unit fails as soon as some single event, such as malfunction of a component, occurs, assuming such events happen independently. In order to gain more insight into failure processes, let us introduce some basic notions in reliability.

Let r.v. T be the time to failure of a component or a system. It is useful to consider a function that gives the probability of failure during a small time increment, assuming that no failure occurred before that time. This function, denoted by $h(t)$, is called the *hazard function* or *failure rate* and is defined by

$$h(t)dt = P(t < T \leq t + dt | T \geq t) \tag{7.64}$$

which gives

$$h(t) = \frac{f_T(t)}{1 - F_T(t)} \tag{7.65}$$

In reliability studies, a hazard function appropriate for many phenomena takes the so-called "bathtub curve" as shown in Figure 7.12. The initial portion of the curve represents "infant mortality," attributable to component defects and manufacturing imperfections. The relative constant portion of the $h(t)$ curve represents the in-usage period in which failure is largely due to chance failure. Wear-out failure near the end of the system life is shown as the increasing portion of the $h(t)$ curve. System reliability can be optimized by initial "burn-in" until time t_1 to avoid premature failure and by part replacement at time t_2 to avoid wear out.

We can now show that the exponential failure law is appropriate during the "in-usage" period of a system's normal life. Substituting

$$f_T(t) = \lambda e^{-\lambda t} \qquad t \geq 0$$

and

$$F_T(t) = 1 - e^{-\lambda t} \qquad t \geq 0$$

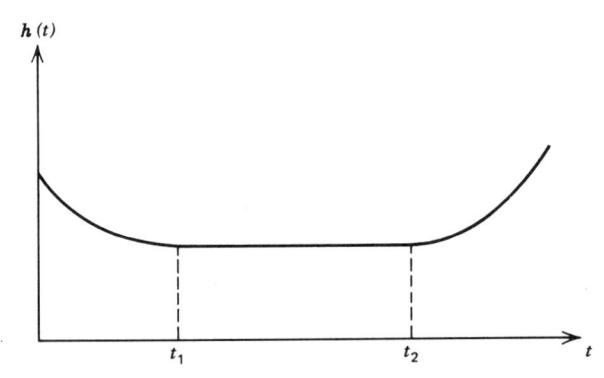

Figure 7.12 Typical shape of a hazard function.

into Equation 7.65, we immediately have

$$h(t) = \lambda \quad (7.66)$$

We see from the above that parameter λ in the exponential distribution plays the role of (constant) failure rate.

We have seen in Example 7.7 that the gamma distribution is appropriate to describe time required for a total of η arrivals. In the context of failure laws, the gamma distribution can be thought of as a generalization of the exponential failure law for systems that fail as soon as exactly η events fail, assuming events take place according to the Poisson law. Thus, the gamma distribution is appropriate as a time-to-failure model for systems having one operating unit and $\eta - 1$ standby units; these standby units go into operation sequentially and each one has an exponential time-to-failure distribution.

VII.4.2 THE CHI-SQUARE (χ^2) DISTRIBUTION

Another important special case of the gamma distribution is the χ^2 distribution, obtained by setting $\lambda = \frac{1}{2}$ and $\eta = n/2$ in Equation 7.52, where n is a positive integer. The χ^2 distribution thus contains one parameter n with probability density function in the form

$$\boxed{\begin{aligned} f_X(x) &= \frac{1}{2^{n/2}\Gamma(n/2)} x^{(n/2)-1} e^{-x/2} \quad x \geq 0 \\ &= 0 \quad \text{elsewhere} \end{aligned}} \quad (7.67)$$

The parameter n is generally referred to as its *degrees of freedom*. The utility of this distribution arises from the fact that a sum of the squares of n independent standardized normal random variables has a χ^2 distribution with n degrees of freedom, that is, if $U_1, U_2, \ldots$, and U_n are independent and distributed $N(0, 1)$, the sum

$$X = U_1^2 + U_2^2 + \cdots + U_n^2 \quad (7.68)$$

has a χ^2 distribution with n degrees of freedom. One can verify this statement by determining the characteristic function of each U_j (see Example 5.7) and use the method of characteristic functions as discussed in Section IV.5 for sums of independent random variables.

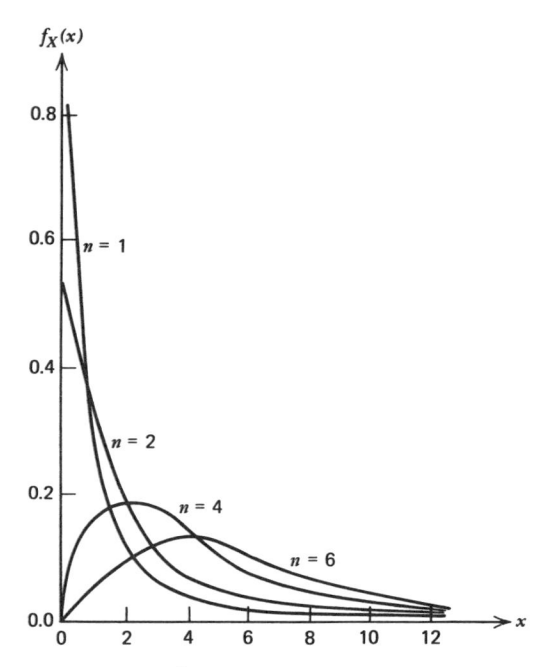

Figure 7.13 χ^2 distribution with several values of n.

Because of this relationship, the χ^2 distribution is one of our main tools in the area of statistical inference and hypothesis testing. These applications are detailed in Chapter X.

The pdf $f_X(x)$ in Equation 7.67 is plotted in Figure 7.13 for several values of n. It is shown that, as n increases, the shape of $f_X(x)$ becomes more symmetric. In view of Equation 7.68, since X can be expressed as a sum of identically distributed random variables, we expect that the χ^2 distribution approaches a normal distribution as $n \to \infty$ on the basis of the central limit theorem.

The mean and variance of r.v. X having a χ^2 distribution are easily obtained from Equation 7.57 to be

$$m_X = n \qquad \sigma_X^2 = 2n \qquad (7.69)$$

VII.5 Beta and Related Distributions

While lognormal and gamma distributions provide a wide diversity of one-sided probability distributions, the beta distribution is rich in providing varied prob-

SOME IMPORTANT CONTINUOUS DISTRIBUTIONS

ability distributions over a finite interval. The beta distribution is characterized by the density function

$$f_X(x) = \frac{\Gamma(\alpha + \beta)}{\Gamma(\alpha)\Gamma(\beta)} x^{\alpha-1}(1-x)^{\beta-1} \quad 0 \leq x \leq 1$$
$$= 0 \quad \text{elsewhere} \tag{7.70}$$

where the parameters α and β take only positive values. The coefficient $\Gamma(\alpha + \beta)/[\Gamma(\alpha)\Gamma(\beta)]$ of $f_X(x)$ can be represented by $1/B(\alpha, \beta)$, where

$$B(\alpha, \beta) = \frac{\Gamma(\alpha)\Gamma(\beta)}{\Gamma(\alpha + \beta)} \tag{7.71}$$

is known as the beta function, hence the name for the distribution given by Equation 7.70.

The parameters α and β are both shape parameters; different combinations of their values permit the density function to take on a wide variety of shapes. When $\alpha, \beta > 1$, the distribution is unimodal with peak at $x = (\alpha - 1)/(\alpha + \beta - 2)$. It becomes U-shaped when $\alpha, \beta < 1$; it is J-shaped when $\alpha \geq 1$ and $\beta < 1$; and it takes the shape of an inverse J when $\alpha < 1$ and $\beta \geq 1$. Finally, as a special case, the uniform distribution over interval $(0, 1)$ results when $\alpha = \beta = 1$. Some of these possible shapes are displayed in Figures 7.14 and 7.15.

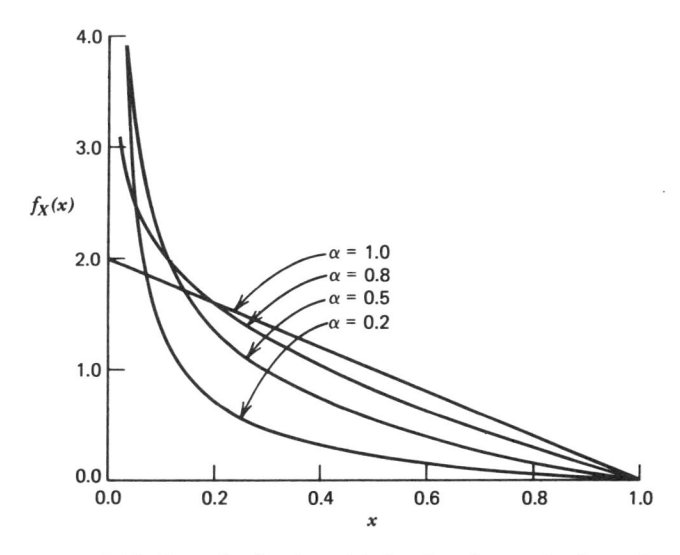

Figure 7.14 Beta distribution with $\beta = 2$ and several values of α.

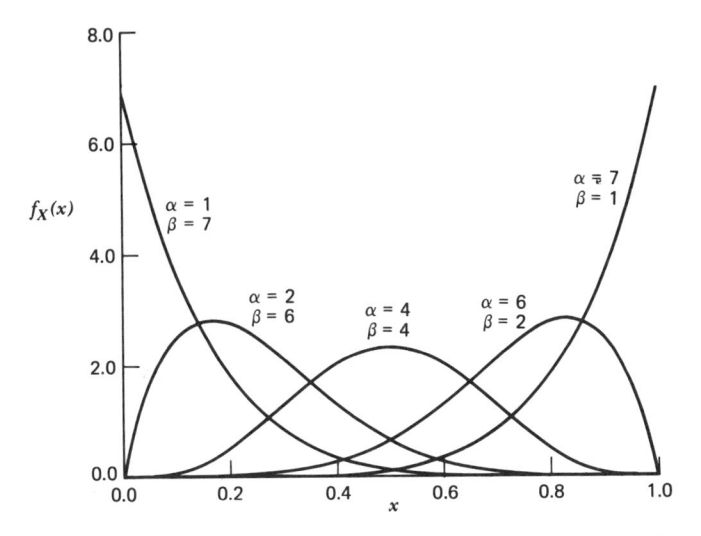

Figure 7.15 Beta distribution with several values of α and β ($\alpha + \beta = 8$).

The mean and variance of a beta-distributed random variable X are, following straightforward integrations,

$$m_X = \frac{\alpha}{\alpha + \beta}$$

$$\sigma_X^2 = \frac{\alpha\beta}{(\alpha + \beta)^2(\alpha + \beta + 1)} \tag{7.72}$$

Because of its versatility as a distribution over a finite interval, the beta distribution is used to represent a large number of physical quantities whose values are restricted to an identifiable interval. Some of the areas of application are tolerance limits, quality control, and reliability.

An interesting situation in which the beta distribution arises is the following. Suppose a random phenomenon Y can be observed independently n times and, after these n independent observations are ranked in order of increasing magnitude, let y_r and y_{n-s+1} be the value of the rth smallest and sth largest observations, respectively. If r.v. X is used to denote the proportion of the original Y taking values between y_r and y_{n-s+1}, it can be shown that X follows a beta distribution with $\alpha = n - r - s + 1$ and $\beta = r + s$, that is,

$$f_X(x) = \frac{\Gamma(n+1)}{\Gamma(n-r-s+1)\Gamma(r+s)} x^{n-r-s}(1-x)^{r+s-1} \quad 0 \leq x \leq 1$$
$$= 0 \quad \text{elsewhere} \tag{7.73}$$

SOME IMPORTANT CONTINUOUS DISTRIBUTIONS 217

This result can be found in Wilks (1942). We will not prove it but only use it in a subsequent example (Example 7.8).

VII.5.1 PROBABILITY TABULATIONS

The probability distribution function associated with the beta distribution is

$$F_X(x) = 0 \qquad x < 0$$
$$= \frac{\Gamma(\alpha + \beta)}{\Gamma(\alpha)\Gamma(\beta)} \int_0^x u^{\alpha-1}(1-u)^{\beta-1}\,du \qquad 0 \le x \le 1 \qquad (7.74)$$
$$= 1 \qquad x > 1$$

which can be integrated directly. It also has the form of an incomplete beta function and can be found in mathematical tables. The incomplete beta function is usually denoted by $I_x(p, q)$. If we write $F_X(x)$ with parameters α and β in the form $F(x; \alpha, \beta)$, the correspondence between $I_x(p, q)$ and $F(x; \alpha, \beta)$ is determined as follows: If $\alpha \ge \beta$, then

$$F(x; \alpha, \beta) = I_x(\alpha, \beta) \qquad (7.75)$$

If $\alpha < \beta$, it is

$$F(x; \alpha, \beta) = 1 - I_{(1-x)}(\beta, \alpha) \qquad (7.76)$$

Another method of evaluating $F_X(x)$ in Equation 7.74 is to note the similarity in form between $f_X(x)$ and $P_Y(k)$ of a binomial distributed r.v. Y for the case where α and β are positive integers. We see from Equation 6.2 that

$$P_Y(k) = \frac{n!}{k!(n-k)!} p^k(1-p)^{n-k} \qquad k = 0, 1, \ldots, n \qquad (7.77)$$

On the other hand, the $f_X(x)$ in Equation 7.70 with α and β being positive integers takes the form

$$f_X(x) = \frac{(\alpha + \beta - 1)!}{(\alpha - 1)!(\beta - 1)!} x^{\alpha-1}(1-x)^{\beta-1} \qquad \alpha, \beta = 1, 2, \ldots \qquad 0 \le x \le 1$$

$$(7.78)$$

and we easily establish the relationship

$$f_X(x) = (\alpha + \beta - 1)p_Y(k) \qquad \alpha, \beta = 1, 2, \ldots \qquad 0 \le x \le 1 \qquad (7.79)$$

where $p_Y(k)$ is evaluated at $k = \alpha - 1$ with $n = \alpha + \beta - 2$ and $p = x$. For example, the value of $f_X(0.5)$ with $\alpha = 2, \beta = 1$ is numerically equal to $2p_Y(1)$ with $n = 1$ and $p = 0.5$, where $p_Y(1)$ can be found from Table A.1 for binomial distributed random variables.

Similarly, the relationship between $F_X(x)$ and $F_Y(k)$ can be established. It takes the form

$$F_X(x) = 1 - F_Y(k) \qquad \alpha, \beta = 1, 2, \ldots \qquad 0 \leq x \leq 1 \qquad (7.80)$$

with $k = \alpha - 1, n = \alpha + \beta - 2$, and $p = x$. The PDF $F_Y(y)$ for a binomial distributed r.v. Y is also widely tabulated and it can be used to advantage here for evaluating $F_X(x)$ associated with the beta distribution.

Example 7.8. In order to establish quality limits for a manufactured item, 10 independent samples are taken at random and the quality limits are established using the lowest and highest sample values. What is the probability that at least 50% of the manufactured items will fall within these limits?

Let X be the proportion of items taking values within the established limits. Its pdf thus takes the form of Equation 7.73 with $n = 10, r = 1$, and $s = 1$. Hence, $\alpha = 10 - 1 - 1 + 1 = 9$ and $\beta = 1 + 1 = 2$ and

$$f_X(x) = \frac{\Gamma(11)}{\Gamma(9)\Gamma(2)} x^8(1 - x)$$

$$= \frac{10!}{8!} x^8(1 - x) \qquad 0 \leq x \leq 1$$

$$= 0 \qquad \text{elsewhere}$$

The desired probability is

$$P(X > 0.50) = 1 - P(X \leq 0.50) = 1 - F_X(0.50) \qquad (7.81)$$

According to Equation 7.80, the value of $F_X(0.50)$ can be found from

$$F_X(0.50) = 1 - F_Y(k) \qquad (7.82)$$

where Y is binomial and $k = \alpha - 1 = 8, n = \alpha + \beta - 2 = 9$, and $p = 0.50$. Using Table A.1, we find that

$$F_Y(8) = 1 - p_Y(9) = 1 - 0.002 = 0.998 \qquad (7.83)$$

Equations 7.82 and 7.81 yield

$$P(X > 0.50) = 1 - F_X(0.50) = 1 - 1 + F_Y(8) = 0.998 \qquad (7.84)$$

VII.5.2 GENERALIZED BETA DISTRIBUTION

The beta distribution can be easily generalized from one restricted to the unit interval $(0, 1)$ to one covering an arbitrary interval (a, b). Let Y be such a generalized random variable. It is clear that the desired transformation is

$$Y = (b - a)X + a \qquad (7.85)$$

where X is beta distributed according to Equation 7.70. Equation 7.85 represents a monotonic transformation from X to Y and the procedure developed in Chapter V can be applied to determine the pdf of Y in a straightforward manner. Following Equation 5.12, we have

$$\boxed{\begin{aligned} f_Y(y) &= \frac{1}{(b-a)^{\alpha+\beta-1}} \frac{\Gamma(\alpha + \beta)}{\Gamma(\alpha)\Gamma(\beta)} (y - a)^{\alpha-1}(b - y)^{\beta-1} \quad & a \le x \le b \\ &= 0 & \text{elsewhere} \end{aligned}} \qquad (7.86)$$

VII.6 Extreme Value Distributions

A structural engineer, concerned with safety of a structure, is often interested in the *maximum* load and *maximum* stress in structural members. In reliability studies, distribution of life of a system having n components in series (system fails when *any* component fails) is a function of the minimum time to failure of these components, whereas for a system with parallel arrangement (system fails when *all* components fail) it is determined by the maximum time to failure distribution. These examples point to our frequent concern with distributions of maximum or minimum values of a number of random variables.

To fix ideas, let $X_j, j = 1, 2, \ldots, n$, denote the jth gust velocity of n gusts occurring in a year and let Y_n denote the annual maximum gust velocity. We are interested in the probability distribution of Y_n in terms of those of X_j. In the following development, attention is given to the case where the r.v.'s X_j, $j = 1, 2, \ldots, n$, are independent and identically distributed with PDF $F_X(x)$ and pdf $f_X(x)$ or pmf $p_X(x)$. Furthermore, asymptotic results for $n \to \infty$ are our primary concern. For the wind gust example given above, these conditions are not unreasonable in determining the distribution of annual maximum gust velocity. We will also determine, under the same conditions, the minimum Z_n of random variables $X_1, X_2, \ldots$, and X_n, which is also of interest in practical applications.

The r.v.'s Y_n and Z_n are defined by

$$Y_n = \max(X_1, X_2, \ldots, X_n) \atop Z_n = \min(X_1, X_2, \ldots, X_n) \qquad (7.87)$$

The PDF of Y_n is

$$F_{Y_n}(y) = P(Y_n \leq y) = P(\text{all } X_j \leq y)$$
$$= P(X_1 \leq y \cap X_2 \leq y \cap \cdots \cap X_n \leq y)$$

Assuming independence, we have

$$F_{Y_n}(y) = F_{X_1}(y) F_{X_2}(y) \cdots F_{X_n}(y) \qquad (7.88)$$

and, if each $F_{X_j}(y) = F_X(y)$, the result is

$$F_{Y_n}(y) = [F_X(y)]^n \qquad (7.89)$$

The pdf of Y_n can be easily derived from the above. When the X_j are continuous, it has the form

$$f_{Y_n}(y) = \frac{dF_{Y_n}(y)}{dy} = n[F_X(y)]^{n-1} f_X(y) \qquad (7.90)$$

The PDF of Z_n is determined in a similar fashion. In this case,

$$F_{Z_n}(z) = P(Z_n \leq z) = P(\text{at least one } X_j \leq z)$$
$$= P(X_1 \leq z \cup X_2 \leq z \cup \cdots \cup X_n \leq z)$$
$$= 1 - P(X_1 > z \cap X_2 > z \cap \cdots \cap X_n > z)$$

When X_j are independent and identically distributed, the foregoing gives

$$F_{Z_n}(z) = 1 - [1 - F_{X_1}(z)][1 - F_{X_2}(z)] \cdots [1 - F_{X_n}(z)]$$
$$= 1 - [1 - F_X(z)]^n \qquad (7.91)$$

If r.v.'s X_j are continuous, the pdf of Z_n is

$$f_{Z_n}(z) = n[1 - F_X(z)]^{n-1} f_X(z) \qquad (7.92)$$

The next step in our development is to determine the forms of $F_{Y_n}(y)$ and $F_{Z_n}(z)$ as expressed by Equations 7.89 and 7.91 as $n \to \infty$. Since the initial distribution $F_X(x)$ of each X_j is sometimes unavailable, we wish to examine whether Equations 7.89 and 7.91 lead to unique distributions for $F_{Y_n}(y)$ and $F_{Z_n}(z)$, respectively, independent of the form of $F_X(x)$. This is not unlike looking for results similar to the powerful ones we obtained for the normal and lognormal distributions via the central limit theorem.

While the distribution functions $F_{Y_n}(y)$ and $F_{Z_n}(z)$ become increasingly insensitive to exact distributional features of X_j as $n \to \infty$, no unique results can be obtained which are completely independent of the form of $F_X(x)$. Some features of the distribution $F_X(x)$ are important and, in what follows, the asymptotic forms of $F_{Y_n}(y)$ and $F_{Z_n}(z)$ are classified into three types based on general features in the distribution tails of X_j. Type I are sometimes referred to as Gumbel's extreme value distributions and included in the Type III is the important Weibull distribution.

VII.6.1 TYPE I ASYMPTOTIC DISTRIBUTIONS OF EXTREME VALUES

Consider first the Type I asymptotic distribution of maximum values. It is the limiting distribution of Y_n (as $n \to \infty$) from an initial distribution $F_X(x)$ whose right tail is unbounded and is of an exponential type, that is, $F_X(x)$ approaches one at least as fast as an exponential distribution. For this case, we can express $F_X(x)$ in the form

$$F_X(x) = 1 - e^{-g(x)} \tag{7.93}$$

where $g(x)$ is an increasing function of x. A number of important distributions fall into this category, such as normal, lognormal, and gamma distributions.

Let

$$\lim_{n \to \infty} Y_n = Y \tag{7.94}$$

We have the following important result.

Theorem

Let r.v.'s $X_1, X_2, \ldots,$ and X_n be independent and identically distributed with the same PDF $F_X(x)$. If $F_X(x)$ is of the form given by Equation 7.93, we have

$$\boxed{F_Y(y) = \exp[-e^{-\alpha(y-u)}] \quad -\infty < y < \infty} \tag{7.95}$$

where $\alpha(>0)$ and u are two parameters of the distribution.

Proof

We shall only sketch the proof here. See Gumbel (1958) for a more comprehensive and rigorous treatment.

Let us first define a quantity u_n known as the characteristic value of Y_n by

$$F_X(u_n) = 1 - \frac{1}{n} \tag{7.96}$$

It is thus the value of X_j, $j = 1, 2, \ldots, n$, at which $P(X_j \leq u_n) = 1 - 1/n$. As n becomes large, $F_X(u_n)$ approaches unity or u_n is in the extreme right tail of the distribution. It can also be shown that u_n is the mode of Y_n that can be verified by, in the case of X_j being continuous, taking the derivative of $f_{Y_n}(y)$ in Equation 7.90 with respect to y and setting it to zero.

If $F_X(x)$ takes the form given by Equation 7.93, we have

$$1 - e^{-g(u_n)} = 1 - \frac{1}{n}$$

or

$$\frac{e^{g(u_n)}}{n} = 1 \tag{7.97}$$

Now consider $F_{Y_n}(y)$ given by Equation 7.89. In view of Equation 7.93, it takes the form

$$F_{Y_n}(y) = [1 - e^{-g(y)}]^n$$

$$= \left[1 - \frac{e^{g(u_n)} e^{-g(y)}}{n}\right]^n$$

$$= \left\{1 - \frac{e^{-[g(y) - g(u_n)]}}{n}\right\}^n \tag{7.98}$$

In the above, we have introduced into the equation the factor $\exp[g(u_n)]/n$, which is unity, as shown by Equation 7.97.

Since u_n is the mode or the "most likely" value of Y_n, the function $g(y)$ in Equation 7.98 can be expanded in powers of $(y - u_n)$ in the form

$$g(y) = g(u_n) + \alpha_n(y - u_n) + \cdots \tag{7.99}$$

SOME IMPORTANT CONTINUOUS DISTRIBUTIONS

where $\alpha_n = dg(y)/dy$ evaluated at $y = u_n$. It is positive since $g(y)$ is an increasing function of y. Retaining only the linear term in Equation 7.99 and substituting it into Equation 7.98, we obtain

$$F_{Y_n}(y) = \left[1 - \frac{e^{-\alpha_n(y-u_n)}}{n}\right]^n \tag{7.100}$$

in which α_n and u_n are only functions of n and not of y. Using the identity

$$\lim_{n \to \infty} \left(1 - \frac{a}{n}\right)^n = e^{-a}$$

for any real a, Equation 7.100 tends to, as $n \to \infty$,

$$F_Y(y) = \exp[-e^{-\alpha(y-u)}] \tag{7.101}$$

which completes the proof. In arriving at Equation 7.101, we have assumed that, as $n \to \infty$, $F_{Y_n}(y)$ converges to $F_Y(y)$ as Y_n converges to Y in some probabilistic sense.

The mean and variance associated with the Type I maximum-value distribution can be obtained through integrations using Equation 7.90. We have noted that u is the mode of the distribution, that is, the value of y at which $f_Y(y)$ is maximum. The mean of Y is

$$m_Y = u + \frac{\gamma}{\alpha} \tag{7.102}$$

where $\gamma \simeq 0.577$ is the Euler's constant, and the variance is given by

$$\sigma_Y^2 = \frac{\pi^2}{6\alpha^2} \tag{7.103}$$

It is seen from the above that u and α are, respectively, location and scale parameters of the distribution. It is interesting to note that the skewness coefficient, defined by Equation 4.11, in this case is

$$\gamma_1 \simeq 1.1396$$

which is independent of α and u. This result indicates that the Type I maximum-value distribution has a fixed shape with a dominant tail to the right. A typical shape of $f_Y(y)$ is shown in Figure 7.16.

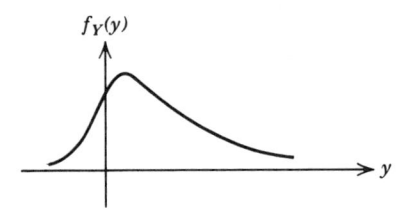

Figure 7.16 A typical plot of type I maximum-value distribution.

The Type I asymptotic distribution for minimum values is the limiting distribution of Z_n in Equation 7.91 as $n \to \infty$ from an initial distribution $F_X(x)$ whose left tail is unbounded and is of exponential type as it decreases to zero on the left. An example of $F_X(x)$ that belongs to this class is the normal distribution.

The distribution of Z_n as $n \to \infty$ can be derived by means of procedures given above for Y_n through a symmetrical argument. Without giving details, if we let

$$\lim_{n \to \infty} Z_n = Z \qquad (7.104)$$

the PDF of Z can be shown to have the form

$$F_Z(z) = 1 - \exp[-e^{\alpha(z-u)}] \qquad -\infty < z < \infty \qquad (7.105)$$

where α and u are again the two parameters of the distribution. It is seen that Type I asymptotic distributions for maximum and minimum values are mirror images of each other. The mode of Z is u and its mean, variance, and skewness coefficients are

$$m_Z = u - \frac{\gamma}{\alpha}$$

$$\sigma_Z^2 = \frac{\pi^2}{6\alpha^2} \qquad (7.106)$$

$$\gamma_1 \simeq -1.1396$$

For probability calculations, tables for probability distribution functions $F_Y(y)$ and $F_Z(z)$ over various ranges of y and z are available. However, with the advent of pocket-size electronic calculators, to carry out these calculations is no longer a significant obstacle.

Example 7.9. The maximum daily gasoline demand Y during the month of May in a given locality follows a Type I asymptotic maximum-value distribution with $m_Y = 2$ and $\sigma_Y = 1$ measured in thousands of gallons. Determine (a) the probability that the demand will exceed 4000 gallons in any day during the month of May and (b) the daily supply level at which 95% of the time will not be exceeded by demand in any given day.

It follows from Equations 7.102 and 7.103 that the parameters α and u are determined from

$$\alpha = \frac{\pi}{\sqrt{6}\,\sigma_Y} = \frac{\pi}{\sqrt{6}} = 1.282$$

$$u = m_Y - \frac{0.577}{\alpha} = 2 - \frac{0.577}{1.282} = 1.55$$

For part (a), the solution is

$$\begin{aligned} P(Y > 4) &= 1 - F_Y(4) \\ &= 1 - \exp[-e^{-1.282(4-1.55)}] \\ &= 1 - 0.958 = 0.042 \end{aligned}$$

For part (b), we need to determine y such that

$$F_Y(y) = P(Y \le y) = 0.95$$

or

$$\exp[-e^{-1.282(y-1.55)}] = 0.95 \tag{7.107}$$

Taking logarithms of Equation 7.107 twice, we obtain

$$y = 3.867$$

or the required supply level is 3867 gallons.

Example 7.10. Consider the problem of estimating floods in design of dams. Let y_T denote the maximum flood associated with return period T. Determine the relationship between y_T and T if the maximum river flow follows the Type I maximum-value distribution. We recall from Example 6.7 that the return period T is defined as the average number of years between floods whose magnitudes are greater than y_T.

Assuming that floods occur independently, the number of years between floods with magnitudes greater than y_T assumes a geometric distribution. Thus

$$T = \frac{1}{P(Y > y_T)} = \frac{1}{1 - F_Y(y_T)} \qquad (7.108)$$

Now,

$$F_Y(y_T) = \exp[-e^{-b}] \qquad (7.109)$$

where $b = \alpha(y_T - u)$. The substitution of Equation 7.109 into Equation 7.108 gives the required relationship.

For values of y_T where $F_Y(y_T) \to 1$, an approximation can be made by noting from Equation 7.109 that

$$e^{-b} = -\ln F_Y(y_T) = -\{[F_Y(y_T) - 1] - \tfrac{1}{2}[F_Y(y_T) - 1]^2 + \cdots\}$$

Since $F_Y(y_T)$ is close to 1, we retain only the first term in the foregoing expansion and obtain

$$1 - F_Y(y_T) \simeq e^{-b}$$

Equation 7.108 thus gives the approximate relationship

$$y_T = u\left(1 + \frac{1}{\alpha u}\ln T\right) \qquad (7.110)$$

where u is the scale factor and the value of αu describes the characteristics of a river; it varies from 1.5 for violent rivers to 10 for stable or mild rivers.

In closing, let us remark again that the Type I maximum-value distribution is valid for initial distributions of such practical importance as normal, lognormal, and gamma distributions. It thus has wide applicability and is sometimes simply called *the extreme value distribution.*

VII.6.2 TYPE II ASYMPTOTIC DISTRIBUTIONS OF EXTREME VALUES

The Type II asymptotic distribution of maximum values arises as the limiting distribution of Y_n as $n \to \infty$ from an initial distribution of the "Pareto" type, that is, the PDF $F_X(x)$ of each X_j is limited on the left at zero and its right tail is unbounded and approaches one according to

$$F_X(x) = 1 - ax^{-k} \qquad a, k > 0 \qquad x \geq 0 \qquad (7.111)$$

SOME IMPORTANT CONTINUOUS DISTRIBUTIONS

For this class, the asymptotic distribution of Y_n, $F_Y(y)$, as $n \to \infty$ takes the form

$$F_Y(y) = \exp\left[-\left(\frac{y}{v}\right)^{-k}\right] \quad v, k > 0 \quad y \geq 0 \quad (7.112)$$

Let us note that, with $F_X(x)$ given by Equation 7.111, each X_j has moments only up to the order r where r is the largest integer less than k. If $k > 1$, the mean of Y is

$$m_Y = v\Gamma\left(1 - \frac{1}{k}\right) \quad (7.113)$$

and, if $k > 2$, the variance has the form

$$\sigma_Y^2 = v^2\left[\Gamma\left(1 - \frac{2}{k}\right) - \Gamma^2\left(1 - \frac{1}{k}\right)\right] \quad (7.114)$$

The derivation of $F_Y(y)$ given by Equation 7.112 follows in broad outline that given for the Type I maximum-value asymptotic distribution and will not be presented here. It has been used as a model in meteorology and hydrology [Gumbel (1958)].

A close relationship exists between the Type I and Type II asymptotic maximum-value distributions. Let Y_I and Y_{II} denote, respectively, these random variables. It can be verified using the techniques of transformations of random variables that they are related by

$$F_{Y_{II}}(y) = F_{Y_I}(\ln y) \quad y \geq 0 \quad (7.115)$$

where the parameters α and u in $F_{Y_I}(y)$ are related to the parameters k and v by

$$u = \ln v \quad \text{and} \quad \alpha = k \quad (7.116)$$

When they are continuous, their pdf's obey the relationship

$$f_{Y_{II}}(y) = \frac{1}{y}f_{Y_I}(\ln y) \quad y \geq 0 \quad (7.117)$$

The Type II asymptotic distribution of minimum values arises under analogous conditions. With the PDF $F_X(x)$ limited on the right at zero and approaching zero on the left in a manner analogous to Equation 7.111, we have

$$F_Z(z) = 1 - \exp\left[-\left|\frac{z}{v}\right|^{-k}\right] \quad v, k > 0 \quad z \leq 0 \quad (7.118)$$

However, it has not been found as useful as its counterparts in Type I and Type III as the required initial distributions are not encountered frequently in practice.

VII.6.3 TYPE III ASYMPTOTIC DISTRIBUTIONS OF EXTREME VALUES

Since the Type III maximum-value asymptotic distribution is of limited practical interest, only the minimum-value distribution will be discussed as follows.

The Type III minimum-value asymptotic distribution is the limiting distribution of Z_n as $n \to \infty$ for an initial distribution $F_X(x)$ whose left tail increases from zero near $x = \varepsilon$ in the manner

$$F_X(x) = c(x - \varepsilon)^k \quad c, k > 0 \quad x \geq \varepsilon \quad (7.119)$$

This class of distributions are bounded on the left at $x = \varepsilon$. The gamma is such a distribution with $\varepsilon = 0$.

Again bypassing derivations, the asymptotic distribution for the minimum-value can be shown to be

$$F_Z(z) = 1 - \exp\left[-\left(\frac{z - \varepsilon}{w - \varepsilon}\right)^k\right] \quad k > 0 \quad w > \varepsilon \quad z \geq \varepsilon \quad (7.120)$$

and, if it is continuous,

$$f_Z(z) = \frac{k}{w - \varepsilon}\left(\frac{z - \varepsilon}{w - \varepsilon}\right)^{k-1} \exp\left[-\left(\frac{z - \varepsilon}{w - \varepsilon}\right)^k\right] \quad k > 0 \quad w > \varepsilon \quad z \geq \varepsilon$$

$$(7.121)$$

The mean and variance of Z are

$$m_Z = \varepsilon + (w - \varepsilon)\Gamma\left(1 + \frac{1}{k}\right)$$

$$\sigma_Z^2 = (w - \varepsilon)^2\left[\Gamma\left(1 + \frac{2}{k}\right) - \Gamma^2\left(1 + \frac{1}{k}\right)\right] \quad (7.122)$$

We have seen in Section VII.4.1 that the exponential distribution is used as a failure law in reliability studies, which corresponds to a constant hazard function (see Equations 7.64 and 7.66). The distribution given by Equations 7.120 and 7.121

SOME IMPORTANT CONTINUOUS DISTRIBUTIONS

is frequently used as a generalized time-to-failure model for cases in which the hazard function varies with time. One can show that the hazard function

$$h(t) = \frac{k}{w}\left(\frac{t}{w}\right)^{k-1} \qquad t \geq 0 \qquad (7.123)$$

is capable of assuming a wide variety of shapes and its associated probability density function for T, time to failure, is given by

$$\boxed{f_T(t) = \frac{k}{w}\left(\frac{t}{w}\right)^{k-1}\exp\left[-\left(\frac{t}{w}\right)^k\right] \qquad w, k > 0 \qquad t \geq 0} \qquad (7.124)$$

It is the so-called *Weibull distribution* after Weibull, who first obtained it heuristically [Weibull (1939)]. Clearly, Equation 7.124 is a special case of Equation 7.121 with $\varepsilon = 0$.

The relationship between Type III and Type I asymptotic distributions can also be established. Let Z_I and Z_{III} be random variables having, respectively, Type I and Type III asymptotic distributions of minimum values. Then

$$F_{Z_{III}}(z) = F_{Z_I}[\ln(z - \varepsilon)] \qquad z \geq \varepsilon \qquad (7.125)$$

with $u = \ln(w - \varepsilon)$ and $\alpha = k$. If they are continuous, the relationship between their pdf's is

$$f_{Z_{III}}(z) = \frac{1}{z - \varepsilon} f_{Z_I}[\ln(z - \varepsilon)] \qquad z \geq \varepsilon \qquad (7.126)$$

One final remark to be made is that asymptotic distributions of maximum and minimum values from the same initial distribution may not be of the same type. For example, for a gamma initial distribution, its asymptotic maximum-value distribution is of Type I whereas the minimum-value distribution falls into Type III. With reference to system time-to-failure models, a system having n components in series with independent gamma life distributions for its components will have a time-to-failure distribution belonging to the Type III asymptotic minimum-value distribution as n becomes large. The corresponding model for a system having n components in parallel is a Type I asymptotic maximum-value distribution.

VII.7 Summary

As in Chapter VI, it is useful to summarize some of the important properties associated with the continuous distributions discussed in this chapter. These are given in Table 7.1.

Table 7.1 Summary of Continuous Distributions

Distribution	Probability Density Function	Parameters	Mean and Variance
Uniform	$f_X(x) = \dfrac{1}{b-a}$, $a \leq x \leq b$	$a, b > a$	$\dfrac{a+b}{2}, \dfrac{(b-a)^2}{12}$
Normal (Gaussian)	$f_X(x) = \dfrac{1}{\sqrt{2\pi}\sigma} \exp\left[-\dfrac{(x-m)^2}{2\sigma^2}\right]$, $-\infty < x < \infty$	$m, \sigma > 0$	m, σ^2
Lognormal	$f_Y(y) = \dfrac{1}{y\sigma_{\ln Y}\sqrt{2\pi}} \exp\left[-\dfrac{1}{2\sigma_{\ln Y}^2}\ln^2\left(\dfrac{y}{\theta_Y}\right)\right]$, $y \geq 0$	$\theta_y > 0, \sigma_{\ln Y} > 0$	$\theta_Y \exp\left(\dfrac{\sigma_{\ln Y}^2}{2}\right), m_Y^2[\exp(\sigma_{\ln Y}^2)-1]$
Gamma	$f_X(x) = \dfrac{\lambda^\eta}{\Gamma(\eta)} x^{\eta-1} e^{-\lambda x}$, $x \geq 0$	$\eta > 0, \lambda > 0$	$\dfrac{\eta}{\lambda}, \dfrac{\eta}{\lambda^2}$
Exponential	$f_X(x) = \lambda e^{-\lambda x}$, $x \geq 0$	$\lambda > 0$	$\dfrac{1}{\lambda}, \dfrac{1}{\lambda^2}$
Chi square	$f_X(x) = \dfrac{1}{2^{\nu/2}\Gamma(\nu/2)} x^{(\nu/2)-1} e^{-x/2}$, $x \geq 0$	ν = positive integer	$\nu, 2\nu$
Beta	$f_X(x) = \dfrac{\Gamma(\alpha+\beta)}{\Gamma(\alpha)\Gamma(\beta)} x^{\alpha-1}(1-x)^{\beta-1}$, $0 \leq x \leq 1$	$\alpha > 0, \beta > 0$	$\dfrac{\alpha}{\alpha+\beta}, \dfrac{\alpha\beta}{(\alpha+\beta)^2(\alpha+\beta+1)}$
Extreme Values			
(a) Type I max	$f_Y(y) = \alpha \exp[-\alpha(y-u) - e^{-\alpha(y-u)}]$, $-\infty < y < \infty$	$\alpha > 0, u$	$u + \dfrac{\gamma}{\alpha}, \dfrac{\pi^2}{6\alpha^2}$
(b) Type I min	$f_Z(z) = \alpha \exp[\alpha(z-u) - e^{\alpha(z-u)}]$, $-\infty < z < \infty$	$\alpha > 0, u$	$u - \dfrac{\gamma}{\alpha}, \dfrac{\pi^2}{6\alpha^2}$
(c) Type II max	$f_Y(y) = \dfrac{k}{v}\left(\dfrac{v}{y}\right)^{k+1} \exp\left[-\left(\dfrac{v}{y}\right)^{-k}\right]$, $y \geq 0$	$v > 0, k > 0$	$v\Gamma\left(1-\dfrac{1}{k}\right)$, $v^2\left[\Gamma\left(1-\dfrac{2}{k}\right) - \Gamma^2\left(1-\dfrac{1}{k}\right)\right]$
(d) Type III min	$f_Z(z) = \dfrac{k}{w-\varepsilon}\left(\dfrac{z-\varepsilon}{w-\varepsilon}\right)^{k-1} \exp\left[-\left(\dfrac{z-\varepsilon}{w-\varepsilon}\right)^k\right]$, $z \geq \varepsilon$	$\varepsilon, w > \varepsilon, k > 0$	$\varepsilon + (w-\varepsilon)\Gamma\left(1+\dfrac{1}{k}\right)$, $(w-\varepsilon)^2\left[\Gamma\left(1+\dfrac{2}{k}\right) - \Gamma^2\left(1+\dfrac{1}{k}\right)\right]$

REFERENCES AND COMMENTS

1. E. J. Gumbel, *Statistics of Extremes*, Columbia University Press, New York, 1958.
2. M. Kramer, "Frequency Surfaces in Two Variables Each of Which is Uniformly Distributed," *The Amer. J. of Hygiene*, **32**, 45–64, 1940.
3. J. W. Lindeberg, "Eine neue Herleitung des Exponentialgesetzes in der Wahrscheinlichkeifsrechnung," *Mathematische Zeitschrift*, **15**, 211–225, 1922.
4. W. Weibull, "A Statistical Theory of the Strength of Materials," *Proc. Royal Swedish Inst. for Engr. Res.*, Stockholm, No. 151, 1939.
5. S. Wilks, "Statistical Prediction with Special Reference to the Problem of Tolerance Limits," *Ann. Math. Stat.*, **13**, 400, 1942.

As we mentioned in Section VII.2.1, the central limit theorem as stated may be generalized in several directions. Extensions since the 1920s include cases in which the r.v. Y in Equation 7.14 is a sum of dependent and not necessarily identically distributed random variables. See, for example, the following two references.

6. M. Loéve, *Probability Theory*, Van Nostrand, New York, 1955.
7. E. Parzen, *Modern Probability Theory and Its Applications*, Wiley, New York, 1960.

Extensive probability tables exist in addition to those given in Appendix A. Probability tables for lognormal, gamma, beta, chi-square, and extreme value distributions can be found in some of the references cited in Chapter VI. In particular, References 3, 4, 7, 10, and 11 are helpful. Additional useful references include the following:

8. J. Aitchison and J. A. C. Brown, *The Log-normal Distribution*, Cambridge University Press, Cambridge, England, 1957.
9. H. L. Harter, *New Tables of the Incomplete Gamma Function Ratio and of Percentage Points of the Chi-square and Beta Distributions*, Aerospace Laboratory, Washington, D.C., GPO, 1964.
10. National Bureau of Standards, *Tables of the Bivariate Normal Distribution and Related Functions* (Applied Mathematics Series 50), Washington, D.C., GPO, 1954.

PROBLEMS

7.1. The r.v.'s X and Y are independent and uniformly distributed in the interval $(0, 1)$. Determine the probability that their product XY is less than $\frac{1}{2}$.

7.2. The characteristic function (Ch.f.) of a r.v. X uniformly distributed in the interval $(-1, 1)$ is

$$\phi_X(t) = \frac{\sin t}{t}$$

(a) Find the Ch.f. of Y, which is uniformly distributed in the interval $(-a, a)$.
(b) Find the Ch.f. of Y if it is uniformly distributed in the interval $(a, a + b)$.

7.3. A machine component consisting of a rod-and-sleeve assembly is shown in the following diagram. Due to machining inaccuracies, the inside diameter of the sleeve is uniformly distributed in the interval (1.98 centimeters, 2.02 centimeters) and the rod diameter is also uniformly distributed in the interval (1.95 centimeters, 2.00 centimeters). Assuming independence of these two distributions, find the probability that:

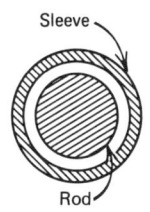

(a) The rod diameter will be smaller than the sleeve diameter.
(b) There is at least a 0.01-centimeter clearance between the rod and the sleeve.

7.4. Repeat Problem 7.3 if the distribution of the rod diameter remains uniform but that of the sleeve inside diameter is $N(2$ centimeters, 0.0004 square centimeters).

7.5. The first mention of the normal distribution was made in the work of de Moivre in 1733 as one method of approximating probabilities of a binomial distribution when n is large. Show that this approximation is valid and give an example showing results of this approximation.

7.6. If the distribution of temperature T of a given volume of gas is $N(400, 1600)$ measured in degrees Fahrenheit, find:

(a) $f_T(450)$
(b) $P(T \leq 450)$
(c) $P(|T - m_T| \leq 20)$
(d) $P(|T - m_T| \leq 20 | T \geq 300)$

SOME IMPORTANT CONTINUOUS DISTRIBUTIONS

7.7. If X is a r.v. and distributed $N(m, \sigma^2)$, show that

$$E\{|X - m|\} = \sqrt{\frac{2}{\pi}}\,\sigma$$

7.8. Let r.v's X and Y be identically and normally distributed. Show that r.v.'s $X + Y$ and $X - Y$ are independent.

7.9. Suppose that the useful lives measured in hours of two electronic devices, say X_1 and X_2, have distributions $N(40, 36)$ and $N(45, 9)$, respectively. If the electronic device is to be used for a 45-hour period, which is to be preferred? Which is preferred if it is to be used for a 48-hour period?

7.10. Verify Equation 7.13 for normal random variables.

7.11. Let r.v.'s $X_1, X_2, \ldots, X_n$ be jointly normal with zero means. Show that

$$E\{X_1 X_2 X_3\} = 0$$

$$E\{X_1 X_2 X_3 X_4\} = E\{X_1 X_2\}E\{X_3 X_4\} + E\{X_1 X_3\}E\{X_2 X_4\}$$
$$+ E\{X_1 X_4\}E\{X_2 X_3\}$$

Generalize the results above and verify Equation 7.35.

7.12. Let r.v.'s X_1, X_2, and X_3 be independent and distributed according to $N(0, 1)$, $N(1, 1)$, and $N(2, 1)$, respectively. Determine the probability $P(X_1 + X_2 + X_3 > 1)$.

7.13. A rope with 100 strands supports a weight of 2100 pounds. If the breaking strength of each strand is random with mean equal to 20 pounds and standard deviation 4 pounds and if breaking strength of the rope is the sum of independent breaking strengths of its strands, determine the probability that the rope will not fail under the load. (Assume no individual strand breakage before rope failure.)

7.14. If $X_1, X_2, \ldots, X_n$ are independent r.v.'s all having the distribution $N(m, \sigma^2)$, determine the conditions that must be imposed on $c_1, c_2, \ldots, c_n$ such that the sum

$$Y = c_1 X_1 + c_2 X_2 + \cdots + c_n X_n$$

is also $N(m, \sigma^2)$. Can all c's be positive?

7.15. The Cauchy distribution has the form

$$f_X(x) = \frac{1}{\pi(1 + x^2)} \qquad -\infty < x < \infty$$

Show that:
(a) It arises from the ratio X_1/X_2, where X_1 and X_2 are independent and distributed $N(0, \sigma^2)$.
(b) The moments of X do not exist.

7.16. Let X_1 and X_2 be independent normal random variables, both with mean 0 and standard deviation 1. Prove that:

$$Y = \arctan \frac{X_2}{X_1}$$

is uniformly distributed from $-\pi$ to π.

7.17. Verify Equations 7.48 for the lognormal distribution.

7.18. The lognormal distribution is found to be a good model for strains in structural members caused by wind loads. Let the strain be represented by X with $m_X = 1$ and $\sigma_X^2 = 0.09$.
(a) Determine the probability $P(X > 1.2)$.
(b) If stress Y in a structural member is related to the strain by $Y = a + bX$ with $b > 0$, determine $f_Y(y)$ and m_Y.

7.19. The life of a power transmission tower is exponentially distributed with mean life equal to 25 years. If three towers, operated independently, are being erected at the same time, what is the probability that at least two will still stand after 35 years?

7.20. For a gamma distributed random variable, show that:
(a) Its mean and variance are those given by Equation 7.57.
(b) It has a positive skewness.

7.21. Show that, if η is a positive integer, the PDF of a gamma distributed r.v. X can be written as

$$F_X(x) = 0 \qquad x \leq 0$$
$$= 1 - \sum_{k=0}^{\eta-1} \frac{(\lambda x)^k e^{-\lambda x}}{k!} \qquad x > 0$$

Recognize that the terms in the sum take the form of Poisson mass function and therefore can be calculated with the aid of probability tables for Poisson distributions.

7.22. The system shown below has three redundant components. Let their operating lives (in hours) be denoted by T_1, T_2, and T_3, respectively. If the redundant parts come into operation only when the on-line component fails (cold redundancy), then the operating life of the system, T, is $T = T_1 + T_2 + T_3$.

Let T_1, T_2, and T_3 be independent r.v.'s and each is distributed as

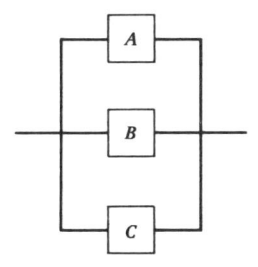

$$f_{T_j}(t_j) = \tfrac{1}{100}e^{-t_j/100} \quad t_j \geq 0 \quad j = 1, 2, 3$$
$$= 0 \quad \text{otherwise}$$

Determine the probability that the system will operate at least 300 hours.

7.23. We showed in Section VII.4.1 that an exponential failure law leads to a constant failure rate. Show that the converse is also true, that is, if $h(t)$ as defined by Equation 7.65 is a constant, then the time to failure T is exponentially distributed.

7.24. A shifted exponential distribution is defined as an exponential distribution shifted to the right an amount a, that is, if r.v. X has an exponential distribution with

$$f_X(x) = \lambda e^{-\lambda x} \quad x \geq 0$$
$$= 0 \quad \text{elsewhere}$$

the r.v. Y has a shifted exponential distribution if $f_Y(y)$ has the same shape as $f_X(x)$ but its nonzero portion starts at a point a rather than zero. Determine the relationship between X and Y and the pdf $f_Y(y)$. What are the mean and variance of Y?

7.25. Let r.v. X be chi square distributed with parameter n. Show that the limiting distribution of $(X - n)/\sqrt{2n}$ as $n \to \infty$ is $N(0, 1)$.

7.26. Let $X_1, X_2, \ldots, X_n$ be independent random variables with common PDF $F_X(x)$ and pdf $f_X(x)$. Equations 7.89 and 7.91 give, respectively, the PDF's of their maximum and minimum values. Let $X_{(j)}$ be the random variable denoting the jth smallest value of $X_1, X_2, \ldots, X_n$. Show that the PDF of $X_{(j)}$ has the form

$$F_{X_{(j)}}(x) = \sum_{k=j}^{n} \binom{n}{k} [F_X(x)]^k [1 - F_X(x)]^{n-k} \quad j = 1, 2, \ldots, n$$

7.27. Ten points are distributed uniformly and independently in the interval (0, 1). Find:
   (a) The probability that the point lying farthest to the right is to the left of $\frac{3}{4}$.
   (b) The probability that the point lying next farthest to the right is to the right of $\frac{1}{2}$.

7.28. Let the number of arrivals in a time interval obey the distribution given in Problem 6.27, which corresponds to a Poisson-type distribution with a time dependent mean rate of arrival. Show that the pdf of time between arrivals is given by

$$f_T(t) = \left(\frac{v}{w}\right) t^{v-1} \exp\left(-\frac{t^v}{w}\right) \qquad t \geq 0$$

$$= 0 \qquad \text{elsewhere}$$

As we see from Equation 7.124, it is the so-called Weibull distribution.

7.29. A multiple-member structure in a parallel arrangement as shown below supports a load s. It is assumed that all members share the load equally, that their resistances are random and identically distributed with common PDF $F_R(r)$, and that they act independently. If a member fails when the load it supports exceeds its resistance, show that the probability that failure will occur to $n - k$ members among n initially existing members is

$$\left[1 - F_R\left(\frac{s}{n}\right)\right]^n \qquad k = n$$

and

$$\sum_{j=1}^{n-k} \binom{n}{j} \left[F_R\left(\frac{s}{n}\right)\right]^j p^n_{(n-j)k}(s) \qquad k = 0, 1, \ldots, n-1$$

where

$$p^n_{kk}(s) = \left[1 - F_R\left(\frac{s}{k}\right)\right]^k$$

$$p^i_{jk}(s) = \sum_{r=1}^{j-k} \binom{j}{r} \left[F_R\left(\frac{s}{j}\right) - F_R\left(\frac{s}{i}\right)\right]^r p^j_{(j-r)k}(s) \qquad n \geq i > j > k$$

SOME IMPORTANT CONTINUOUS DISTRIBUTIONS

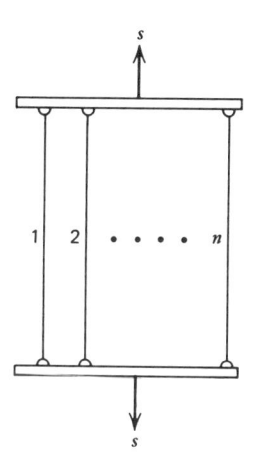

7.30. What is the probability sought in Problem 7.29 if the load is also a random variable S with pdf $f_S(s)$?

7.31. Let $n = 3$ in Problem 7.29. Determine the probabilities of zero-, one-, two-, and three-member failures in Problem 7.29 if R follows a uniform distribution over the interval (80, 100) and $s = 270$. Is partial failure (one- or two-member failure) possible in this case?

7.32. To show that, as a time-to-failure model, the Weibull distribution corresponds to a wide variety of shapes of the hazard function, graph the hazard function in Equation 7.123 and the corresponding Weibull distribution in Equation 7.124 for the following combinations of parameter values: $k = 0.5, 1, 2, 3$ and $w = 1, 2$.

7.33. The ranges of n independent test flights of a supersonic aircraft are assumed to be identically distributed with PDF $F_X(x)$ and pdf $f_X(x)$. If *range span* is defined as the distance between the maximum and minimum ranges of these n values, determine the pdf of the range span in terms of $F_X(x)$ or $f_X(x)$. Expressing it mathematically, the pdf of interest is that of S where

$$S = Y - Z$$

with

$$Y = \max(X_1, X_2, \ldots, X_n)$$

and

$$Z = \min(X_1, X_2, \ldots, X_n)$$

Note that r.v.'s Y and Z are *not* independent.

Part B

Statistical Inference, Parameter Estimation, and Model Verification

VIII
Observed Data and Graphical Representation

Referring to Figure 1.1 in Chapter I, we are concerned in this and subsequent chapters with the step $D \to E$ of the basic cycle in probabilistic modeling, that is, model verification and parameter estimation on the basis of observed data. In the preceding chapters, our major concern has been the selection of an appropriate model (probability distribution) to represent a physical or natural phenomenon based on our understanding of its underlying properties. In order to specify the model completely, however, it is required that the parameters in the distribution be assigned. We now consider this problem of parameter estimation using available data. Included in this discussion are techniques for assessing the reasonableness of a selected model and the problem of selecting a model from among a number of contending distributions when no single one is preferred on the basis of underlying physical characteristics of a given phenomenon.

Let us emphasize at the outset that, owing to probabilistic nature of the situation, the problem of parameter estimation is precisely that—an estimation problem. A sequence of observations, say n in number, is a *sample* of observed values of the underlying random variable. If we were to repeat the sequence of n observations, the random nature of the experiment would produce a different sample of observed values. Any reasonable rule for extracting parameter estimates from a set of n observations will thus give different estimates for different sets of observations. In other words, no single sequence of observations finite in number

VIII.1 Histograms

Given a set of independent observations $x_1, x_2, \ldots,$ and x_n of a r.v. X, a useful first step is to organize and present them properly so that they can be easily interpreted and evaluated. When there are a large number of observed data, a *histogram* is an excellent graphical representation of the data that facilitates (a) an evaluation of adequacy of the assumed model, (b) estimation of percentiles of the distribution, and (c) estimation of the distribution parameters.

Let us consider, for example, a chemical process that is producing batches of desired material. Two hundred observed values of the percent yield X, representing a large sample size, are given in Table 8.1. The sample values vary from 64 to 76. Dividing this range into 12 equal intervals and plotting the total number of observed yields in each interval as the height of a rectangle over the interval results in the histogram as shown in Figure 8.1. A *frequency diagram* is obtained if the ordinate of the histogram is divided by the total number of observations, 200 in this case. We see that the histogram or the frequency diagram gives an immediate impression of the range, relative frequency, and scatter associated with observed data.

Table 8.1. *Chemical Yield Data.*

Batch No.	Yield (%)	Batch No.	Yield (%)	Batch No.	Yield (%)	Batch No.	Yield (%)	Batch No.	Yield (%)
1	68.4	41	68.7	81	68.5	121	73.3	161	70.5
2	69.1	42	69.1	82	71.4	122	75.8	162	68.8
3	71.0	43	69.3	83	68.9	123	70.4	163	72.9
4	69.3	44	69.4	84	67.6	124	69.0	164	69.0
5	72.9	45	71.1	85	72.2	125	72.2	165	68.1
6	72.5	46	69.4	86	69.0	126	69.8	166	67.7
7	71.1	47	75.6	87	69.4	127	68.3	167	67.1
8	68.6	48	70.1	88	73.0	128	68.4	168	68.1
9	70.6	49	69.0	89	71.9	129	70.0	169	71.7
10	70.9	50	71.8	90	70.7	130	70.9	170	69.0

OBSERVED DATA AND GRAPHICAL REPRESENTATION

Table 8.1. (*continued*)

Batch No.	Yield (%)	Batch No.	Yield (%)	Batch No.	Yield (%)	Batch No.	Yield (%)	Batch No.	Yield (%)
11	68.7	51	70.1	91	67.0	131	72.6	171	72.0
12	69.5	52	64.7	92	71.1	132	70.1	172	71.5
13	72.6	53	68.2	93	71.8	133	68.9	173	74.9
14	70.5	54	71.3	94	67.3	134	64.6	174	78.7
15	68.5	55	71.6	95	71.9	135	72.5	175	69.0
16	71.0	56	70.1	96	70.3	136	73.5	176	70.8
17	74.4	57	71.8	97	70.0	137	68.6	177	70.0
18	68.8	58	72.5	98	70.3	138	68.6	178	70.3
19	72.4	59	71.1	99	72.9	139	64.7	179	67.5
20	69.2	60	67.1	100	68.5	140	65.9	180	71.7
21	69.5	61	70.6	101	69.8	141	69.3	181	74.0
22	69.8	62	68.0	102	67.9	142	70.3	182	67.6
23	70.3	63	69.1	103	69.8	143	70.7	183	71.1
24	69.0	64	71.7	104	66.5	144	65.7	184	64.6
25	66.4	65	72.2	105	67.5	145	71.1	185	74.0
26	72.3	66	69.7	106	71.0	146	70.4	186	67.9
27	74.4	67	68.3	107	72.8	147	69.2	187	68.5
28	69.2	68	68.7	108	68.1	148	73.7	188	73.4
29	71.0	69	73.1	109	73.6	149	68.5	189	70.4
30	66.5	70	69.0	110	68.0	150	68.5	190	70.7
31	69.2	71	69.8	111	69.6	151	70.7	191	71.6
32	69.0	72	69.6	112	70.6	152	72.3	192	66.9
33	69.4	73	70.2	113	70.0	153	71.4	193	72.6
34	71.5	74	68.4	114	68.5	154	69.2	194	72.2
35	68.0	75	68.7	115	68.0	155	73.9	195	69.1
36	68.2	76	72.0	116	70.0	156	70.2	196	71.3
37	71.1	77	71.9	117	69.2	157	69.6	197	67.9
38	72.0	78	74.1	118	70.3	158	71.6	198	66.1
39	68.3	79	69.3	119	67.2	159	69.7	199	70.8
40	70.6	80	69.0	120	70.7	160	71.2	200	69.5

Taken from Hill (1975). Reprinted with permission.

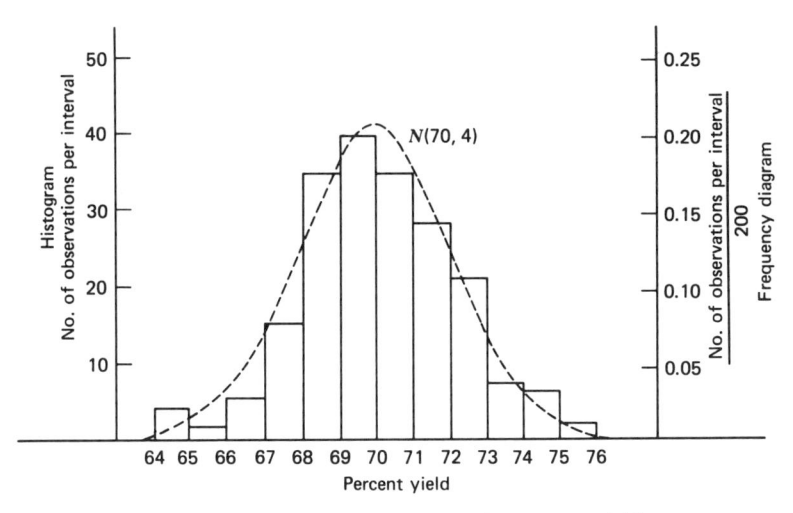

244   PARAMETER ESTIMATION AND MODEL VERIFICATION

Figure 8.1 Histogram and frequency diagram for percent yield.

In the case of a discrete random variable, the histogram and frequency diagram as obtained from data take the shape of a bar chart as opposed to connected rectangles in the continuous case. Consider, for example, the distribution of the number of accidents per driver during a six-year span in California. The data given in Table 8.2 are six-year accident records of 7842 California drivers [Burg (1967 and 1968)]. Based upon this set of observations, the histogram has the form given in Figure 8.2.

Returning now to the chemical yield example, the frequency diagram as shown in Figure 8.1 has the familiar properties of a probability density function if the height of each interval is divided by the interval width (which happens to be one in this example). Hence, probabilities associated with various events can be estimated.

Table 8.2. *Six-year Accident Record for 7842 California Drivers*

No. of Accidents	No. of Drivers
0	5147
1	1859
2	595
3	167
4	54
5	14
5+	6

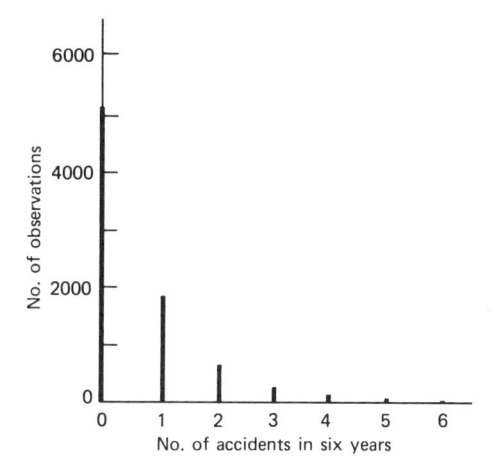

Figure 8.2 Histogram from six-year accident data.

For example, the probability of a batch having yield less than 68% can be read off from the frequency diagram by summing over the areas, with each interval height divided by its interval width, to the left of 68%. It is 0.02 + 0.01 + 0.025 + 0.075 = 0.13. Similarly, the probability of a batch having yields greater than 72% is 0.105 + 0.035 + 0.03 + 0.01 = 0.18. Let us remember, however, these are probabilities calculated based on the histogram that is constructed from the observed data. A different set of data obtained from the same chemical process would in general lead to a different histogram and hence different values for these probabilities. Consequently, they are, at best, estimates of the probabilities $P(X < 68)$ and $P(X > 72)$ associated with the underlying random variable X.

A remark on the choice of the number of intervals for histogram plotting is in order. For this example, the choice of 12 intervals is convenient on account of the range of values spanned by the observations and of the fact that the resulting resolution is adequate for calculations of probabilities carried out earlier. In Figure 8.3, a histogram is constructed using 4 intervals instead of 12 for the same example. It is easy to see that it projects quite a different visual impression of the data behavior. It is thus important to choose the number of intervals *consistent* with the information one wishes to extract from the mathematical model. As a practical guide, Sturges (1926) suggests that an approximate value for the number of intervals, k, be determined from

$$k = 1 + 3.3 \log_{10} n \qquad (8.1)$$

where n is the sample size.

From the modeling point of view, it is reasonable to select a normal distribution as the probabilistic model for percent yield X by observing that its random variations are the resultant of numerous independent random sources in the

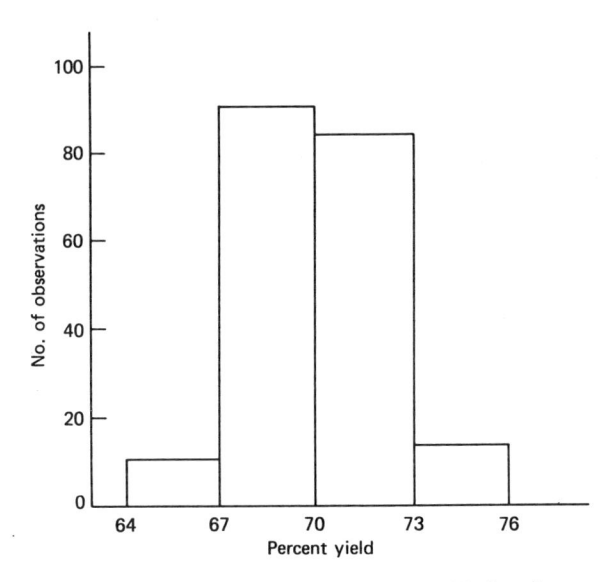

Figure 8.3 Histogram for percent yield with four intervals.

chemical manufacturing process. Whether or not this is a reasonable selection can be evaluated in a subjective way using the histogram given in Figure 8.1. The normal density function with mean 70 and variance 4 is superimposed on the histogram in Figure 8.1, which shows a reasonable match. Based on this normal distribution, we can calculate the probabilities given above, giving a further assessment of adequacy of the model. For example, with the aid of Table A.3,

$$P(X < 68) = F_U\left(\frac{68 - 70}{2}\right) = F_U(-1)$$
$$= 1 - F_U(1) = 0.159$$

which compares with 0.13 using the histogram.

In the above, the choice of 70 and 4, respectively, as estimates of the mean and variance of X is made by observing that the mean of the distribution should be close to the arithmetic mean of the sample, that is,

$$m_X \cong \frac{1}{n} \sum_{j=1}^{n} x_j \qquad (8.2)$$

and the variance can be approximated by

$$\sigma_X^2 \cong \frac{1}{n} \sum_{j=1}^{n} (x_j - m_X)^2 \qquad (8.3)$$

which gives the arithmetic average of the squares of sample values with respect to their arithmetic mean.

Let us emphasize that our use of Equations 8.2 and 8.3 is guided largely by intuition. It is clear that we need to address the problem of estimating the parameter values in an objective and systematic fashion. In addition, procedures need to be developed that permit us to assess the adequacy of the normal model chosen for this example. These are subjects of discussion in the chapters to follow.

REFERENCES AND COMMENTS

1. J. R. Benjamin and C. A. Cornell, *Probability, Statistics, and Decision for Civil Engineers*, McGraw-Hill, New York, 1970.
2. A. Burg, *The Relationship between Vision Test Scores and Driving Record*, Department of Engineering, UCLA, Los Angeles, 1967 and 1968.
3. K. K. Chen and R. B. Krieger, "A Statistical Analysis of the Influence of Cyclic Variation on the Formation of Nitric Oxide in Spark Ignition Engines," *Combustion Sci. Tech.*, **12**, 125–134, 1976.
4. J. W. Dunham, G. N. Brekke, and G. N. Thompson, *Live Loads on Floors in Buildings*, Building Materials and Structures Report 133, National Bureau of Standards, 1952.
5. J. Ferreira, Jr., "The Long-term Effects of Merit-rating Plans for Individual Motorists," *Oper. Research*, **22**, 954–978, 1974.
6. W. J. Hill, *Statistical Analysis for Physical Scientists*, Class Notes, SUNY/Buffalo, Buffalo, New York, 1975.
7. R. W. Jelliffe, J. Buell, R. Kalaba, R. Sridhar, and R. Rockwell, "A Mathematical Study of the Metabolic Conversion of Digitoxin to Digoxin in Man," *Math. Biosci.*, **6**, 387–403, 1970.
8. V. F. Link, *Statistical Analysis of Blemishes in a SEC Image Tube*, M.S. Thesis, SUNY/Buffalo, Buffalo, New York, 1972.
9. H. A. Sturges, "The Choice of a Class Interval," *J. Am. Stat. Assoc.*, **21**, 65–66, 1926.

PROBLEMS

8.1. It has been shown that the frequency diagram, with interval heights divided by interval widths, gives a graphical representation of the probability density function. Use the data given in Table 8.1 and construct a diagram which approximates the probability distribution function of the percent yield X.

248 PARAMETER ESTIMATION AND MODEL VERIFICATION

In each of the problems given below, observations or sample values of size n are given for a random phenomenon.

(a) *If not already given, plot histogram and frequency diagram associated with the designated r.v. X.*

(b) *Based on the shape of these diagrams and on your understanding of the underlying physical situation, suggest one probability distribution (e.g., normal, Poisson, gamma, etc.), which may be appropriate for X. Estimate parameter values by means of Equations 8.2 and 8.3 and, for the purpose of comparison, plot the proposed pdf or pmf and superimpose it on the frequency diagram.*

8.2. $X =$ maximum annual flood flow of the Feather River at Oroville, California. Data given below are records of maximum flood flows in 1000 cfs for years 1902–1960. [Source: Benjamin and Cornell (1970).]

Year	Flood	Year	Flood	Year	Flood
		1921	62	1941	84
1902	41		36		110
	102		22		108
	118		42		25
	81		64		60
	128		56		54
	230		94		46
	16		185		37
	140		14		17
	31		80		46
	75		12		92
	16		23		13
	17		9		59
	122		20		113
	81		59		55
	42		85		203
	80		19		83
	28		185		102
	66		8		35
1920	23	1940	152	1960	135

OBSERVED DATA AND GRAPHICAL REPRESENTATION

8.3. X = number of accidents per driver during six-year span in California. Data are given in Table 8.2 for 7842 drivers.

8.4. X = time gap in seconds between cars on a stretch of highway. The following gives measurements of time gaps in seconds between successive vehicles at a given location ($n = 100$).

4.1	3.5	2.2	2.7	2.7	4.1	3.4	1.8	3.1	2.1
2.1	1.7	2.3	3.0	4.1	3.2	2.2	2.3	1.5	1.1
2.5	4.7	1.8	4.8	1.8	4.0	4.9	3.1	5.7	5.7
3.1	2.0	2.9	5.9	2.1	3.0	4.4	2.1	2.6	2.7
3.2	2.5	1.7	2.0	2.7	1.2	9.0	1.8	2.1	5.4
2.1	3.8	4.5	3.3	2.1	2.1	7.1	4.7	3.1	1.7
2.2	3.1	1.7	3.1	2.3	8.1	5.7	2.2	4.0	2.7
1.5	1.7	4.0	6.4	1.5	2.2	1.2	5.1	2.7	2.4
1.7	1.2	2.7	7.0	3.9	5.2	2.7	3.5	2.9	1.2
1.5	2.7	2.9	4.1	3.1	1.9	4.8	4.0	3.0	2.7

8.5. X = sum of two successive gaps in Problem 8.4.

8.6. X = number of vehicles arriving per minute at a toll booth on New York Thruway. Measurements of 105 one-minute arrivals are given as follows.

9	9	11	15	6	11	9	6	11	8	10
3	9	8	5	7	15	7	14	6	6	16
6	8	10	6	10	11	9	7	7	11	10
3	8	4	7	15	6	7	7	8	7	5
13	12	11	10	8	14	3	15	13	5	7
12	7	10	4	16		7	11	11	13	10
9	10	11	6	6		8	9	5	5	5
11	6	7	9	5		12	12	4	13	4
12	16	10	14	15		16	10	8	10	6
18	13	6	9	4		13	14	6	10	10

8.7. X = number of five-minute arrivals in Problem 8.6.

8.8. X = amount of yearly snowfall in inches in Buffalo, New York. Given below are recorded snowfalls in inches from 1909 to 1979.

Year	Snowfall	Year	Snowfall	Year	Snowfall
1909–1910	126.4	1934–1935	49.1	1959–1960	115.6
	82.4		103.9		102.4
	78.1		51.6		101.4
	51.1		82.4		89.8
	90.9		83.6		71.5
	76.2		77.8		70.9
	104.5		79.3		98.3
	87.4		89.6		55.5
	110.5		85.5		66.1
	25.0		58.0		78.4
	69.3		120.7		120.5
	53.5		110.5		97.0
	39.8		65.4		109.9
	63.6		39.9		78.8
	46.7		40.1		88.7
	72.9		88.7		95.6
	74.6		71.4		82.5
	83.6		83.0		199.4
	80.7		55.9		154.3
	60.3		89.9	1978–1979	97.3
	79.0		84.6		
	74.4		105.2		
	49.6		113.7		
	54.7		124.7		
1933–1934	71.8	1958–1959	114.5		

8.9. X = peak combustion pressure in k Pa per cycle. In spark ignition engines, cylinder pressure during combustion varies from cycle to cycle. The histogram of peak combustion pressure in k Pa is shown below for 280 samples. [Source: Chen and Krieger (1976).]

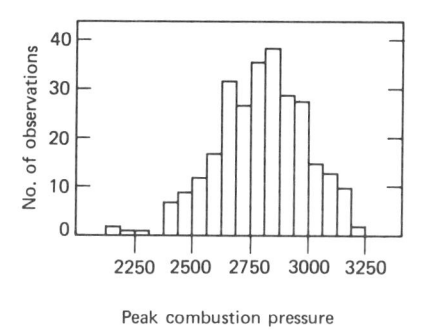

OBSERVED DATA AND GRAPHICAL REPRESENTATION 251

[Histogram: No. of observations vs Peak combustion pressure, x-axis 2250 to 3250]

Peak combustion pressure

8.10. X_1, X_2, X_3 = annual premiums paid by low-, medium-, and high-risk drivers. Frequency diagram for each group is given below. [Source: Simulated results over 50 years due to Ferreira (1974).]

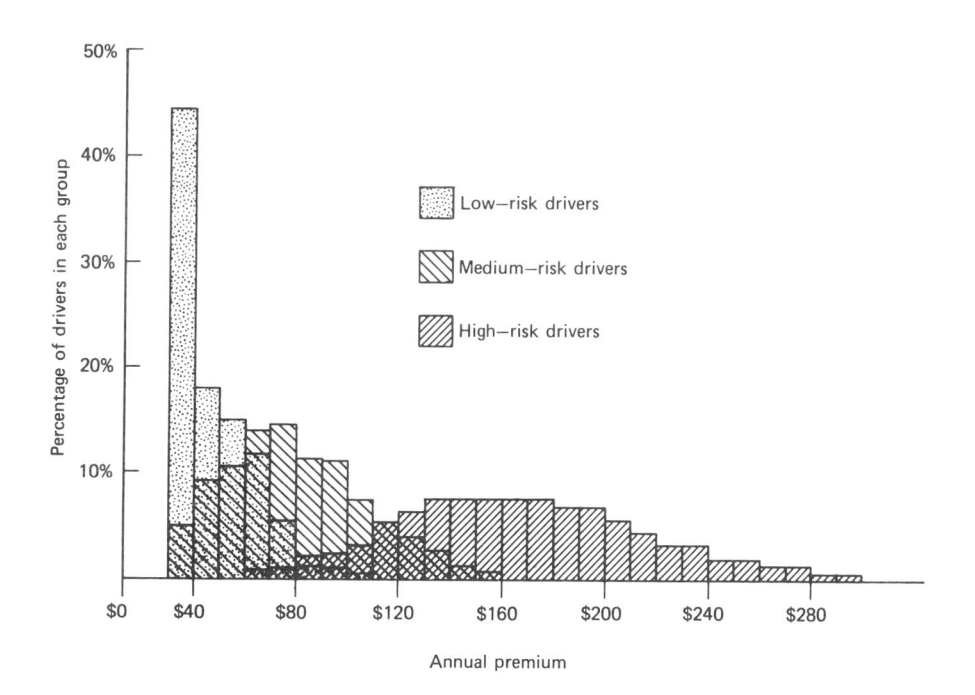

[Stacked histogram: Percentage of drivers in each group vs Annual premium, with Low-risk drivers, Medium-risk drivers, and High-risk drivers]

Annual premium

8.11. X = number of blemishes in a certain type of image tubes for television. Fifty-eight data points are used for construction of the histogram shown below. [Source: Link (1972).]

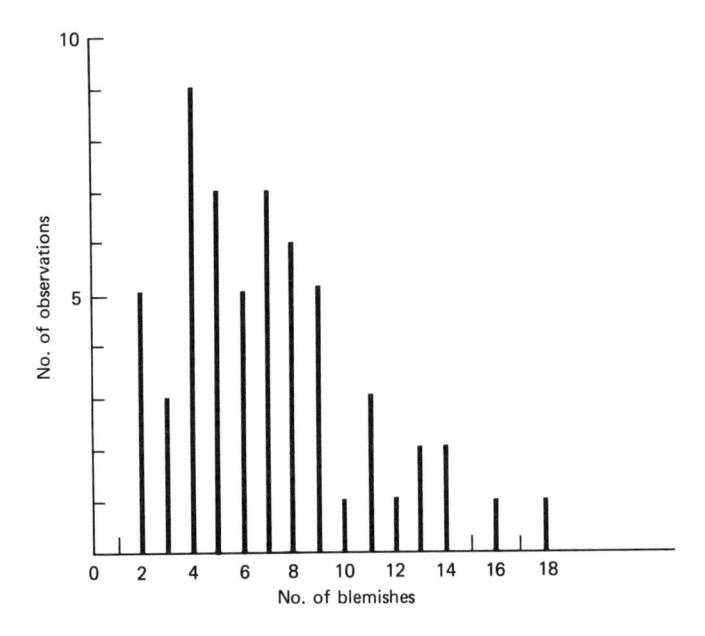

8.12. X = difference between observed and computed urinary digitoxin excretion in micrograms per day. In a study of metabolism of digitoxin to digoxin in patients, long-term studies of urinary digitoxin excretion were carried out on four patients. A histogram of the difference between observed and computed urinary digitoxin excretion in micrograms per day is given as follows ($n = 110$). [Source: Jelliffe et al. (1970).]

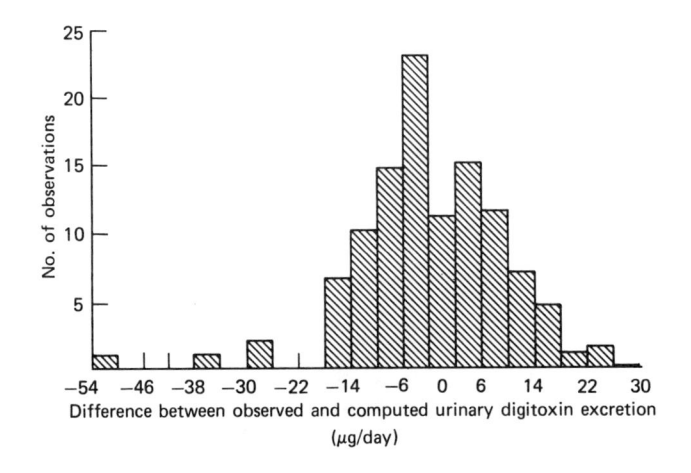

8.13. X = live load in psf in warehouses. Histogram given below represents 220 measurements of live loads on different floors of a warehouse over bays of areas of approximately 400 square feet. [Source: Dunham (1952).]

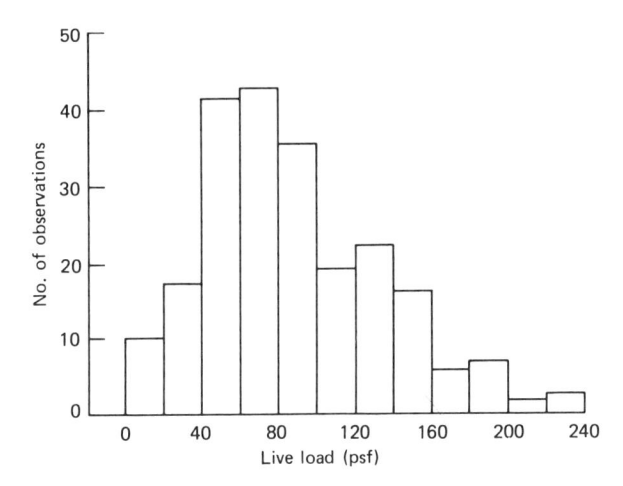

IX
Parameter Estimation

Suppose that a probabilistic model, represented by the probability density function $f(x)$, has been chosen for a physical or natural phenomenon whose parameters $\theta_1, \theta_2, \ldots$ are to be estimated from observed data $x_1, x_2, \ldots, x_n$. Let us consider for a moment a single parameter θ for simplicity and write $f(x; \theta)$ to mean a specified probability distribution where θ is the unknown parameter to be estimated. The parameter estimation problem is then one of determining an appropriate function of $x_1, x_2, \ldots, x_n$, say $h(x_1, x_2, \ldots, x_n)$, which gives the "best" estimate for θ. In order to develop systematic estimation procedures, we need to make more precise terms that were defined rather loosely in the preceding chapter and introduce some new concepts needed for this development.

IX.1 Samples and Statistics

Given a data set $x_1, x_2, \ldots, x_n$, let

$$\hat{\theta} = h(x_1, x_2, \ldots, x_n) \qquad (9.1)$$

be an estimate of the parameter θ. In order to ascertain its general properties, it is recognized that, if the experiment that yielded the data set were to be repeated, we would obtain different values for $x_1, x_2, \ldots, x_n$. The function $h(x_1, x_2, \ldots, x_n)$ when applied to the new data set would yield a different value for $\hat{\theta}$. We thus see that the estimate $\hat{\theta}$ is itself a random variable possessing a probability distribution, which depends both on the functional form defined by h and on the distribution of the underlying r.v. X. The appropriate representation of $\hat{\theta}$ is thus

$$\hat{\Theta} = h(X_1, X_2, \ldots, X_n) \tag{9.2}$$

where $X_1, X_2, \ldots, X_n$ are random variables, representing a *sample* from the r.v. X, which is sometimes referred to as the *population*. In practically all applications, we shall assume that the sample $X_1, X_2, \ldots, X_n$ possesses the following properties:

1. $X_1, X_2, \ldots, X_n$ are independent.
2. $f_{X_j}(x) = f_X(x)$ for all $x, j = 1, 2, \ldots, n$.

The r.v.'s $X_1, \ldots, X_n$ satisfying these conditions are called a *random sample* of size n. The word "random" in this definition is usually omitted for the sake of brevity. If X is a random variable of the discrete type with pmf $p_X(x)$, then $p_{X_j}(x) = p_X(x)$ for each j.

A specific set of observed values $(x_1, x_2, \ldots, x_n)$ is a set of *sample values* assumed by the sample. The problem of parameter estimation is one class in the broader topic of statistical inference in which our object is to make inferences about various aspects of the underlying population distribution on the basis of observed sample values. For the purpose of clarification, the interrelationships among X, $(X_1, X_2, \ldots, X_n)$, and $(x_1, x_2, \ldots, x_n)$ are schematically shown in Figure 9.1.

Let us note that the properties of a sample as given above imply that certain conditions are imposed on the manner in which observed data are obtained. Each data point must be observed from the population independently and under essentially identical conditions. In sampling a population of percent yield as discussed in Chapter VIII, for example, one would avoid taking adjacent batches if correlation between them is to be expected.

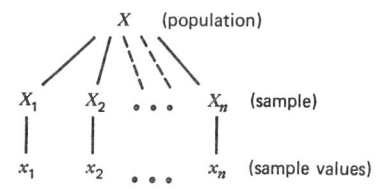

Figure 9.1 Population, sample, and sample values.

A *statistic* is any function of a given sample $X_1, X_2, \ldots, X_n$ which does not depend on the unknown parameter. The function $h(X_1, X_2, \ldots, X_n)$ in Equation 9.2 is thus a statistic whose value can be determined once the sample values have been observed. It is important to note that a statistic, being a function of random variables, is a random variable. When used to estimate a distribution parameter, its statistical properties, such as mean, variance, and distribution, give information concerning the quality of this particular estimation procedure. Certain statistics play an important role in statistical estimation theory; these include sample mean, sample variance, order statistics, and other sample moments. Some properties of these important statistics are discussed below.

IX.1.1 SAMPLE MEAN

The statistic

$$\boxed{\bar{X} = \frac{1}{n} \sum_{i=1}^{n} X_i} \tag{9.3}$$

is called the *sample mean* of the population X. Let the population mean and variance be

$$E\{X\} = m \quad \text{and} \quad \text{var}\{X\} = \sigma^2 \tag{9.4}$$

The mean and variance of $\bar{X}$, the sample mean, are easily found to be

$$E\{\bar{X}\} = \frac{1}{n} \sum_{i=1}^{n} E\{X_i\} = \frac{1}{n}(nm) = m \tag{9.5}$$

and, due to independence,

$$\text{var}\{\bar{X}\} = E\{(\bar{X} - m)^2\} = E\left\{\left[\frac{1}{n}\sum_{i=1}^{n}(X_i - m)\right]^2\right\}$$
$$= \frac{1}{n^2}(n\sigma^2) = \frac{\sigma^2}{n} \tag{9.6}$$

which is inversely proportional to the sample size n. As n increases, the variance of $\bar{X}$ decreases and the distribution of $\bar{X}$ becomes sharply peaked at $E\{\bar{X}\} = m$. Hence, it is intuitively clear that the statistic $\bar{X}$ is a good estimate of the population mean m. This is another statement of the law of large numbers that was discussed in Examples 4.12 and 4.13.

Since $\bar{X}$ is a sum of independent random variables, its distribution can also be determined either by the use of techniques developed in Chapter V or by means of the method of characteristic functions given in Section IV.5. We further observe that, on the basis of the central limit theorem, the sample mean $\bar{X}$ approaches a normal distribution as $n \to \infty$. More precisely, the r.v. $(\bar{X} - m)/(\sigma/\sqrt{n})$ approaches $N(0, 1)$ as $n \to \infty$.

IX.1.2 SAMPLE VARIANCE

The statistic

$$S^2 = \frac{1}{n-1} \sum_{i=1}^{n} (X_i - \bar{X})^2 \qquad (9.7)$$

is called the *sample variance* of the population X. The mean of S^2 can be found by expanding the squares in the sum and taking termwise expectations. We first write Equation 9.7 as

$$S^2 = \frac{1}{n-1} \sum_{i=1}^{n} [(X_i - m) - (\bar{X} - m)]^2$$

$$= \frac{1}{n-1} \sum_{i=1}^{n} \left[(X_i - m) - \frac{1}{n} \sum_{j=1}^{n} (X_j - m) \right]^2$$

$$= \frac{1}{n} \sum_{i=1}^{n} (X_i - m)^2 - \frac{1}{n(n-1)} \sum_{\substack{i,j=1 \\ i \neq j}}^{n} (X_i - m)(X_j - m)$$

Taking termwise expectations and noting mutual independence, we have

$$E\{S^2\} = \sigma^2 \qquad (9.8)$$

where m and σ^2 are defined in Equation 9.4. We remark at this point that the reason for using $1/(n-1)$ rather than $1/n$ in Equation 9.7 is to make the mean of S^2 equal to σ^2. As we shall see in the next section, this is a desirable property for S^2 if it is to be used to estimate σ^2, the true variance of X.

The variance of S^2 is found from

$$\text{var}\{S^2\} = E\{(S^2 - \sigma^2)^2\} \qquad (9.9)$$

Upon expanding the right-hand side and carrying out expectations term by term, we find that

$$\operatorname{var}\{S^2\} = \frac{1}{n}\left[\mu_4 - \frac{n-3}{n-1}\sigma^4\right] \tag{9.10}$$

where μ_4 is the fourth central moment of X, that is,

$$\mu_4 = E\{(X - m)^4\} \tag{9.11}$$

Equation 9.10 again shows that the variance of S^2 is an inverse function of n.

In principle, the distribution of S^2 can be derived using techniques advanced in Chapter V. It is, however, a tedious process because of complex nature of the expression for S^2 as defined by Equation 9.7. For the case in which the population X is distributed according to $N(m, \sigma^2)$, we have the following result.

Theorem

Let S^2 be the sample variance of size n from the normal population $N(m, \sigma^2)$, then $(n - 1)S^2/\sigma^2$ has a chi-square distribution with $(n - 1)$ degrees of freedom.

Proof

The chi-square distribution is given in Section VII.4.2. In order to sketch a proof for this theorem, let us note from Section VII.4.2 that the random variable

$$Y = \frac{1}{\sigma^2} \sum_{j=1}^{n}(X_j - m)^2 \tag{9.12}$$

has a chi-square distribution of n degrees of freedom since each term in the sum is a squared normal random variable and is independent of other random variables in the sum. Now, we can show that the difference between Y and $(n - 1)S^2/\sigma^2$ is

$$Y - \frac{(n-1)S^2}{\sigma^2} = \left(\frac{\overline{X} - m}{\sigma/\sqrt{n}}\right)^2 \tag{9.13}$$

Since the right-hand side of Equation 9.13 is a random variable having a chi-square distribution with one degree of freedom, Equation 9.13 would lead to the result that $(n - 1)S^2/\sigma^2$ is chi-square distributed with $(n - 1)$ degrees of freedom provided that independence exists between $(n - 1)S^2/\sigma^2$ and

PARAMETER ESTIMATION

$[(\bar{X} - m)/(\sigma/\sqrt{n})]^2$. The proof of this independence is not given here but can be found in more advanced texts [e.g., Anderson and Bancroft (1952)].

IX.1.3 SAMPLE MOMENTS

The kth sample moment is

$$M_k = \frac{1}{n} \sum_{j=1}^{n} X_j^k \qquad (9.14)$$

Following similar procedures as given above, we can show that

$$E\{M_K\} = \alpha_k$$
$$\text{var}\{M_k\} = \frac{1}{n}(\alpha_{2k} - \alpha_k^2) \qquad (9.15)$$

where α_k is the kth moment of the population X.

IX.1.4 ORDER STATISTICS

A sample $X_1, X_2, \ldots, X_n$ can be ranked in order of increasing numerical magnitude. Let $X_{(1)}, X_{(2)}, \ldots, X_{(n)}$ be such a rearranged sample where $X_{(1)}$ is the smallest and $X_{(n)}$ the largest. Then $X_{(k)}$ is called the *k*th-order statistic. The extreme values $X_{(1)}$ and $X_{(n)}$ are of particular importance in applications and their properties have been discussed in Section VII.6.

In terms of the PDF of population X, $F_X(x)$, it follows from Equations 7.89 and 7.91 that the PDF's of $X_{(1)}$ and $X_{(n)}$ are

$$F_{X_{(1)}}(x) = 1 - [1 - F_X(x)]^n \qquad (9.16)$$
$$F_{X_{(n)}}(x) = F_X^n(x) \qquad (9.17)$$

If X is continuous, the pdf's of $X_{(1)}$ and $X_{(n)}$ are of the forms (see Equations 7.90 and 7.92)

$$f_{X_{(1)}}(x) = n[1 - F_X(x)]^{n-1} f_X(x) \qquad (9.18)$$
$$f_{X_{(n)}}(x) = n F_X^{n-1}(x) f_X(x) \qquad (9.19)$$

The means and variances of order statistics can be obtained through integration but they are not expressible as simple functions of the moments of the population X.

IX.2 Quality Criteria for Estimates

We are now in a position to propose a number of criteria under which the quality of an estimate can be evaluated. These criteria define generally desirable properties for an estimate to have as well as provide a guide by which the quality of one estimate can be compared to that of another.

Before proceeding, a remark is in order regarding the notation to be used. As seen in Equation 9.2, our objective in parameter estimation is to determine a statistic

$$\hat{\Theta} = h(X_1, X_2, \ldots, X_n) \tag{9.20}$$

which gives a good estimate of the parameter θ. This statistic will be called an *estimator* for θ, whose properties, such as mean, variance, or distribution, provide a measure of quality of this estimator. Once we have observed sample values $x_1, x_2, \ldots, x_n$, the observed estimator

$$\hat{\theta} = h(x_1, x_2, \ldots, x_n) \tag{9.21}$$

has a numerical value and will be called an *estimate* of the parameter θ.

IX.2.1 UNBIASEDNESS

An estimator $\hat{\Theta}$ is said to be an *unbiased* estimator for θ if

$$\boxed{E\{\hat{\Theta}\} = \theta} \tag{9.22}$$

for all θ. This is clearly a desirable property for $\hat{\Theta}$, which states that, on the average, we expect $\hat{\Theta}$ to be close to the true parameter value θ. Let us note here that the requirement of unbiasedness may lead to other undesirable consequences. Hence, the overall quality of an estimator does not rest on any single criterion but on a set of criteria.

We have studied two statistics, $\bar{X}$ and S^2, in Sections IX.1.1 and IX.1.2. It is seen from Equations 9.5 and 9.8 that, if $\bar{X}$ and S^2 are used as estimators for the population mean m and population variance σ^2, respectively, they are unbiased estimators. This nice property for S^2 suggests that the sample variance defined by

PARAMETER ESTIMATION

Equation 9.7 is preferred over the more natural choice obtained by replacing $1/(n-1)$ by $1/n$ in Equation 9.7. Indeed, if we let

$$S^{2*} = \frac{1}{n} \sum_{i=1}^{n} (X_i - \bar{X})^2 \qquad (9.23)$$

its mean is

$$E\{S^{2*}\} = \frac{n-1}{n} \sigma^2$$

and the estimator S^{2*} has a bias indicated by the coefficient $(n-1)/n$.

IX.2.2 MINIMUM VARIANCE

It seems natural that, if $\hat{\Theta} = h(X_1, X_2, \ldots, X_n)$ is to qualify as a good estimator for θ, not only its mean should be close to the true value θ but also there should be a good probability that any of its observed values $\hat{\theta}$ will be close to θ. This can be achieved by selecting a statistic in such a way that not only is $\hat{\Theta}$ unbiased but also its variance is as small as possible. Hence, the second desirable property is that of minimum variance.

Definition. Let $\hat{\Theta}$ be an unbiased estimator for θ. It is an *unbiased minimum-variance* estimator for θ if, for all other unbiased estimators Θ^* for θ from the same sample,

$$\boxed{\operatorname{var}\{\hat{\Theta}\} \leq \operatorname{var}\{\Theta^*\}} \qquad (9.24)$$

for all θ.

Given two unbiased estimators for a given parameter, the one with smaller variance is preferred because smaller variance implies that observed values of the estimator tend to be closer to its mean, the true parameter value.

Example 9.1. We have seen that $\bar{X}$ obtained from a sample of size n is an unbiased estimator for the population mean m. Does the quality of $\bar{X}$ improve as n increases?

We easily see from Equation 9.5 that the mean of $\bar{X}$ is independent of the sample size; it thus remains unbiased as n increases. Its variance, on the other hand, as given by Equation 9.6 is

$$\operatorname{var}\{\bar{X}\} = \frac{\sigma^2}{n} \qquad (9.25)$$

which decreases as n increases. Thus, based on the minimum variance criterion, the quality of $\bar{X}$ as an estimator for m improves as n increases.

Example 9.2. Based on a fixed sample size n, is $\bar{X}$ the best estimator for m in terms of unbiasedness and minimum variance?

In order to answer this question, it is necessary to show that the variance of $\bar{X}$ as given by Equation 9.25 is the smallest among all unbiased estimators, which can be constructed from the sample. This is certainly difficult to do. However, a powerful theorem given below shows that it is possible to determine the minimum achievable variance of any unbiased estimator obtained from a given sample. This lower bound on the variance thus permits us to answer questions such as the one just posed.

Theorem (The Cramér–Rao Inequality)

Let $X_1, X_2, \ldots, X_n$ denote a sample of size n from a population X with pdf $f(x;\theta)$, where θ is the unknown parameter, and let $\hat{\Theta} = h(X_1, X_2, \ldots, X_n)$ be an unbiased estimator for θ. Then the variance of $\hat{\Theta}$ satisfies the inequality

$$\text{var}\{\hat{\Theta}\} \geq \frac{1}{nE\{[\partial \ln f(X;\theta)/\partial \theta]^2\}} \tag{9.26}$$

if the indicated expectation and differentiation exist. An analogous result with $p(X;\theta)$ replacing $f(X;\theta)$ is obtained when X is discrete.

Proof

The joint probability density function of $X_1, X_2, \ldots,$ and X_n is, because of their mutual independence, $f(x_1;\theta)f(x_2;\theta)\cdots f(x_n;\theta)$. The mean of the statistic $\hat{\Theta} = h(X_1, X_2, \ldots, X_n)$ is

$$E\{\hat{\Theta}\} = E\{h(X_1, X_2, \ldots, X_n)\}$$

and, since $\hat{\Theta}$ is unbiased, it gives

$$\theta = \int_{-\infty}^{\infty} \cdots \int_{-\infty}^{\infty} h(x_1, \ldots, x_n) f(x_1;\theta) \cdots f(x_n;\theta) dx_1 \cdots dx_n \tag{9.27}$$

Another relation we need is the identity

$$1 = \int_{-\infty}^{\infty} f(x_i;\theta) dx_i \qquad i = 1, 2, \ldots, n \tag{9.28}$$

PARAMETER ESTIMATION

Upon differentiating both sides of each of Equations 9.27 and 9.28 with respect to θ, we have

$$1 = \int_{-\infty}^{\infty} \cdots \int_{-\infty}^{\infty} h(x_1, \ldots, x_n) \left[\sum_{j=1}^{n} \frac{1}{f(x_j; \theta)} \frac{\partial f(x_j; \theta)}{\partial \theta} \right]$$
$$\times f(x_1; \theta) \cdots f(x_n; \theta) dx_1 \cdots dx_n$$
$$= \int_{-\infty}^{\infty} \cdots \int_{-\infty}^{\infty} h(x_1, \ldots, x_n) \left[\sum_{j=1}^{n} \frac{\partial \ln f(x_j; \theta)}{\partial \theta} \right]$$
$$\times f(x_1; \theta) \cdots f(x_n; \theta) dx_1 \cdots dx_n \qquad (9.29)$$

$$0 = \int_{-\infty}^{\infty} \frac{\partial f(x_i; \theta)}{\partial \theta} dx_i$$
$$= \int_{-\infty}^{\infty} \frac{\partial \ln f(x_i; \theta)}{\partial \theta} f(x_i; \theta) dx_i \qquad i = 1, 2, \ldots, n \qquad (9.30)$$

Let us define a new r.v. Y by

$$Y = \sum_{j=1}^{n} \frac{\partial \ln f(X_j; \theta)}{\partial \theta} \qquad (9.31)$$

Equation 9.30 shows that

$$E\{Y\} = 0$$

Moreover, since Y is a sum of n independent random variables each with mean zero and variance $E\{[\partial \ln f(X; \theta)/\partial \theta]^2\}$, the variance of Y is the sum of the n variances and has the form

$$\sigma_Y^2 = nE\left\{ \left[\frac{\partial \ln f(X; \theta)}{\partial \theta} \right]^2 \right\} \qquad (9.32)$$

Now, it follows from Equation 9.29 that

$$1 = E\{\hat{\Theta} Y\} \qquad (9.33)$$

and recall that

$$E\{\hat{\Theta} Y\} = E\{\hat{\Theta}\} E\{Y\} + \rho_{\hat{\Theta} Y} \sigma_{\hat{\Theta}} \sigma_Y$$

or

$$1 = \theta(0) + \rho_{\hat{\Theta}Y}\sigma_{\hat{\Theta}}\sigma_Y \qquad (9.34)$$

As a consequence of the property $\rho^2 \leq 1$, we finally have

$$\frac{1}{\sigma_{\hat{\Theta}}^2 \sigma_Y^2} \leq 1$$

or, using Equation 9.32,

$$\sigma_{\hat{\Theta}}^2 \geq \frac{1}{\sigma_Y^2} = \frac{1}{nE\left\{\left[\dfrac{\partial \ln f(X;\theta)}{\partial \theta}\right]^2\right\}} \qquad (9.35)$$

The proof is now complete.

In the above, we have assumed that differentiations with respect to θ under an integral or sum sign is permissible. Equation 9.26 gives a lower bound on the variance of any unbiased estimator and it expresses a fundamental limitation on the accuracy with which a parameter can be estimated. We also note that this lower bound is in general a function of θ, the true parameter value.

Several remarks in connection with the Cramér–Rao lower bound (CRLB) are now in order.

1. The expectation in Equation 9.26 is equivalent to $-E\{\partial^2 \ln f(X;\theta)/\partial \theta^2\}$ or

$$\sigma_{\hat{\Theta}}^2 \geq \frac{-1}{nE\{\partial^2 \ln f(X;\theta)/\partial \theta^2\}} \qquad (9.36)$$

   This alternate expression offers computational advantages in some cases.

2. The result can be extended easily to multiple parameter cases. Let $\theta_1, \theta_2, \ldots,$ and θ_m $(m \leq n)$ be the unknown parameters in $f(x;\theta_1,\ldots,\theta_m)$, which are to be estimated on the basis of a sample of size n. In vector notation, we can write

$$\boldsymbol{\theta}^T = [\theta_1 \quad \theta_2 \quad \cdots \quad \theta_m] \qquad (9.37)$$

with the corresponding vector unbiased estimator

$$\hat{\boldsymbol{\Theta}}^T = [\hat{\Theta}_1 \quad \hat{\Theta}_2 \quad \cdots \quad \hat{\Theta}_m] \qquad (9.38)$$

PARAMETER ESTIMATION

Following similar steps in the derivation of Equation 9.26, we can show that the Cramér–Rao inequality for multiple parameters is of the form

$$\operatorname{cov}\{\hat{\Theta}\} \geq \frac{\Lambda^{-1}}{n} \qquad (9.39)$$

where Λ^{-1} is the inverse of the matrix Λ whose elements are

$$\Lambda_{ij} = E\left\{\frac{\partial \ln f(X;\theta)}{\partial \theta_i} \frac{\partial \ln f(X;\theta)}{\partial \theta_j}\right\} \quad i,j = 1, 2, \ldots, m \qquad (9.40)$$

Equation 9.39 implies that

$$\operatorname{var}\{\hat{\Theta}_j\} \geq \frac{(\Lambda^{-1})_{jj}}{n} \geq \frac{1}{n\Lambda_{jj}} \quad j = 1, 2, \ldots, m \qquad (9.41)$$

where $(\Lambda^{-1})_{jj}$ is the jjth element of Λ^{-1}.

3. The CRLB can be transformed easily under a transformation of the parameter. Suppose that, instead of θ, the parameter $\phi = g(\theta)$ is one to one and differentiable with respect to θ, then

$$\text{CRLB for } \operatorname{var}\{\hat{\Phi}\} = \left[\frac{dg(\theta)}{d\theta}\right]^2 [\text{CRLB for } \operatorname{var}(\hat{\Theta})] \qquad (9.42)$$

where $\hat{\Phi}$ is an unbiased estimator for ϕ.

4. Given an unbiased estimator $\hat{\Theta}$ for the parameter θ, the ratio of its CRLB to its variance is called the *efficiency* of $\hat{\Theta}$. The efficiency of any unbiased estimator is thus always less than or equal to 1. An unbiased estimator with efficiency equal to 1 is said to be *efficient*. We must point out, however, efficient estimators exist only under certain conditions.

We are finally in the position to answer the question posed in Example 9.2. First, we note that, in order to apply the CRLB, the pdf $f(x;\theta)$ of the population X must be known. Suppose that $f(x; m)$ in Example 9.2 is according to $N(m, \sigma^2)$ where m is the only unknown parameter. We have

$$\ln f(X; m) = \ln\left[\frac{1}{\sqrt{2\pi}\sigma} e^{-(X-m)^2/2\sigma^2}\right]$$

$$= \ln\left[\frac{1}{\sqrt{2\pi}\sigma}\right] - \frac{(X-m)^2}{2\sigma^2}$$

and

$$\frac{\partial \ln f(X;m)}{\partial m} = \frac{X-m}{\sigma^2}$$

Thus,

$$E\left\{\left[\frac{\partial \ln f(X;m)}{\partial m}\right]^2\right\} = \frac{1}{\sigma^4} E\{(X-m)^2\} = \frac{1}{\sigma^2}$$

Equation 9.26 then shows that the CRLB for the variance of any unbiased estimator for m is σ^2/n. Since the variance of $\bar{X}$ is σ^2/n, it has the minimum variance among all unbiased estimators for m when the population X is distributed normally with m as the only unknown parameter.

Example 9.3. Consider a population X having a normal distribution $N(0, \sigma^2)$ where σ^2 is an unknown parameter to be estimated from a sample of size $n > 1$.

(a) Determine the CRLB for the variance of any unbiased estimator for σ^2.
(b) Is the sample variance S^2 an efficient estimator for σ^2?

Let us denote σ^2 by θ. Then

$$f(X;\theta) = \frac{1}{\sqrt{2\pi\theta}} e^{-X^2/2\theta}$$

and

$$\ln f(X;\theta) = -\frac{X^2}{2\theta} - \tfrac{1}{2}\ln 2\pi\theta$$

$$\frac{\partial \ln f(X;\theta)}{\partial \theta} = \frac{X^2}{2\theta^2} - \frac{1}{2\theta}$$

$$\frac{\partial^2 \ln f(X;\theta)}{\partial \theta^2} = -\frac{X^2}{\theta^3} + \frac{1}{2\theta^2}$$

$$E\left\{\frac{\partial^2 \ln f(X;\theta)}{\partial \theta^2}\right\} = -\frac{\theta}{\theta^3} + \frac{1}{2\theta^2} = -\frac{1}{2\theta^2}$$

Hence, according to Equation 9.36, the CRLB for the variance of any unbiased estimator for θ is $2\theta^2/n$.

PARAMETER ESTIMATION

For S^2, it has been shown in Section IX.1.2 that it is an unbiased estimator for θ and its variance is (see Equation 9.10)

$$\text{var}\{S^2\} = \frac{1}{n}\left(\mu_4 - \frac{n-3}{n-1}\sigma^4\right)$$

$$= \frac{1}{n}\left(3\sigma^4 - \frac{n-3}{n-1}\sigma^4\right)$$

$$= \frac{2\sigma^4}{n-1} = \frac{2\theta^2}{n-1}$$

since $\mu_4 = 3\sigma^4$ when X is normally distributed. The efficiency of S^2, denoted by $e(S^2)$, is thus

$$e(S^2) = \frac{\text{CRLB}}{\text{var}(S^2)} = \frac{n-1}{n}$$

We see that the sample variance is not an efficient estimator for θ in this case. It is, however, *asymptotically efficient* in the sense that $e(S^2) \to 1$ as $n \to \infty$.

Example 9.4. Determine the CRLB for the variance of any unbiased estimator for θ in the lognormal distribution

$$f(x;\theta) = \frac{1}{x\sqrt{2\pi\theta}}\exp\left(-\frac{1}{2\theta}\ln^2 x\right) \quad x \geq 0 \quad \theta > 0$$

$$= 0 \quad \text{elsewhere}$$

We have

$$\frac{\partial \ln f(X;\theta)}{\partial \theta} = -\frac{1}{2\theta} + \frac{\ln^2 X}{2\theta^2}$$

$$\frac{\partial^2 \ln f(X;\theta)}{\partial \theta^2} = \frac{1}{2\theta^2} - \frac{\ln^2 X}{\theta^3}$$

$$E\left\{\frac{\partial^2 \ln f(X;\theta)}{\partial \theta^2}\right\} = \frac{1}{2\theta^2} - \frac{\theta}{\theta^3} = -\frac{1}{2\theta^2}$$

It thus follows from Equation 9.36 that the desired CRLB is $2\theta^2/n$.

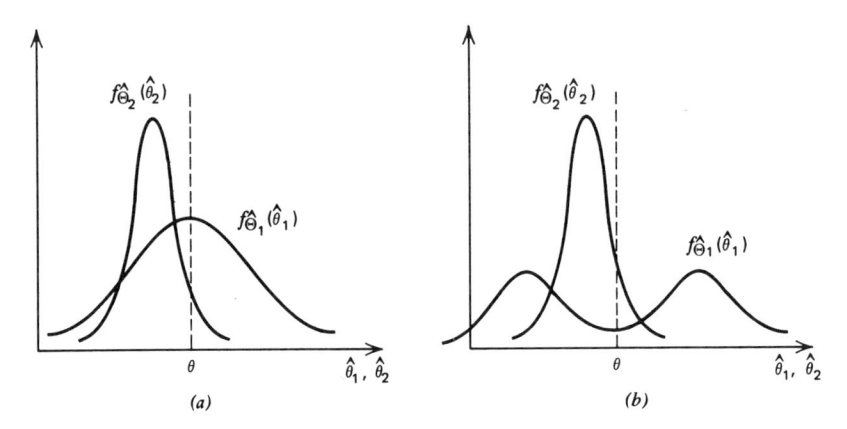

Figure 9.2 Probability density functions of $\hat{\Theta}_1$ and $\hat{\Theta}_2$.

Before going to the next criterion, it is worth mentioning again that, while unbiasedness as well as small variance is desirable, it does not mean that we should discard all biased estimators as inferior. Consider two estimators for a parameter θ, $\hat{\Theta}_1$ and $\hat{\Theta}_2$, whose pdf's are depicted in Figure 9.2a. Although $\hat{\Theta}_2$ is biased, because of its smaller variance, the probability of an observed value of $\hat{\Theta}_2$ being closer to the true value θ can well be higher than that associated with an observed value of $\hat{\Theta}_1$. Hence, one can argue convincingly that $\hat{\Theta}_2$ is the better estimator of the two. A more dramatic situation is shown in Figure 9.2b. Clearly, based on a particular sample of size n, an observed value of $\hat{\Theta}_2$ will likely be closer to the true value θ than that of $\hat{\Theta}_1$ although $\hat{\Theta}_1$ is again unbiased. It is worthwhile for us to reiterate our remark advanced in Section IX.2.1 that the quality of an estimator does not rest on any single criterion but on a combination of them.

Example 9.5. To illustrate the point that unbiasedness can be outweighed by other considerations, consider the problem of estimating the parameter θ in the binomial distribution

$$p_X(k) = \theta^k(1-\theta)^{1-k} \quad k = 0, 1 \tag{9.43}$$

Let us propose two estimators $\hat{\Theta}_1$ and $\hat{\Theta}_2$ for θ given by

$$\hat{\Theta}_1 = \overline{X}$$
$$\hat{\Theta}_2 = \frac{n\overline{X} + 1}{n + 2} \tag{9.44}$$

where $\bar{X}$ is the sample mean based on a sample of size n. The choice of $\hat{\Theta}_1$ is intuitively obvious and the choice of $\hat{\Theta}_2$ is based on a prior probability argument that is not our concern at this point.

Since

$$E\{\bar{X}\} = \theta \quad \text{and} \quad \sigma_{\bar{X}}^2 = \frac{\theta(1-\theta)}{n}$$

We have

$$E\{\hat{\Theta}_1\} = \theta$$

$$E\{\hat{\Theta}_2\} = \frac{n\theta + 1}{n+2}$$

(9.45)

and

$$\sigma_{\hat{\Theta}_1}^2 = \frac{\theta(1-\theta)}{n}$$

$$\sigma_{\hat{\Theta}_2}^2 = \frac{n^2}{(n+2)^2}\sigma_{\bar{X}}^2 = \frac{n\theta(1-\theta)}{(n+2)^2}$$

(9.46)

We see from the above that, although $\hat{\Theta}_2$ is a biased estimator, its variance is smaller than that of $\hat{\Theta}_1$, particularly when n is of a moderate value. This is a valid reason for choosing $\hat{\Theta}_2$ as a better estimator as compared with $\hat{\Theta}_1$ for θ in certain cases.

IX.2.3 CONSISTENCY

An estimator $\hat{\Theta}$ is said to be a *consistent* estimator for θ if, as sample size n increases,

$$\boxed{\lim_{n \to \infty} P[|\hat{\Theta} - \theta| \geq \varepsilon] = 0} \qquad (9.47)$$

for all $\varepsilon > 0$. The consistency condition states that the estimator $\hat{\Theta}$ converges in the sense above to the true value θ as sample size increases. It is thus a large-sample concept and is a good quality for an estimator to have.

Example 9.6. Show that the estimator S^2 in Example 9.3 is a consistent estimator for σ^2.

Using Chebyshev inequality defined in Section IV.2, we can write

$$P\{|S^2 - \sigma^2| \geq \varepsilon\} \leq \frac{1}{\varepsilon^2} E\{(S^2 - \sigma^2)^2\}$$

We have shown that $E\{S^2\} = \sigma^2$ and $\text{var}\{S^2\} = 2\theta^2/(n-1)$. Hence,

$$\lim_{n \to \infty} P\{|S^2 - \sigma^2| \geq \varepsilon\} \leq \lim_{n \to \infty} \frac{1}{\varepsilon^2}\left(\frac{2\theta^2}{n-1}\right) = 0$$

Thus S^2 is a consistent estimator for σ^2.

The example above gives an expedient procedure for checking whether an estimator is consistent. We shall state this procedure as a theorem below. It is important to note that this theorem gives a *sufficient*, but not *necessary*, condition for consistency.

Theorem

Let $\hat{\Theta}$ be an estimator for θ based on a sample of size n. Then, if

$$\lim_{n \to \infty} E\{\hat{\Theta}\} = \theta \quad \text{and} \quad \lim_{n \to \infty} \text{var}\{\hat{\Theta}\} = 0 \quad (9.48)$$

the estimator $\hat{\Theta}$ is a consistent estimator for θ.

The proof is essentially given in Example 9.6 and will not be repeated here.

IX.2.4 SUFFICIENCY

Let $X_1, X_2, \ldots, X_n$ be a sample of a population X whose distribution depends on unknown parameter θ. If $Y = h(X_1, X_2, \ldots, X_n)$ is a statistic such that, for any other statistic $Z = g(X_1, X_2, \ldots, X_n)$, the conditional distribution of Z given $Y = y$ does not depend on θ, then Y is called a *sufficient statistic* for θ. If also $E\{Y\} = \theta$, then Y is said to be a *sufficient estimator* for θ.

In words, the definition for sufficiency states that, if Y is a sufficient statistic for θ, all the sample information concerning θ is contained in Y. A sufficient statistic is thus of interest in that, if it can be found for a parameter, then an estimator based on this statistic is able to make use of all the information that the sample contains regarding the value of the unknown parameter. Moreover, an important property of a sufficient estimator is that, starting with any unbiased estimator of a parameter θ which is not a function of the sufficient estimator, it is possible to find an unbiased estimator based on the sufficient statistic which has a smaller variance

PARAMETER ESTIMATION 271

than that of the initial estimator. Sufficient estimators thus have variances that are smaller than any other unbiased estimators that do not depend on sufficient statistics.

If a sufficient statistic for a parameter θ exists, the theorem stated below without proof provides an easy way of finding it.

Theorem (Fisher–Neyman Factorization Criterion)

Let $Y = h(X_1, X_2, \ldots, X_n)$ be a statistic based on a sample of size n. Then Y is a sufficient statistic for θ if and only if the joint probability density function of $X_1, X_2, \ldots,$ and $X_n, f_X(x_1; \theta) \cdots f_X(x_n; \theta)$, can be factorized in the form

$$\prod_{j=1}^{n} f_X(x_j; \theta) = g_1[h(x_1, \ldots, x_n), \theta]g_2(x_1, \ldots, x_n) \qquad (9.49)$$

If X is discrete, we have

$$\prod_{j=1}^{n} p_X(x_j; \theta) = g_1[h(x_1, \ldots, x_n), \theta]g_2(x_1, \ldots, x_n) \qquad (9.50)$$

The sufficiency of the factorization criterion was first pointed out by Fisher (1922). Neyman (1935) showed that it is also necessary.

The foregoing results can be extended to the multiple parameter case. Let $\boldsymbol{\theta}^T = [\theta_1 \cdots \theta_m]$, $m \leq n$, be the parameter vector. Then $Y_1 = h_1(X_1, \ldots, X_n), \ldots, Y_r = h_r(X_1, \ldots, X_n)$, $r \geq m$, is a set of sufficient statistics for $\boldsymbol{\theta}$ if and only if

$$\prod_{j=1}^{n} f_X(x_j; \boldsymbol{\theta}) = g_1[\mathbf{h}(x_1, \ldots, x_n), \boldsymbol{\theta}]g_2(x_1, \ldots, x_n) \qquad (9.51)$$

where $\mathbf{h}^T = [h_1 \cdots h_r]$. Similar expression holds when X is discrete.

Example 9.7. Let us show that the statistic $\overline{X}$ is a sufficient statistic for θ in Example 9.5. In this case,

$$\prod_{j=1}^{n} p_X(x_j; \theta) = \prod_{j=1}^{n} \theta^{x_j}(1-\theta)^{1-x_j}$$
$$= \theta^{\sum x_j}(1-\theta)^{n-\sum x_j} \qquad (9.52)$$

We see that the jpmf is a function of $\sum x_j$ and θ. If we let $Y = \sum_{j=1}^{n} X_j$, the jpmf of $X_1, \ldots,$ and X_n takes the form given by Equation 9.50 with

$$g_1 = \theta^{\sum x_j}(1-\theta)^{n-\sum x_j} \quad \text{and} \quad g_2 = 1$$

In this example, $\sum_{j=1}^{n} X_j$ is thus a sufficient statistic for θ. We have seen in Example 9.5 that both $\hat{\Theta}_1 = \bar{X}$ and $\hat{\Theta}_2 = (n\bar{X} + 1)/(n + 2)$ are based on this sufficient statistic. Furthermore $\hat{\Theta}_1$, being unbiased, is a sufficient estimator for θ.

Example 9.8. Suppose $X_1, X_2, \ldots,$ and X_n are a sample taken from a Poisson distribution

$$p_X(k; \theta) = \frac{\theta^k e^{-\theta}}{k!} \qquad k = 0, 1, 2, \ldots \tag{9.53}$$

where θ is the unknown parameter. We have

$$\prod_{j=1}^{n} p_X(x_j; \theta) = \frac{\theta^{\sum x_j} e^{-n\theta}}{\prod x_j!} \tag{9.54}$$

which can be factorized in the form of Equation 9.50 by letting

$$g_1 = \theta^{\sum x_j} e^{-n\theta} \quad \text{and} \quad g_2 = \frac{1}{\prod x_j!}$$

It is seen that $Y = \sum_{j=1}^{n} X_j$ is a sufficient statistic for θ.

IX.3 Methods of Estimation

Based on the estimation criteria defined in Section IX.2, some estimation techniques that yield "good," and sometimes "best," estimates of distribution parameters are now developed.

Two approaches to the parameter estimation problem are discussed in what follows: point estimation and interval estimation. In point estimation, we use certain prescribed methods to arrive at a number $\hat{\theta}$ as a function of observed data that we accept as a "good" estimate of θ, good in terms of unbiasedness, minimum variance, etc., as defined by the estimation criteria.

In many scientific studies it is more useful to obtain information about a parameter beyond a single number as its estimate. The interval estimation is a procedure by which bounds of the parameter value are obtained that not only give information on the numerical value of the parameter but also an indication of the level of confidence one can place on the possible numerical value of the parameter on the basis of a sample. Point estimation will be discussed first, followed by the development of methods of interval estimation.

IX.4 Point Estimation

We now proceed to present two general methods of finding point estimators for distribution parameters on the basis of a sample from a population.

IX.4.1 THE METHOD OF MOMENTS

The oldest systematic method of point estimation was proposed by Pearson (1894) and was extensively used by him and his co-workers. It was neglected for a number of years because of its general lack of optimum properties and because of the popularity and universal appeal associated with the method of maximum likelihood. The moment method, however, appears to be regaining its acceptance primarily due to its expediency in terms of computational labor and to the fact that it can be improved upon easily in certain cases.

The method of moments is simple in concept. Consider a selected probability density function $f(x; \theta_1, \theta_2, \ldots, \theta_m)$ whose parameters $\theta_j, j = 1, 2, \ldots, m$, are to be estimated based on the sample $X_1, X_2, \ldots, X_n$ of X. The theoretical or population moments of X are

$$\alpha_i = \int_{-\infty}^{\infty} x^i f(x; \theta_1, \ldots, \theta_m) dx \qquad i = 1, 2, \ldots \qquad (9.55)$$

They are, in general, functions of the unknown parameters, that is,

$$\alpha_i = \alpha_i(\theta_1, \theta_2, \ldots, \theta_m) \qquad (9.56)$$

On the other hand, sample moments of various orders can be found from the sample by (see Equation 9.14)

$$M_i = \frac{1}{n} \sum_{j=1}^{n} X_j^i \qquad i = 1, 2, \ldots \qquad (9.57)$$

The method of moments suggests that, in order to determine the estimators $\hat{\Theta}_1, \ldots,$ and $\hat{\Theta}_m$ from the sample, we equate a sufficient number of the sample moments to the corresponding population moments. By establishing and solving as many resulting moment equations as there are parameters to be estimated, estimators for the parameters are obtained. Hence, the procedure for determining $\hat{\Theta}_1, \hat{\Theta}_2, \ldots,$ and $\hat{\Theta}_m$ consists of the following steps.

1. Let

$$\boxed{\alpha_i(\hat{\Theta}_1, \ldots, \hat{\Theta}_m) = M_i \qquad i = 1, 2, \ldots, m} \qquad (9.58)$$

These yield m moment equations in m unknowns $\hat{\Theta}_j, j = 1, \ldots, m$.

2. Solve for $\hat{\Theta}_j, j = 1, \ldots, m$, from this system of equations. These are called the *moment estimators* for $\theta_1, \ldots,$ and θ_m.

Let us remark that it is not necessary to consider m consecutive moment equations as indicated by Equations 9.58; any convenient set of m equations that lead to the solutions for $\hat{\Theta}_j, j = 1, \ldots, m$, is sufficient. Lower-order moment equations are preferred, however, since they require less manipulation of observed data.

An attractive feature of the method of moments is that the moment equations are straightforward to establish, and there is seldom any difficulty in solving them. On the other hand, a shortcoming is that such desirable properties as unbiasedness or efficiency are not generally guaranteed for estimators so obtained.

However, consistency of moment estimators can be established under general conditions. In order to show this, let us consider a single parameter θ whose moment estimator $\hat{\Theta}$ satisfies the moment equation

$$\alpha_i(\hat{\Theta}) = M_i \tag{9.59}$$

for some i. The solution of Equation 9.59 for $\hat{\Theta}$ can be represented by $\hat{\Theta} = \hat{\Theta}(M_i)$, whose Taylor's expansion about $\alpha_i(\theta)$ gives

$$\hat{\Theta} = \theta + \hat{\Theta}^{(1)}[\alpha_i(\theta)][M_i - \alpha_i(\theta)]$$
$$+ \frac{\hat{\Theta}^{(2)}[\alpha_i(\theta)]}{2!}[M_i - \alpha_i(\theta)]^2 + \cdots \tag{9.60}$$

where the superscript (k) denotes the kth derivative with respect to M_i. Upon performing successive differentiation of Equation 9.59 with respect to M_i, Equation 9.60 becomes

$$\hat{\Theta} - \theta = [M_i - \alpha_i(\theta)]\left[\frac{d\alpha_i(\theta)}{d\theta}\right]^{-1}$$
$$- \tfrac{1}{2}[M_i - \alpha_i(\theta)]^2\left[\frac{d^2\alpha_i(\theta)}{d\theta^2}\right]\left[\frac{d\alpha_i(\theta)}{d\theta}\right]^{-3} + \cdots \tag{9.61}$$

The bias and variance of $\hat{\Theta}$ can be found by taking the expectation of Equation 9.61 and the expectation of the square of Equation 9.61, respectively. Up to the order of $1/n$, we find

$$E\{\hat{\Theta}\} - \theta = -\frac{1}{2n}(\alpha_{2i} - \alpha_i^2)\left(\frac{d^2\alpha_i}{d\theta^2}\right)\left(\frac{d\alpha_i}{d\theta}\right)^{-3}$$
$$\text{var}\{\hat{\Theta}\} = \frac{1}{n}(\alpha_{2i} - \alpha_i^2)\left(\frac{d\alpha_i}{d\theta}\right)^{-2} \tag{9.62}$$

PARAMETER ESTIMATION

Assuming that all the indicated moments and their derivatives exist, Equations 9.62 show that

$$\lim_{n \to \infty} E\{\hat{\Theta}\} = \theta \quad \text{and} \quad \lim_{n \to \infty} \text{var}\{\hat{\Theta}\} = 0$$

and hence $\hat{\Theta}$ is consistent.

Example 9.9. Let us select normal distribution as a model for the percent yield discussed in Chapter VIII, that is,

$$f(x; m, \sigma^2) = \frac{1}{\sqrt{2\pi}\sigma} \exp\left[-\frac{(x-m)^2}{2\sigma^2}\right] \quad -\infty < x < \infty \quad (9.63)$$

We wish to estimate the parameters $\theta_1 = m$ and $\theta_2 = \sigma^2$ based on the 200 sample values given in Table 8.1.

Following the method of moments, we need two moment equations and the most convenient ones are obviously

$$\alpha_1 = M_1 = \bar{X} \quad \text{and} \quad \alpha_2 = M_2$$

Now,

$$\alpha_1 = \theta_1$$

Hence, the first of these moment equations gives

$$\hat{\Theta}_1 = \bar{X} = \frac{1}{n} \sum_{j=1}^{n} X_j \quad (9.64)$$

The properties of this estimator have already been discussed in Example 9.2. It is unbiased and has minimum variance among all unbiased estimators for m. We see that the method of moments produces desirable results in this case.

The second moment equation gives

$$\hat{\Theta}_1^2 + \hat{\Theta}_2 = M_2 = \frac{1}{n} \sum_{j=1}^{n} X_j^2$$

or

$$\hat{\Theta}_2 = M_2 - M_1^2 = \frac{1}{n} \sum_{j=1}^{n} (X_j - \bar{X})^2 \quad (9.65)$$

This, as we have shown, is a biased estimator for σ^2.

The estimates $\hat{\theta}_1$ and $\hat{\theta}_2$ of $\theta_1 = m$ and $\theta_2 = \sigma^2$ based on the sample values given by Table 8.1 are, following Equations 9.64 and 9.65,

$$\hat{\theta}_1 = \frac{1}{200} \sum_{j=1}^{200} x_j \cong 70$$

$$\hat{\theta}_2 = \frac{1}{200} \sum_{j=1}^{200} (x_j - \hat{\theta}_1)^2 \cong 4$$

where $x_j, j = 1, 2, \ldots, 200$, are sample values given in Table 8.1.

Example 9.10. Consider the binomial distribution

$$p_X(k; p) = p^k(1 - p)^{1-k} \qquad k = 0, 1 \tag{9.66}$$

We wish to estimate the parameter p based on a sample of size n.

The method of moments suggests that we determine the estimator for p, $\hat{P}$, by equating α_1 to $M_1 = \overline{X}$. Since

$$\alpha_1 = E\{X\} = p$$

we have

$$\hat{P} = \overline{X} \tag{9.67}$$

The mean of $\hat{P}$ is

$$E\{\hat{P}\} = \frac{1}{n} \sum_{j=1}^{n} E\{X_j\} = p \tag{9.68}$$

Hence it is an unbiased estimator. Its variance is given by

$$\text{var}\{\hat{P}\} = \text{var}\{\overline{X}\} = \frac{\sigma^2}{n} = \frac{p(1-p)}{n} \tag{9.69}$$

It is easy to derive the CRLB for this case and show that the $\hat{P}$ defined by Equation 9.67 is also efficient.

Example 9.11. A set of 214 observed gaps in traffic on a section of Arroyo Seco Freeway is given in Table 9.1. If the exponential density function

$$f(t; \lambda) = \lambda e^{-\lambda t} \qquad t \geq 0 \tag{9.70}$$

is proposed for the gap, we wish to determine parameter λ from the data.

Table 9.1. *Observed Traffic Gaps on Arroyo Seco Freeway*

Gap Length (sec)	No. of Gaps	Gap Length (sec)	No. of Gaps
0–1	18	16–17	6
1–2	25	17–18	4
2–3	21	18–19	3
3–4	13	19–20	3
4–5	11	20–21	1
5–6	15	21–22	1
6–7	16	22–23	1
7–8	12	23–24	0
8–9	11	24–25	1
9–10	11	25–26	0
10–11	8	26–27	1
11–12	12	27–28	1
12–13	6	28–29	1
13–14	3	29–30	2
14–15	3	30–31	1
15–16	3		

Taken from Gerlough (1955).

In this case,

$$\alpha_1 = \frac{1}{\lambda}$$

and, following the method of moments, the simplest estimator $\hat{\Lambda}$ for λ is obtained from

$$\alpha_1 = \overline{X} \quad \text{or} \quad \hat{\Lambda} = \frac{1}{\overline{X}} \tag{9.71}$$

Hence, the desired estimate is

$$\hat{\lambda} = \left(\frac{1}{214} \sum_{j=1}^{214} x_j\right)^{-1}$$

$$= \frac{214}{18(0.5) + 25(1.5) + \cdots + 1(30.5)}$$

$$= 0.13 \text{ sec}^{-1} \tag{9.72}$$

Let us note that, although $\overline{X}$ is an unbiased estimator for α_1, the estimator for λ obtained above is not unbiased since

$$E\left\{\frac{1}{\overline{X}}\right\} \neq \frac{1}{E\{\overline{X}\}}$$

Example 9.12. Suppose that population X has a uniform distribution over the range $(0, \theta)$ and we wish to estimate the parameter θ from a sample of size n. The density function is

$$f(x; \theta) = \frac{1}{\theta} \quad 0 \leq x \leq \theta \quad (9.73)$$
$$= 0 \quad \text{elsewhere}$$

and the first moment is

$$\alpha_1 = \frac{\theta}{2} \quad (9.74)$$

If follows from the method of moments that, on letting $\alpha_1 = \overline{X}$, we obtain

$$\hat{\Theta} = 2\overline{X} = \frac{2}{n}\sum_{j=1}^{n} X_j \quad (9.75)$$

Upon little reflection, the validity of this estimator is somewhat questionable because, by definition, all values assumed by X are supposed to lie within the interval $(0, \theta)$. However, we see from Equation 9.75 that it is possible that some of the sample are greater than $\hat{\Theta}$. Intuitively, a better estimator might be

$$\hat{\Theta} = X_{(n)} \quad (9.76)$$

where $X_{(n)}$ is the nth-order statistic. As we will see, this would be the outcome following the method of maximum likelihood.

Since the method of moments requires only α_i, the moments of population X, the knowledge of its probability density function is not necessary. This advantage is demonstrated in Example 9.13.

Example 9.13. Consider measurement of the length r of an object using a sensing instrument. Due to inherent inaccuracies in the instrument, what is actually measured is X as shown in Figure 9.3 where X_1 and X_2 are identically and normally

PARAMETER ESTIMATION 279

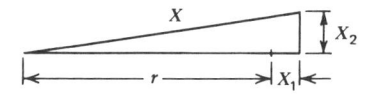

Figure 9.3 Measurement X in Example 9.13.

distributed with mean zero and unknown variance σ^2. Determine a moment estimator $\hat{\Theta}$ for $\theta = r^2$ on the basis of a sample of size n from X.

Now the r.v. X is

$$X = [(r + X_1)^2 + X_2^2]^{1/2} \tag{9.77}$$

The pdf of X with unknown parameters θ and σ^2 can be found using techniques developed in Chapter V. It is, however, unnecessary here since some moments of X can be directly generated from Equation 9.77. We remark that, although an estimator for σ^2 is not required, it is nevertheless an unknown parameter and must be considered together with θ. In applied literature, an unknown parameter whose value is of no interest is sometimes referred to as a *nuisance parameter*.

Two moment equations are needed in this case. However, we see from Equation 9.77 that the odd-order moments of X are quite complicated. For simplicity, the second-order and fourth-order moment equations will be used. We easily obtain from Equation 9.77

$$\begin{aligned}\alpha_2 &= \theta + 2\sigma^2 \\ \alpha_4 &= \theta^2 + 8\theta\sigma^2 + 8\sigma^4\end{aligned} \tag{9.78}$$

The two moment equations are

$$\begin{aligned}\hat{\Theta} + 2\widehat{\Sigma^2} &= M_2 \\ \hat{\Theta}^2 + 8\hat{\Theta}\widehat{\Sigma^2} + 8\widehat{\Sigma^2}^2 &= M_4\end{aligned} \tag{9.79}$$

Solving for $\hat{\Theta}$, we have

$$\hat{\Theta} = (2M_2^2 - M_4)^{1/2} \tag{9.80}$$

Incidentally, a moment estimator $\widehat{\Sigma^2}$ for σ^2, if needed, is obtained from Equations 9.79 to be

$$\widehat{\Sigma^2} = \tfrac{1}{2}(M_2 - \hat{\Theta}) \tag{9.81}$$

Combined Moment Estimators. Let us take another look at Example 9.11 for the purpose of motivating the following development. In this example, an estimator for λ has been obtained by using the first-order moment equation. Based on the same sample, one can obtain additional moment estimators for λ by using higher-order moment equations. For example, since $\alpha_2 = 2/\lambda^2$, the second-order moment equation

$$\alpha_2 = M_2$$

produces a moment estimator $\hat{\Lambda}$ for λ in the form

$$\hat{\Lambda} = \sqrt{\frac{2}{M_2}} \qquad (9.82)$$

While this estimator may be inferior to $1/\bar{X}$ in terms of the quality criteria we have established, an interesting question arises: Given two or more moment estimators, can they be combined to yield an estimator superior to any of the individual moment estimators?

In what follows, we consider a combined moment estimator derived from an optimal linear combination of a set of moment estimators. Let $\hat{\Theta}^{(1)}, \hat{\Theta}^{(2)}, \ldots, \hat{\Theta}^{(p)}$ be p moment estimators for the same parameter θ. We seek a combined estimator Θ_p^* in the form

$$\Theta_p^* = w_1 \hat{\Theta}^{(1)} + \cdots + w_p \hat{\Theta}^{(p)} \qquad (9.83)$$

where the coefficients $w_1, \ldots,$ and w_p are to be chosen in such a way that it is unbiased if $\hat{\Theta}^{(j)}, j = 1, 2, \ldots, p$, are unbiased and the variance of Θ_p^* is minimized.

The unbiasedness condition requires that

$$w_1 + \cdots + w_p = 1 \qquad (9.84)$$

We thus wish to determine the coefficients w_j by minimizing

$$Q = \text{var}\{\Theta_p^*\} = \text{var}\left\{\sum_{j=1}^{p} w_j \hat{\Theta}^{(j)}\right\} \qquad (9.85)$$

subject to Equation 9.84.

Let $\mathbf{u}^T = [1 \cdots 1], \hat{\mathbf{\Theta}}^T = [\hat{\Theta}^{(1)} \cdots \hat{\Theta}^{(p)}]$, and $\mathbf{w}^T = [w_1 \cdots w_p]$. Equations 9.84 and 9.85 can be written in the vector-matrix notation

$$\mathbf{w}^T \mathbf{u} = 1 \qquad (9.86)$$

PARAMETER ESTIMATION

and

$$Q(\mathbf{w}) = \text{var}\left\{\sum_{j=1}^{p} w_j \hat{\Theta}^{(j)}\right\} = \mathbf{w}^T \Lambda \mathbf{w} \qquad (9.87)$$

where $\Lambda = [\lambda_{ij}]$ with $\lambda_{ij} = \text{cov}\{\hat{\Theta}^{(i)}, \hat{\Theta}^{(j)}\}$.

In order to minimize Equation 9.87 subject to Equation 9.86, we consider

$$Q_1(\mathbf{w}) = \mathbf{w}^T \Lambda \mathbf{w} - \mathbf{w}^T \mathbf{u} \lambda - \lambda \mathbf{u}^T \mathbf{w} \qquad (9.88)$$

where λ is the Lagrange multiplier. Taking the variation of Equation 9.88 we obtain

$$\delta Q_1(\mathbf{w}) = \delta \mathbf{w}^T (\Lambda \mathbf{w} - \mathbf{u}\lambda) + (\mathbf{w}^T \Lambda - \lambda \mathbf{u}^T)\delta \mathbf{w} = 0$$

as a condition of extreme. Since $\delta \mathbf{w}$ and $\delta \mathbf{w}^T$ are arbitrary, we require that

$$\Lambda \mathbf{w} - \mathbf{u}\lambda = 0 \quad \text{and} \quad \mathbf{w}^T \Lambda - \lambda \mathbf{u}^T = 0 \qquad (9.89)$$

and either of these two relations gives

$$\mathbf{w}^T = \lambda \mathbf{u}^T \Lambda^{-1} \qquad (9.90)$$

The constraint Equation 9.86 is now used to determine λ. It implies that

$$\mathbf{w}^T \mathbf{u} = \lambda \mathbf{u}^T \Lambda^{-1} \mathbf{u} = 1$$

or

$$\lambda = \frac{1}{\mathbf{u}^T \Lambda^{-1} \mathbf{u}} \qquad (9.91)$$

Hence, we have from Equations 9.90 and 9.91

$$\mathbf{w}^T = \frac{\mathbf{u}^T \Lambda^{-1}}{\mathbf{u}^T \Lambda^{-1} \mathbf{u}} \qquad (9.92)$$

The variance of Θ_p^* is

$$\text{var}\{\Theta_p^*\} = \mathbf{w}^T \Lambda \mathbf{w} = \frac{1}{\mathbf{u}^T \Lambda^{-1} \mathbf{u}} \qquad (9.93)$$

in view of Equation 9.92.

Several attractive features are possessed by Θ_p^*. For example, we can show that its variance is smaller than or equal to that of any of the simple moment estimators $\hat{\Theta}^{(j)}, j = 1, 2, \ldots, p$, and furthermore [see Soong (1969)],

$$\text{var}\{\Theta_p^*\} \leq \text{var}\{\Theta_q^*\} \tag{9.94}$$

if $p \geq q$.

Example 9.14. Consider the problem of estimating parameter θ in the log-normal distribution

$$f(x;\theta) = \frac{1}{x\sqrt{2\pi\theta}} \exp\left[-\frac{1}{2\theta}\ln^2 x\right] \quad x \geq 0 \quad \theta > 0 \tag{9.95}$$

from a sample of size n.

Three moment estimators for θ ($\hat{\Theta}^{(1)}$, $\hat{\Theta}^{(2)}$, $\hat{\Theta}^{(3)}$) can be found by means of establishing and solving the first three moment equations. Let Θ_2^* be the combined moment estimator of $\hat{\Theta}^{(1)}$ and $\hat{\Theta}^{(2)}$, and Θ_3^* the combined estimator of all three. As we have obtained the CRLB for the variance of any unbiased estimator for θ in Example 9.4, the efficiency of each of the above estimators can be calculated. Figure 9.4 shows these efficiencies as $n \to \infty$. As we can see, a significant increase

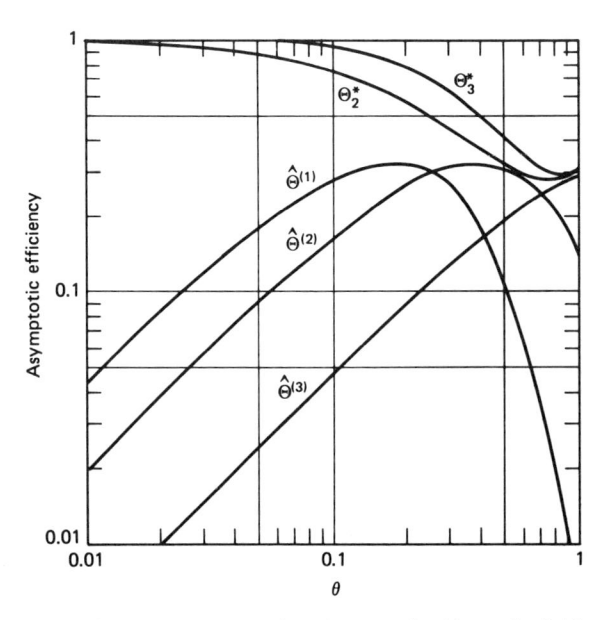

Figure 9.4 Efficiencies of estimators in Example 9.14 as $n \to \infty$.

IX.4.2 THE METHOD OF MAXIMUM LIKELIHOOD

First introduced by Fisher in 1922, the method of maximum likelihood has become the most important general method of estimation from theoretical point of view. Its greatest appeal stems from the fact that some very general properties associated with this procedure can be derived and, in the case of large samples, they are optimal properties in terms of the criteria set forth in Section IX.2.

Let $f(x;\theta)$ be the density function of population X where, for simplicity, θ is the only parameter to be estimated from a set of sample values $x_1, x_2, \ldots, x_n$. The joint density function of the corresponding sample $X_1, X_2, \ldots, X_n$ has the form

$$f(x_1;\theta)f(x_2;\theta)\cdots f(x_n;\theta)$$

We define the *likelihood function L* of a set of n sample values from the population by

$$\boxed{L(x_1, x_2, \ldots, x_n;\theta) = f(x_1;\theta)f(x_2;\theta)\cdots f(x_n;\theta)} \qquad (9.96)$$

In the case when X is discrete, we write

$$\boxed{L(x_1, x_2, \ldots, x_n;\theta) = p(x_1;\theta)p(x_2;\theta)\cdots p(x_n;\theta)} \qquad (9.97)$$

When the sample values are given, the likelihood function L becomes a function of a single variable θ. The estimation procedure for θ based on the method of maximum likelihood consists of choosing, as an estimate of θ, the particular value of θ that maximizes L. The maximum of $L(\theta)$ occurs in most cases at the value of θ where $dL(\theta)/d\theta = 0$. Hence, in a large number of cases, the *maximum likelihood estimate* (MLE) $\hat{\theta}$ of θ based on sample values $x_1, x_2, \ldots,$ and x_n can be determined from

$$\frac{dL(x_1, x_2, \ldots, x_n;\hat{\theta})}{d\hat{\theta}} = 0 \qquad (9.98)$$

As we see from Equations 9.96 and 9.97, the function L is in the form of a product of many functions of θ. Since L is always nonnegative and attains its maximum for the same value of $\hat{\theta}$ as ln L. It is generally easier to obtain the MLE $\hat{\theta}$ by solving

$$\boxed{\frac{d \ln L(x_1, \ldots, x_n; \hat{\theta})}{d\hat{\theta}} = 0} \qquad (9.99)$$

because ln L is in the form of a sum rather than a product.

Equation 9.99 is referred to as the *likelihood equation*. The desired solution is one where the root $\hat{\theta}$ is a function of x_j, $j = 1, 2, \ldots, n$, if such a root exists. When several roots of Equation 9.99 exist, the MLE is the root corresponding to the global maximum of L or ln L.

To see that this procedure is plausible, we observe that the quantity

$$L(x_1, x_2, \ldots, x_n; \theta) dx_1 \, dx_2 \cdots dx_n$$

is the probability that the sample $X_1, X_2, \ldots, X_n$ takes values in the region defined by $(x_1 + dx_1, x_2 + dx_2, \ldots, x_n + dx_n)$. Given the sample values, this probability gives a measure of likelihood that they are from the population. By choosing a value of θ that maximizes L, or ln L, we in fact say that we prefer the value of θ that makes as probable as possible the event that the sample values indeed come from the population.

The extension to the case of several parameters is straightforward. In the case of m parameters, the likelihood function becomes

$$L(x_1, \ldots, x_n; \theta_1, \ldots, \theta_m)$$

and the MLE's of θ_j, $j = 1, \ldots, m$, are obtained by solving simultaneously the system of likelihood equations

$$\frac{\partial \ln L}{\partial \hat{\theta}_j} = 0 \qquad j = 1, 2, \ldots, m \qquad (9.100)$$

A discussion of some of the important properties associated with a maximum likelihood estimator is now in order. Let us represent the solution of the likelihood equation (9.99) by

$$\hat{\theta} = h(x_1, x_2, \ldots, x_n) \qquad (9.101)$$

The *maximum likelihood estimator* $\hat{\Theta}$ for θ is then

$$\hat{\Theta} = h(X_1, X_2, \ldots, X_n) \qquad (9.102)$$

The universal appeal enjoyed by maximum likelihood estimators stems from the optimal properties they possess when the sample size becomes large. Under mild regularity conditions imposed on the pdf or pmf of population X, two notable properties are given below without proof.

Consistency and Asymptotic Efficiency. Let $\hat{\Theta}$ be the maximum likelihood estimator for θ in the density function $f(x;\theta)$ on the basis of a sample of size n. Then, as $n \to \infty$,

$$E\{\hat{\Theta}\} \to \theta \qquad (9.103)$$

and

$$\text{var}\{\hat{\Theta}\} \to \frac{1}{nE\{[\partial \ln f(X;\theta)/\partial \theta]^2\}} \qquad (9.104)$$

Analogous results are obtained when the population X is discrete. Furthermore, the distribution of $\hat{\Theta}$ tends to a normal distribution as n becomes large.

This important result shows that MLE $\hat{\Theta}$ is consistent. Since the variance given by Equation 9.104 is equal to the Cramér–Rao lower bound, it is efficient as n becomes large, or *asymptotically efficient*. The fact that MLE $\hat{\Theta}$ is normally distributed as $n \to \infty$ is also of considerable practical interest since probability statements can be made regarding any observed value of a maximum likelihood estimator as n becomes large.

Let us remark, however, these important properties are *large sample* properties. Unfortunately, very little can be said in cases of small sample size; it may be biased and nonefficient. This lack of reasonable small-sample properties can be explained in part by the fact that MLE is based on finding the mode of a distribution by attempting to select the true parameter value. Estimators, on the other hand, are generally designed to approach the true value rather than to give an exact hit. Modes are therefore not as desirable as mean or median when the sample size is small.

Invariant Property. It can be shown that, if $\hat{\Theta}$ is the MLE of θ, then the MLE of a function of θ, say $g(\theta)$, is $g(\hat{\Theta})$, where $g(\theta)$ is assumed to represent a one-to-one transformation and be differentiable with respect to θ. This important invariance property implies that, for example, if $\hat{\Sigma}$ is the MLE of the standard deviation σ in a distribution, then the MLE of the variance σ^2, $\widehat{\Sigma^2}$, is $\hat{\Sigma}^2$.

Let us also make an observation on the solution procedure for solving likelihood equations. While it is fairly simple to establish Equation 9.99 or Equations 9.100, they are frequently highly nonlinear in the unknown estimates and close-form solutions for the MLE are sometimes difficult, if not impossible, to achieve. In many cases, iterations or numerical schemes are necessary.

Example 9.15. Let us consider Example 9.9 again and determine the MLE's of m and σ^2. The logarithm of the likelihood function is

$$\ln L = -\frac{1}{2\sigma^2} \sum_{j=1}^{n} (x_j - m)^2 - \tfrac{1}{2} n \ln \sigma^2 - \tfrac{1}{2} n \ln 2\pi \qquad (9.105)$$

Let $\theta_1 = m$ and $\theta_2 = \sigma^2$ as before, the likelihood equations are

$$\frac{\partial \ln L}{\partial \hat{\theta}_1} = \frac{1}{\hat{\theta}_2} \sum_{j=1}^{n} (x_j - \hat{\theta}_1) = 0$$

$$\frac{\partial \ln L}{\partial \hat{\theta}_2} = \frac{1}{2\hat{\theta}_2^2} \sum_{j=1}^{n} (x_j - \hat{\theta}_1)^2 - \frac{n}{2\hat{\theta}_2} = 0$$

Solving the above equations simultaneously, the maximum likelihood estimates of m and σ^2 are found to be

$$\hat{\theta}_1 = \frac{1}{n} \sum_{j=1}^{n} x_j$$

and

$$\hat{\theta}_2 = \frac{1}{n} \sum_{j=1}^{n} (x_j - \hat{\theta}_1)^2$$

The maximum likelihood estimators for m and σ^2 are, therefore,

$$\hat{\Theta}_1 = \frac{1}{n} \sum_{j=1}^{n} X_j = \bar{X}$$

$$\hat{\Theta}_2 = \frac{1}{n} \sum_{j=1}^{n} (X_j - \bar{X})^2 = \frac{n-1}{n} S^2 \qquad (9.106)$$

which coincide with their moment estimators in this case. Although $\hat{\Theta}_2$ is biased, consistency and asymptotic efficiency for both $\hat{\Theta}_1$ and $\hat{\Theta}_2$ can be easily verified.

Example 9.16. Let us determine the MLE of θ considered in Example 9.12. Now,

$$f(x; \theta) = \frac{1}{\theta} \quad 0 \le x \le \theta$$
$$= 0 \quad \text{elsewhere} \qquad (9.107)$$

PARAMETER ESTIMATION

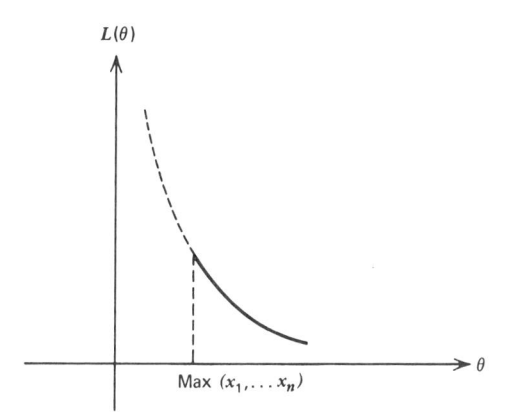

Figure 9.5 Likelihood function in Example 9.16.

The likelihood function becomes

$$L(x_1, x_2, \ldots, x_n; \theta) = \left(\frac{1}{\theta}\right)^n \quad 0 \leq x_i \leq \theta \text{ for all } i \tag{9.108}$$

A plot of L is given in Figure 9.5. However, we note from the condition associated with Equation 9.108 that all sample values x_i must be smaller than or equal to θ, this implying that only the portion of the curve to the right of $\max(x_1, \ldots, x_n)$ is applicable. Hence, the maximum of L occurs at $\theta = \max(x_1, x_2, \ldots, x_n)$ or the MLE for θ is

$$\hat{\theta} = \max(x_1, x_2, \ldots, x_n) \tag{9.109}$$

and the maximum likelihood estimator for θ is

$$\hat{\Theta} = \max(X_1, X_2, \ldots, X_n) = X_{(n)} \tag{9.110}$$

This estimator is seen to be different from that obtained using the moment method (Equation 9.75) and, as we already commented in Example 9.12, it appears to be a more logical choice.

Let us also note that we did not obtain Equation 9.109 by solving the likelihood equation. The likelihood equation does not apply in this case as the maximum of L occurs at the boundary and the derivative is not zero there.

It is instructive to study some of the properties of $\hat{\Theta}$ given by Equation 9.110. The pdf of $\hat{\Theta}$ is given by (see Equation 9.19)

$$f_{\hat{\Theta}}(x) = nF_X^{n-1}(x)f_X(x) \tag{9.111}$$

With $f_X(x)$ given by Equation 9.107 and

$$\begin{aligned} F_X(x) &= 0 & x &< 0 \\ &= \frac{x}{\theta} & 0 &\leq x \leq \theta \\ &= 1 & x &> \theta \end{aligned} \tag{9.112}$$

we have

$$\begin{aligned} f_{\hat{\Theta}}(x) &= \frac{nx^{n-1}}{\theta^n} & 0 \leq x \leq \theta \\ &= 0 & \text{elsewhere} \end{aligned} \tag{9.113}$$

The mean and variance of $\hat{\Theta}$ are

$$E\{\hat{\Theta}\} = \int_0^\theta x f_{\hat{\Theta}}(x) dx = \frac{n}{n+1} \theta \tag{9.114}$$

$$\text{var}\{\hat{\Theta}\} = \int_0^\theta \left(x - \frac{n}{n+1}\theta\right)^2 f_{\hat{\Theta}}(x) dx = \left[\frac{n}{(n+1)^2(n+2)}\right] \theta^2 \tag{9.115}$$

We see that $\hat{\Theta}$ is biased but consistent.

Example 9.17. Let us now determine the MLE of $\theta = r^2$ in Example 9.13. To carry out this estimation procedure, it is now necessary to determine the pdf of X given by Equation 9.77. Applying techniques developed in Chapter V, we can show that X is characterized by the *Rice distribution* with pdf given by [see Rice (1959)]

$$\begin{aligned} f_X(x; \theta, \sigma^2) &= \frac{x}{\sigma^2} I_0\left(\frac{\sqrt{\theta}x}{\sigma^2}\right) \exp\left(-\frac{x^2 + \theta}{2\sigma^2}\right) & x \geq 0 \\ &= 0 & \text{elsewhere} \end{aligned} \tag{9.116}$$

where I_0 is the modified zeroth-order Bessel function of the first kind.

Given a sample of size n from population X, the likelihood function takes the form

$$L = \prod_{j=1}^{n} f_X(x_j; \theta, \sigma^2) \qquad (9.117)$$

The MLE's of θ and σ^2, $\hat{\theta}$ and $\widehat{\sigma^2}$, satisfy the likelihood equations

$$\frac{\partial \ln L}{\partial \hat{\theta}} = 0 \quad \text{and} \quad \frac{\partial \ln L}{\partial \widehat{\sigma^2}} = 0 \qquad (9.118)$$

which, upon simplifying, can be written as

$$\frac{1}{n\sqrt{\hat{\theta}}} \sum_{j=1}^{n} \frac{x_j I_1(y_j)}{I_0(y_j)} - 1 = 0 \qquad (9.119)$$

$$\widehat{\sigma^2} = \frac{1}{2}\left(\frac{1}{n}\sum_{j=1}^{n} x_j^2 - \hat{\theta}\right) \qquad (9.120)$$

where I_1 is the first-order Bessel function of the first kind and

$$y_j = \frac{x_j\sqrt{\hat{\theta}}}{\widehat{\sigma^2}} \qquad (9.121)$$

As we can see, while likelihood equations can be established, they are complicated functions of $\hat{\theta}$ and $\widehat{\sigma^2}$ and we must resort to numerical means for their solutions. As we have pointed out earlier, this difficulty is often encountered in using the method of maximum likelihood. Indeed, Example 9.13 shows that the method of moments offers considerable computational advantage in this case.

The variances of the maximum likelihood estimators for θ and σ^2 can be obtained by simulation using Equations 9.119 and 9.120. We can also show that their variances can be larger than those associated with the moment estimators obtained in Example 9.13 for moderate sample sizes [see Benedict and Soong (1967)]. This observation serves to remind us again that, while maximum likelihood estimators possess optimal asymptotic properties, they may perform poorly when sample size is small.

IX.5 Interval Estimation

We now examine another approach to the problem of parameter estimation. As stated in Section IX.3, the interval estimation provides, on the basis of a sample from a population, not only information on the parameter values to be estimated but also an indication of the level of confidence that can be placed on possible numerical values of the parameters. Before describing the theory of interval estimation, an example will be used to demonstrate that a method that appears to be almost intuitively obvious could lead to conceptual difficulties.

Suppose that five sample values, 3, 2, 1.5, 0.5, and 2.1, are observed from a normal distribution having an unknown mean m and a known variance $\sigma^2 = 9$. From Example 9.15 we see that the MLE of m is the sample mean $\bar{X}$ and thus

$$\hat{m} = \tfrac{1}{5}(3 + 2 + 1.5 + 0.5 + 2.1) = 1.82 \tag{9.122}$$

Our additional task is to determine the upper and lower limits of an interval such that, with a specified level of confidence, the true mean m will lie in this interval.

The maximum likelihood estimator for m is $\bar{X}$ which, being a sum of normal random variables, is normal with mean m and variance $\sigma^2/n = \tfrac{9}{5}$. The standardized r.v. U defined by

$$U = \frac{\sqrt{5}(\bar{X} - m)}{3} \tag{9.123}$$

is then $N(0, 1)$ and it has the pdf

$$f_U(u) = \frac{1}{\sqrt{2\pi}} e^{-u^2/2} \qquad -\infty < u < \infty \tag{9.124}$$

Suppose we specify that the probability of U being in the interval $(-u_1, u_1)$ is equal to 0.95. From Table A.3 we find that $u_1 = 1.96$ and

$$P(-1.96 < U < 1.96) = \int_{-1.96}^{1.96} f_U(u)\,du = 0.95 \tag{9.125}$$

or, on substituting Equation 9.123 into Equation 9.125,

$$P(\bar{X} - 2.63 < m < \bar{X} + 2.63) = 0.95 \tag{9.126}$$

and, using Equation 9.122, the observed interval is

$$P(-0.81 < m < 4.45) = 0.95 \tag{9.127}$$

Equation 9.127 gives the desired result but it must be interpreted carefully. The mean m, although unknown, is nevertheless deterministic; and it either lies in an interval or it does not. On the other hand, we see from Equation 9.126 that the interval is a function of the statistic $\bar{X}$. Hence, the proper way to interpret Equations 9.126 and 9.127 is that the probability of the *random interval* ($\bar{X} - 2.63$, $\bar{X} + 2.63$) covering the distribution's true mean m is 0.95, and Equation 9.127 gives the observed interval based upon the given sample values.

Let us place the concept illustrated by the example above in a more general and precise setting.

Definition. Supoose that a sample $X_1, X_2, \ldots, X_n$ is drawn from a population having pdf $f(x; \theta)$, θ being the parameter to be estimated. Further suppose that $L_1(X_1, \ldots, X_n)$ and $L_2(X_1, \ldots, X_n)$ are two statistics such that $L_1 < L_2$ with probability 1. The interval (L_1, L_2) is called a $100(1 - \alpha)\%$ *confidence interval* for θ if L_1 and L_2 can be selected such that

$$P(L_1 < \theta < L_2) = 1 - \alpha \qquad (9.128)$$

The limits L_1 and L_2 are called, respectively, *lower* and *upper confidence limits* for θ, and $1 - \alpha$ is called the *confidence coefficient*. The value of $1 - \alpha$ is generally taken as 0.80, 0.90, 0.95, etc.

We now make several remarks concerning the foregoing definition.

1. As we see from Equation 9.126, confidence limits are functions of a given sample. The confidence interval thus will generally vary in position and width from sample to sample.

2. For a given sample, confidence limits are not unique. In other words, many pairs of statistics L_1 and L_2 exist that satisfy Equation 9.128. For example, in addition to the pair $(-1.96, 1.96)$, there are many other pairs of values (not symmetric about zero) that could give the probability 0.95 in Equation 9.125. However, it is easy to see that this particular pair gives the minimum-width interval.

3. In view of the above, it is thus desirable to define a set of quality criteria for interval estimators so that the "best" interval can be obtained. Intuitively, the "best" interval is the shortest interval. Moreover, since the interval width $L = L_2 - L_1$ is a random variable, we may like to choose "minimum expected interval width" as a good criterion. Unfortunately, there may not exist statistics L_1 and L_2 that give rise to an expected interval width that is minimum for all values of θ.

4. Just as in point estimation, sufficient statistics also play an important role in interval estimation as the following theorem demonstrates.

Theorem

Let L_1 and L_2 be two statistics based on a sample $X_1, \ldots, X_n$ from a population X with pdf $f(x;\theta)$ such that $P(L_1 < \theta < L_2) = 1 - \alpha$. Let $Y = h(X_1, \ldots, X_n)$ be a sufficient statistic. Then there exist two functions R_1 and R_2 of Y such that $P(R_1 < \theta < R_2) = 1 - \alpha$ and such that the two interval widths $L = L_2 - L_1$ and $R = R_2 - R_1$ have the same distribution.

This theorem shows that, if minimum interval width exists, it can be obtained by using functions of sufficient statistics as confidence limits.

The construction of confidence intervals for some important cases will be carried out in the following sections. The method consists essentially of finding an appropriate random variable whose values can be calculated on the basis of observed sample values and the parameter value, but whose distribution does not depend on the parameter. More general methods for obtaining confidence intervals are discussed in Mood (1950) and Wilks (1962).

IX.5.1 CONFIDENCE INTERVAL FOR m IN $N(m, \sigma^2)$ WITH KNOWN σ^2

The confidence interval given by Equation 9.126 is designed to estimate the mean of a normal population with known variance. In general terms, the procedure shows that we first determine an (symmetric) interval in U to achieve a confidence coefficient of $1 - \alpha$. Writing $u_{\alpha/2}$ for the value of U above which the area under $f_U(u)$ is $\alpha/2$, that is, $P(U > u_{\alpha/2}) = \alpha/2$ (see Figure 9.6), we have

$$P(-u_{\alpha/2} < U < u_{\alpha/2}) = 1 - \alpha \qquad (9.129)$$

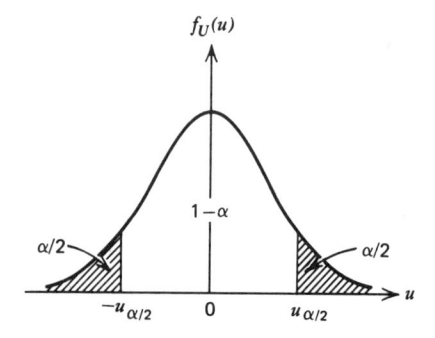

Figure 9.6 $100(1 - \alpha)\%$ confidence limits for U.

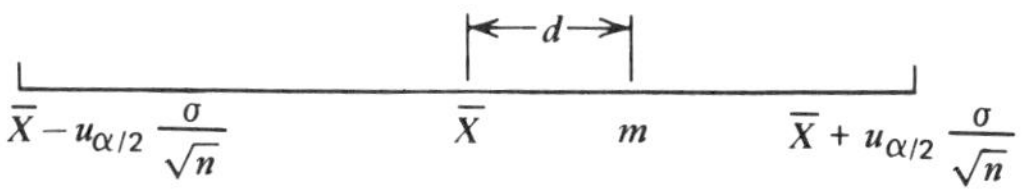

```
                |←─d─→|
└────────┬─────────┴────────┴─────────┘
X̄ - u_{α/2} σ/√n    X̄    m    X̄ + u_{α/2} σ/√n
```

Figure 9.7 Error in point estimators $\bar{X}$ for m.

Hence, using the transformation given by Equation 9.123 we have the general result

$$P\left(\bar{X} - \frac{u_{\alpha/2}\sigma}{\sqrt{n}} < m < \bar{X} + \frac{u_{\alpha/2}\sigma}{\sqrt{n}}\right) = 1 - \alpha \qquad (9.130)$$

This result can also be used to estimate means of nonnormal populations with known variances if the sample size is large enough to justify use of the central limit theorem.

It is noteworthy that, in this case, the position of the interval is a function of $\bar{X}$ and therefore is a function of the sample. The width of the interval, on the other hand, is only a function of the sample size n, being inversely proportional to $\sqrt{n}$.

The $100(1 - \alpha)\%$ confidence interval for m given in Equation 9.130 also provides an estimate of the accuracy of our point estimator $\bar{X}$ for m. As we see from Figure 9.7, the true mean m lies within the indicated interval with $100(1 - \alpha)\%$ confidence. Since $\bar{X}$ is at the center of the interval, the distance between $\bar{X}$ and m can be at most equal to one-half of the interval width. We thus have the following result.

Theorem

Let $\bar{X}$ be an estimator for m. Then, with $100(1 - \alpha)\%$ confidence, the error of using this estimator for m is less than $u_{\alpha/2}\sigma/\sqrt{n}$.

Example 9.18. Let the population X be normally distributed with known variance σ^2. If $\bar{X}$ is used as an estimator for mean m, determine the sample size n needed so that the estimation error will be less than a specified amount ε with $100(1 - \alpha)\%$ confidence.

Using the theorem given above, the minimum sample size n must satisfy

$$\varepsilon = \frac{u_{\alpha/2}\sigma}{\sqrt{n}}$$

Hence, the solution for n is

$$n = \left(\frac{u_{\alpha/2}\sigma}{\varepsilon}\right)^2 \qquad (9.131)$$

IX.5.2 CONFIDENCE INTERVAL FOR m IN $N(m, \sigma^2)$ WITH UNKNOWN σ^2

The difference between this problem and the preceding one is that, since σ is not known, we can no longer use

$$U = \frac{\bar{X} - m}{\sigma/\sqrt{n}}$$

as the random variable for confidence limit calculations regarding the mean m. Let us then use the sample variance S^2 as an unbiased estimator for σ^2 and consider the random variable

$$Y = \frac{\bar{X} - m}{S/\sqrt{n}} \quad (9.132)$$

The r.v. Y is now a function of r.v.'s $\bar{X}$ and S. In order to determine its distribution, we first state the following theorem.

Theorem (Student's t-Distribution)
Consider a r.v. T defined by

$$T = \frac{U}{\sqrt{V/n}} \quad (9.133)$$

If U is $N(0, 1)$, V is χ^2-distributed with n degrees of freedom, and U and V are independent, then the pdf of T has the form

$$\boxed{f_T(t) = \frac{\Gamma[(n+1)/2]}{\Gamma(n/2)\sqrt{n\pi}} \left(1 + \frac{t^2}{n}\right)^{-(n+1)/2} \quad -\infty < t < \infty} \quad (9.134)$$

This distribution is known as the *student's t-distribution* with n degrees of freedom; it is named after W.S. Gosset, who used the pseudonym "student" in his research publications.

Proof
The proof is straightforward following methods given in Chapter V. Since U and V are independent, their jpdf is

$$f_{UV}(u, v) = \left(\frac{1}{\sqrt{2\pi}} e^{-u^2/2}\right) \left[\frac{1}{2^{n/2}\Gamma(n/2)} v^{(n/2)-1} e^{-v/2}\right] \quad \begin{matrix} -\infty < u < \infty \\ v > 0 \end{matrix} \quad (9.135)$$
$$= 0 \quad \text{elsewhere}$$

PARAMETER ESTIMATION

Consider the transformation from U and V to T and V. The method discussed in Section V.3 leads to

$$f_{TV}(t, v) = f_{UV}[g_1^{-1}(t, v), g_2^{-1}(t, v)]|J| \qquad (9.136)$$

where

$$g_1^{-1}(t, v) = t\sqrt{\frac{v}{n}} \qquad g_2^{-1}(t, v) = v \qquad (9.137)$$

and the Jacobian is

$$J = \begin{vmatrix} \dfrac{\partial g_1^{-1}}{\partial t} & \dfrac{\partial g_1^{-1}}{\partial v} \\ \dfrac{\partial g_2^{-1}}{\partial t} & \dfrac{\partial g_2^{-1}}{\partial v} \end{vmatrix} = \begin{vmatrix} \sqrt{\dfrac{v}{n}} & \dfrac{t}{2n\sqrt{v/n}} \\ 0 & 1 \end{vmatrix} = \sqrt{\dfrac{v}{n}} \qquad (9.138)$$

The substitution of Equations 9.135, 9.137, and 9.138 into Equation 9.136 gives the jpdf $f_{TV}(t, v)$ of T and V. The pdf of T as given by Equation 9.134 is obtained by integrating $f_{TV}(t, v)$ with respect to v.

It is seen from Equation 9.134 that the t-distribution is symmetrical about the origin. As n increases, it approaches that of a standardized normal random variable.

Returning to r.v. Y defined by Equation 9.132, let

$$U = \frac{\overline{X} - m}{\sigma/\sqrt{n}} \qquad \text{and} \qquad V = \frac{(n-1)S^2}{\sigma^2}$$

Then

$$Y = \frac{U}{\sqrt{V/(n-1)}} \qquad (9.139)$$

where U is clearly distributed according to $N(0, 1)$. We also see from Section IX.1.2 that $(n-1)S^2/\sigma^2$ has a χ^2 distribution with $(n-1)$ degrees of freedom. Furthermore, although we will not verify it here, it can be shown that $\overline{X}$ and S^2 are independent. In accordance with the theorem given above, the r.v. Y thus has a t-distribution with $(n-1)$ degrees of freedom.

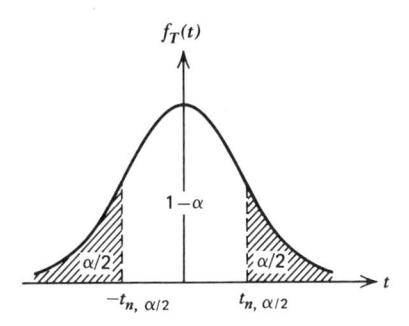

Figure 9.8 $100(1 - \alpha)\%$ confidence limits for T with n degrees of freedom.

The r.v. Y can now be used to establish confidence intervals for mean m. We note that the value of Y depends on the unknown mean m but its distribution does not.

The t-distribution is tabulated in Table A.4 in Appendix A. Let $t_{n,\alpha/2}$ be the value such that $P(T > t_{n,\alpha/2}) = \alpha/2$ with n representing the number of degrees of freedom (see Figure 9.8). We have the result

$$P(-t_{n-1,\alpha/2} < Y < t_{n-1,\alpha/2}) = 1 - \alpha \qquad (9.140)$$

Upon substituting Equation 9.132 into Equation 9.140, a $100(1 - \alpha)\%$ confidence interval for mean m is thus given by

$$\boxed{P\left(\bar{X} - \frac{t_{n-1,\alpha/2} S}{\sqrt{n}} < m < \bar{X} + \frac{t_{n-1,\alpha/2} S}{\sqrt{n}}\right) = 1 - \alpha} \qquad (9.141)$$

Since both $\bar{X}$ and S are functions of the sample, both position and width of the confidence interval given above will vary from sample to sample.

Example 9.19. Let us assume that annual snowfall in the Buffalo area is normally distributed. Using the snowfall record from 1970–1979 as given in Problem 8.8, determine a 95% confidence interval for the mean m.

For this example, $\alpha = 0.05$, $n = 10$, the observed sample mean is

$$\bar{x} = \tfrac{1}{10}(120.5 + 97.0 + \cdots + 97.3) = 112.4$$

and the observed sample variance is

$$s^2 = \tfrac{1}{9}[(120.5 - 112.4)^2 + (97.0 - 112.4)^2 + \cdots + (97.3 - 112.4)^2]$$
$$= 1414.3$$

Using Table A.4, we find that $t_{9, 0.025} = 2.262$. Substituting all the values given above into Equation 9.141 gives

$$P(85.5 < m < 139.3) = 0.95$$

It is clear that this interval would be different if we had incorporated more observations into our calculations or if we had chosen a different set of yearly snowfall data.

IX.5.3 CONFIDENCE INTERVAL FOR σ^2 IN $N(m, \sigma^2)$

An unbiased point estimator for the population variance σ^2 is S^2. For the construction of confidence intervals for σ^2, let us use the random variable

$$D = \frac{(n-1)S^2}{\sigma^2} \qquad (9.142)$$

which has been shown in Section IX.1.2 to have a χ^2 distribution with $(n-1)$ degrees of freedom. Letting $\chi^2_{n, \alpha/2}$ be the value such that $P(D > \chi^2_{n, \alpha/2}) = \alpha/2$ with n degrees of freedom, we can write (see Figure 9.9)

$$P(\chi^2_{n-1, 1-\alpha/2} < D < \chi^2_{n-1, \alpha/2}) = 1 - \alpha \qquad (9.143)$$

which gives, upon substituting Equation 9.142 for D,

$$\boxed{P\left[\frac{(n-1)S^2}{\chi^2_{n-1, \alpha/2}} < \sigma^2 < \frac{(n-1)S^2}{\chi^2_{n-1, 1-\alpha/2}}\right] = 1 - \alpha} \qquad (9.144)$$

Let us note that the $100(1 - \alpha)\%$ confidence interval for σ^2 as defined by Equation 9.144 is not the minimum-width interval on the basis of a given sample.

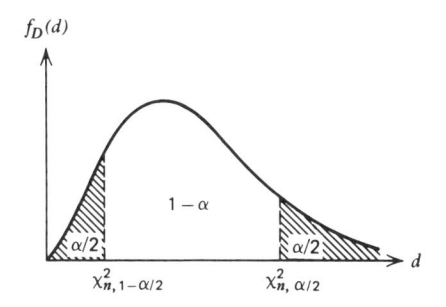

Figure 9.9 $100(1 - \alpha)\%$ confidence limits for D with n degrees of freedom.

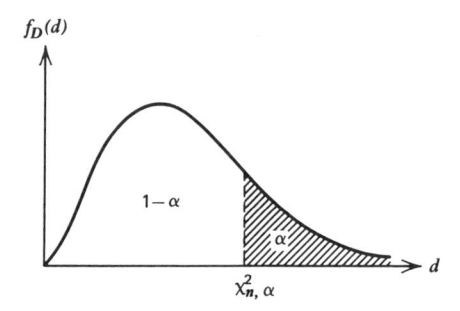

Figure 9.10 One-sided $100(1 - \alpha)\%$ confidence limit for D with n degrees of freedom.

As we see in Figure 9.9, a shift to the left leaving area $\alpha/2 - \varepsilon$ to the left and area $\alpha/2 + \varepsilon$ to the right under the $f_D(d)$-curve, where ε is an appropriate amount, will result in a smaller confidence interval. This is because the width needed at the left to give an increase ε in the area is less than the corresponding width eliminated at the right. The minimum interval width for a given number of degrees of freedom can be determined by interpolation from tabulated values of the PDF of χ^2 distribution.

Table A.5 in Appendix A gives selected values of $\chi^2_{n,\alpha}$ for various values of n and α. For convenience, Equation 9.144 is commonly used for constructing two-sided confidence intervals for σ^2 of a normal population. If a one-sided confidence interval is desired, it is then given by (see Figure 9.10)

$$P\left[\sigma^2 > \frac{(n-1)S^2}{\chi^2_{n-1,\alpha}}\right] = 1 - \alpha \tag{9.145}$$

Example 9.20. Consider Example 9.19 again and let us determine both two-sided and one-sided 95% confidence intervals for σ^2.

As seen from Example 9.19, the observed sample variance is

$$s^2 = 1414.3$$

The values of $\chi^2_{9,0.975}$, $\chi^2_{9,0.025}$, and $\chi^2_{9,0.05}$ are obtained from Table A.5 to be

$$\chi^2_{9,0.975} = 2.700 \qquad \chi^2_{9,0.025} = 19.023 \qquad \chi^2_{9,0.05} = 16.919$$

Equations 9.144 and 9.145 thus lead to, with $n = 10$ and $\alpha = 0.05$,

$$P(669.12 < \sigma^2 < 4714.33) = 0.95$$

and

$$P(\sigma^2 > 752.3) = 0.95$$

IX.5.4 CONFIDENCE INTERVAL FOR A PROPORTION

Consider now the construction of confidence intervals for p in the binomial distribution

$$p_X(k) = p^k(1-p)^{1-k} \qquad k = 0, 1$$

The parameter p represents the proportion in a binomial experiment. Given a sample of size n from population X, we see from Example 9.10 that an unbiased and efficient estimator for p is $\bar{X}$. For large n, the r.v. $\bar{X}$ is approximately normal with mean p and variance $p(1-p)/n$.

Defining

$$U = \frac{\bar{X} - p}{\sqrt{p(1-p)/n}} \tag{9.146}$$

the r.v. U tends to $N(0, 1)$ as n becomes large. In terms of U, we have the same situation as in Section IX.5.1 and Equation 9.129 gives

$$P(-u_{\alpha/2} < U < u_{\alpha/2}) = 1 - \alpha \tag{9.147}$$

The substitution of Equation 9.146 into Equation 9.147 gives

$$P\left[-u_{\alpha/2} < \frac{\bar{X} - p}{\sqrt{p(1-p)/n}} < u_{\alpha/2}\right] = 1 - \alpha \tag{9.148}$$

In order to determine confidence limits for p, we need to solve for p satisfying the equation

$$\frac{|\bar{X} - p|}{\sqrt{p(1-p)/n}} \leq u_{\alpha/2}$$

or, equivalently,

$$(\bar{X} - p)^2 \leq \frac{u_{\alpha/2}^2 p(1-p)}{n} \tag{9.149}$$

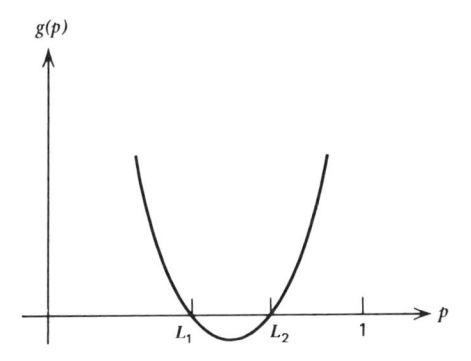

Figure 9.11 Parabola $g(p) = p^2(1 + u_{\alpha/2}^2/n) - p(2\bar{X} + u_{\alpha/2}^2/n) + \bar{X}^2$.

Upon transposing the right-hand side, we have

$$p^2\left(1 + \frac{u_{\alpha/2}^2}{n}\right) - p\left(2\bar{X} + \frac{u_{\alpha/2}^2}{n}\right) + \bar{X}^2 \leq 0 \tag{9.150}$$

In Equation 9.150, the left-hand side defines a parabola as shown in Figure 9.11 and the two roots L_1 and L_2 of Equation 9.150 with the equal sign define the interval within which the parabola is negative. Hence, solving the quadratic equation defined by Equation 9.150, we have

$$\boxed{L_{1,2} = \frac{(\bar{X} + u_{\alpha/2}^2/2n) \mp (u_{\alpha/2}/\sqrt{n})\sqrt{\bar{X}(1 - \bar{X}) + u_{\alpha/2}^2/4n}}{1 + u_{\alpha/2}^2/n}} \tag{9.151}$$

For large n, they can be approximated by

$$L_{1,2} = \bar{X} \mp u_{\alpha/2}\sqrt{\frac{\bar{X}(1 - \bar{X})}{n}} \tag{9.152}$$

An approximate $100(1 - \alpha)\%$ confidence interval for p is thus given by, for large n,

$$\boxed{P\left(\bar{X} - u_{\alpha/2}\sqrt{\frac{\bar{X}(1 - \bar{X})}{n}} < p < \bar{X} + u_{\alpha/2}\sqrt{\frac{\bar{X}(1 - \bar{X})}{n}}\right) = 1 - \alpha} \tag{9.153}$$

In this approximation, the sample mean $\bar{X}$ is at center of the interval whose width is a function of the sample and the sample size.

Example 9.21. In a random sample of 500 persons in the city of Los Angeles, it was found that 372 did not approve of the U.S. energy policy. Determine a 95% confidence interval for p, the actual proportion of Los Angeles' population registering their disapproval.

In this example, $n = 500$, $\alpha = 0.05$, and the observed sample mean is $\bar{x} = \frac{372}{500} = 0.74$. Table A.3 gives $u_{0.025} = 1.96$. Substituting these values into Equation 9.153 then yields

$$P(0.74 - 0.04 < p < 0.74 + 0.04) = P(0.70 < p < 0.78) = 0.95$$

REFERENCES AND COMMENTS

1. R. L. Anderson and T. A. Bancroft, *Statistical Theory in Research*, McGraw-Hill, New York, 1952.
2. T. R. Benedict and T. T. Soong, "The Joint Estimation of Signal and Noise from the Sum Envelope," *IEEE Trans. Information Theory*, **IT-13**, 447–454, 1967.
3. R. A. Fisher, "On the Mathematical Foundation of Theoretical Statistics," *Phil. Trans. Roy. Soc. London*, Ser. A, **222**, 309–368, 1922.
4. D. L. Gerlough, "The Use of Poisson Distribution in Traffic," in *Poisson and Traffic*, The Eno Foundation for Highway Trafic Control, Saugatuk, Conn., 1955.
5. A. M. Mood, *Introduction to the Theory of Statistics*, McGraw-Hill, New York, 1950 (Chapter 11).
6. J. Neyman, "Su un Teorema Concernente le Cosiddeti Statistiche Sufficienti," *Giorn. Inst. Ital. Attuari.*, **6**, 320–334, 1935.
7. K. Pearson, "Contributions to the Mathematical Theory of Evolution," *Phil. Trans. Roy. Soc. London*, Ser. A, **185**, 71–78, 1894.
8. T. T. Soong, "An Extension of the Moment Method in Statistical Estimation," *SIAM J. Appl. Math.*, **17**, 560–568, 1969.
9. S. S. Wilks, *Mathematical Statistics*, John Wiley, New York, 1962 (Chapter 12).

The Cramér–Rao inequality is named after two well-known statisticians, H. Cramér and C. R. Rao, who independently established this result in the following references. However, this inequality was first stated by Fisher in Reference 2 cited above. In fact, much of the foundation of parameter estimation and statistical inference in general, such as concepts of consistency, efficiency, and sufficiency, was laid by Fisher in a series of publications given below.

10. H. Cramér, *Mathematical Methods of Statistics*, Princeton University Press, Princeton, N.J., 1946.

11. R. A. Fisher, "On a Distribution Yielding the Error Functions of Several Well-Known Statistics," *Proc. Int. Math. Congress*, Vol. II, Toronto, 805–813, 1924.
12. R. A. Fisher, *Statistical Methods for Research Workers*, 14th Ed., Hafner Press, New York, 1970 (First Edition, 1925).
13. R. A. Fisher, "Theory of statistical Estimation," *Proc. Camb. Phil. Soc.*, **22**, 700–725, 1925.
14. C. R. Rao, "Information and Accuracy Attainable in the Estimation of Statistical Parameters," *Bull. Calcutta Math. Soc.*, **37**, 81–91, 1945.

PROBLEMS

The following notations and abbreviations are used in some statements of the problems:

$\bar{X}$ = sample mean, $\bar{x}$ = observed sample mean
S^2 = sample variance, s^2 = observed sample variance
ME = moment estimator or moment estimate
MLE = maximum likelihood estimator or maximum likelihood estimate
CRLB = Cramér–Rao lower bound

9.1. In order to make enrollment projections, a survey is made of numbers of children in 100 families in a school district and the result is given below. Determine $\bar{x}$, the observed sample mean, and s^2, the observed sample variance, on the basis of these 100 sample values.

No. of Children	No. of Families
0	21
1	24
2	30
3	16
4	4
5	4
6	0
7	1
	$n = 100$

PARAMETER ESTIMATION

9.2. Verify that the variance of sample variance S^2 as defined by Equation 9.7 is given by Equation 9.10.

9.3. Verify that the mean and variance of the kth sample moment M_k as defined by Equation 9.14 are given by Equations 9.15.

9.4. Let $X_1, X_2, \ldots, X_{10}$ be a sample of size 10 from the standardized normal distribution $N(0, 1)$. Determine the probability $P(\overline{X} \leq 1)$.

9.5. Let $X_1, X_2, \ldots, X_{10}$ be a sample of size 10 from a uniformly distributed random variable in the interval $(0, 1)$.

(a) Determine the pdf's of $X_{(1)}$ and $X_{(10)}$.
(b) Find the probabilities $P(X_{(1)} > 0.5)$ and $P(X_{(10)} \leq 0.5)$.
(c) Determine $E\{X_{(1)}\}$ and $E\{X_{(10)}\}$.

9.6. A sample of size n is taken from a population X with pdf

$$f_X(x) = e^{-x} \quad x \geq 0$$
$$= 0 \quad \text{elsewhere}$$

Determine the probability density function of the statistic $\overline{X}$. (Hint: Use the method of characteristic functions discussed in Chapter IV.)

9.7. Two samples X_1 and X_2 are taken from an exponential r.v. X with unknown parameter θ, that is,

$$f_X(x; \theta) = \frac{1}{\theta} e^{-x/\theta} \quad x \geq 0$$

We propose two estimators for θ in the forms

$$\hat{\Theta}_1 = \overline{X} = \frac{X_1 + X_2}{2}$$

$$\hat{\Theta}_2 = \frac{4}{\pi} \sqrt{X_1 X_2}$$

In terms of unbiasedness and minimum variance, which one is the better of the two?

9.8. Let X_1 and X_2 be a sample of size two from a population X with mean m and variance σ^2.

(a) Two estimators for m are proposed to be

$$\hat{M}_1 = \bar{X} = \frac{X_1 + X_2}{2}$$

$$\hat{M}_2 = \frac{X_1 + 2X_2}{3}$$

Which is the better estimator?

(b) Consider an estimator for m in the form

$$\hat{M} = aX_1 + (1-a)X_2 \qquad 0 \leq a \leq 1$$

Determine the value a, which gives the best estimator in this form.

9.9. The geometrical mean $(X_1 X_2 \cdots X_n)^{1/n}$ is proposed as an estimator for the unknown median of a lognormally distributed random variable X. Is it unbiased? Is it unbiased as $n \to \infty$?

9.10. Let X_1, X_2, X_3 be a sample of size three from a uniform distribution whose pdf is

$$f_X(x; \theta) = \frac{1}{\theta} \qquad 0 \leq x \leq \theta$$

$$= 0 \qquad \text{elsewhere}$$

Suppose that $aX_{(1)}$ and $bX_{(3)}$ are proposed as two possible estimators for θ.

(a) Determine a and b such that these estimators are unbiased.
(b) Which one is the better of the two? In the above, $X_{(j)}$ is the jth order statistic.

9.11. Let $X_1, \ldots, X_n$ be a sample from a population whose kth moment $\alpha_k = E\{X^k\}$ exists. Show that the kth sample moment

$$M_k = \frac{1}{n} \sum_{j=1}^{n} X_j^k$$

is a consistent estimator for α_k.

PARAMETER ESTIMATION

9.12. Let θ be the parameter to be estimated in each of the distributions given below. For each case, determine the CRLB for the variance of any unbiased estimator for θ.

(a) $f(x;\theta) = \dfrac{1}{\theta} e^{-x/\theta}$ $x \geq 0$

(b) $f(x;\theta) = \theta x^{\theta-1}$ $0 \leq x \leq 1$ $\theta > 0$

(c) $p(x;\theta) = \theta^x (1-\theta)^{1-x}$ $x = 0, 1$

(d) $p(x;\theta) = \dfrac{\theta^x e^{-\theta}}{x!}$ $x = 0, 1, 2, \ldots$

9.13. Determine CRLB for the variances of $\widehat{M}$ and $\widehat{\Sigma^2}$, which are, respectively, unbiased estimators for m and σ^2 in the normal distribution $N(m, \sigma^2)$.

9.14. Consider each distribution in Problem 9.12.
(a) Determine a ME for θ on the basis of a sample of size n by using the first-order moment equation. Determine its asymptotic efficiency (efficiency as $n \to \infty$) (hint: Use the second of Equations 9.62 for asymptotic variance of ME).
(b) Determine the MLE for θ.

9.15. Electronic components are tested for reliability. Let p be the probability of an electronic component being successful and $1 - p$ be the probability of component failure. If X is the number of trials at which the first failure occurs, then it has the geometric distribution

$$p_X(k;p) = (1-p)p^{k-1} \quad k = 1, 2, \ldots$$

Suppose that a sample $X_1, \ldots, X_n$ is taken from the population X, each X_j consisting of testing X_j components when the first failure occurs.
(a) Determine the MLE of p.
(b) Determine the MLE of $P(X > 9)$, the probability that the component will not fail in nine trials. Note: $P(X > 9) = 1 - \sum_{k=1}^{9}(1-p)p^{k-1}$.

9.16. The pdf of a population is given by

$$f_X(x;\theta) = \dfrac{2x}{\theta^2} \quad 0 \leq x \leq \theta$$

$$= 0 \quad \text{elsewhere}$$

Based on a sample of size n:
(a) Determine the MLE and ME for θ.
(b) Which one of the two is the better estimator?

9.17. Assume that X is a continuous r.v. having a shifted exponential distribution with

$$f_X(x; a) = e^{a-x} \qquad x \geq a$$

On the basis of a sample of size n from X, determine the MLE and ME for a.

9.18. Let $X_1, X_2, \ldots, X_n$ be a sample of size n from a uniform distribution

$$f(x; \theta) = 1 \qquad \theta - \tfrac{1}{2} \leq x \leq \theta + \tfrac{1}{2}$$
$$ = 0 \qquad \text{elsewhere}$$

Show that every statistic $h(X_1, \ldots, X_n)$ satisfying

$$X_{(n)} - \tfrac{1}{2} \leq h(X_1, \ldots, X_n) \leq X_{(1)} + \tfrac{1}{2}$$

is a MLE for θ, where $X_{(j)}$ is the jth order statistic. Determine a maximum likelihood estimate for θ when observed sample values are (1.5, 1.4, 2.1, 2.0, 1.9, 2.0, 2.3) with $n = 7$.

9.19. Using the 214 measurements given in Example 9.11, determine the MLE for λ in the exponential distribution given by Equation 9.70.

9.20. Let us assume that the r.v. X in Problem 8.11 is Poisson distributed. Using the 58 sample values given in Problem 8.11, determine the MLE and ME for the mean number of blemishes.

9.21. The time-to-failure T of a certain device has a shifted exponential distribution, that is,

$$f_T(t; t_0, \lambda) = \lambda e^{-\lambda(t-t_0)} \qquad t \geq t_0$$
$$ = 0 \qquad \text{elsewhere}$$

Let $T_1, T_2, \ldots, T_n$ be a sample from T.
(a) Determine the MLE and ME for λ assuming t_0 is known.
(b) Determine the MLE and ME for t_0 assuming λ is known.
(c) Determine the MLE and ME for both λ and t_0 assuming both are unknown.

9.22. If $X_1, X_2, \ldots, X_n$ is a sample from the gamma distribution

$$f(x; r, \lambda) = \frac{\lambda^r x^{r-1}}{\Gamma(r)} e^{-\lambda x} \qquad x \geq 0 \qquad r, \lambda > 0$$

PARAMETER ESTIMATION   307

show that:
(a) If r is known and λ is the parameter to be estimated, both the MLE and ME for λ are $\hat{\Lambda} = r/\overline{X}$.
(b) If both r and λ are to be estimated, then the method of moments and the method of maximum likelihood lead to different estimators for r and λ. (It is not necessary to determine these estimators.)

9.23. Consider Buffalo yearly snowfall data as given in Problem 8.8 and assume that a normal distribution is appropriate.
(a) Find estimates for the parameters by means of the moment method and the method of maximum likelihood.
(b) Estimate from the model the probability of having another blizzard of 1977 $[P(X > 199.4)]$.

9.24. Recorded annual flow Y (in cfs) of a river at a given point are 141, 146, 166, 209, 228, 234, 260, 278, 319, 351, 383, 500, 522, 589, 696, 833, 888, 1173, 1200, 1258, 1340, 1390, 1420, 1423, 1443, 1561, 1650, 1810, 2004, 2013, 2016, 2080, 2090, 2143, 2185, 2316, 2582, 3050, 3186, 3222, 3660, 3799, 3824, 4099, 6634. Assuming that Y follows a lognormal distribution, determine the MLE's of the distribution parameters.

9.25. Let X_1 and X_2 be a sample of size two from a uniform distribution with pdf

$$f(x;\theta) = \frac{1}{\theta} \quad 0 \leq x \leq \theta$$
$$= 0 \quad \text{elsewhere}$$

Determine the constant c so that the interval

$$0 < \theta < c(X_1 + X_2)$$

is a $100(1 - \alpha)\%$ confidence interval for θ.

9.26. The fuel consumption of a certain type of vehicle is approximately normal with standard deviation of 3 miles per gallon. If a sample of 64 vehicles has an average fuel consumption of 16 miles per gallon:
(a) Determine a 95% confidence interval for the mean fuel consumption of all vehicles of this type.
(b) With 95% confidence, what is the possible error if the mean fuel consumption is taken to be 16 miles per gallon?
(c) How large a sample is needed if we wish to be 95% confident that the mean will be within 0.5 miles per gallon of the true mean?

9.27. Seventy yearly Buffalo snowfall measurements are given in Problem 8.8. Assume that it is approximately normal with standard deviation $\sigma = 25$ inches. Determine 95% confidence intervals for the mean using measurements of (a) 1909 to 1939, (b) 1909 to 1959, and (c) 1909 to 1979. Display these intervals graphically.

9.28. Let $\bar{X}_1$ and $\bar{X}_2$ be independent sample means from two normal populations $N(m_1, \sigma_1^2)$ and $N(m_2, \sigma_2^2)$, respectively. If σ_1^2 and σ_2^2 are known, show that a $100(1 - \alpha)\%$ confidence interval for $m_1 - m_2$ is

$$P\left[(\bar{X}_1 - \bar{X}_2) - u_{\alpha/2}\sqrt{\frac{\sigma_1^2}{n_1} + \frac{\sigma_2^2}{n_2}} < m_1 - m_2 < (\bar{X}_1 - \bar{X}_2) + u_{\alpha/2}\sqrt{\frac{\sigma_1^2}{n_1} + \frac{\sigma_2^2}{n_2}}\right] = 1 - \alpha$$

where n_1 and n_2 are, respectively, the sample sizes from $N(m_1, \sigma_1^2)$ and $N(m_2, \sigma_2^2)$, and $u_{\alpha/2}$ is the value of the standardized normal r.v. U such that $P(U > u_{\alpha/2}) = \alpha/2$.

9.29. Let us assume that r.v. X in Problem 8.6 has a Poisson distribution with pmf

$$p_X(k; \lambda) = \frac{\lambda^k e^{-\lambda}}{k!} \quad k = 0, 1, 2, \ldots$$

Use the sample values of X given in Problem 8.6 and:
(a) Determine the MLE $\hat{\lambda}$ for λ.
(b) Determine a 95% confidence interval for λ using asymptotic properties of the MLE $\hat{\lambda}$.

9.30. Assume that the life span of American males is normally distributed with unknown mean m and unknown variance σ^2. A sample of 30 mortality history of American males shows that

$$\bar{x} = \frac{1}{30}\sum_{i=1}^{30} x_i = 71.3 \text{ years}$$

$$s^2 = \frac{1}{29}\sum_{i=1}^{30}(x_i - \bar{x})^2 = 128 \text{ years}^2$$

Determine the observed values of 95% confidence intervals for m and σ^2.

9.31. The life of light bulbs manufactured in a certain plant can be assumed to be normally distributed. A sample of 15 light bulbs gives the observed sample

mean $\bar{x} = 1100$ hours and observed sample standard deviation $s = 50$ hours.
- (a) Determine a 95% confidence interval for the average life.
- (b) Determine both two-sided and one-sided 95% confidence intervals for its variance.

9.32. Twelve of 100 manufactured items examined are found to be defective.
- (a) Find a 99% confidence interval for the proportion of defective items in the manufacturing process.
- (b) With 99% confidence, what is the possible error if the proportion is estimated to be $\frac{12}{100} = 0.12$?

9.33. In a public opinion poll such as the one described in Example 9.21, determine the minimum sample size needed for the poll so that with 95% confidence the sample mean will be within 0.05 of the true proportion (hint: Use the fact that $\bar{X}(1 - \bar{X}) \leq \frac{1}{4}$ in Equation 9.153).

X
Model Verification

The parameter estimation procedures developed in the preceding chapter presume a distribution for the population. To a large extent, the validity of the model building process based on this approach thus hinges on the substantiability of this hypothesized distribution. Indeed, if the hypothesized distribution is off the mark, the resulting probabilistic model with parameters estimated by any, however elegant, procedure would still be illusive and, at best, give a poor representation of the underlying physical or natural phenomenon.

In this chapter, we wish to develop methods of testing or verifying a hypothesized distribution for a population on the basis of a sample taken from the population. Some aspects of this problem have been addressed in Chapter VIII in which, by means of histograms and frequency diagrams, a graphical comparison between the hypothesized distribution and observed data can be made. In the chemical-yield example, for instance, a comparison between the shape of a normal distribution and the histogram constructed from the data as shown in Figure 8.1 suggests that the normal model is reasonable in that case.

However, the graphical procedure described above is clearly subjective and nonquantitative. On a more objective and quantitative basis, the problem of model verification on the basis of sample information falls within the framework of testing of hypotheses. Some basic concepts in this area of statistical inference are now introduced.

X.1 Preliminaries

In our development, statistical hypotheses concern functional forms of the assumed distributions; these distributions may be specified completely with given values for their parameters or they may be specified with parameters yet to be estimated from the sample.

Let $X_1, X_2, \ldots, X_n$ be a sample of size n from a population X with a hypothesized pdf $f(x; \boldsymbol{\theta})$, where $\boldsymbol{\theta}$ may be specified or unspecified. We denote as *hypothesis H* the hypothesis that the sample represents n values of a random variable with pdf $f(x; \boldsymbol{\theta})$. The hypothesis is called a *simple hypothesis* when the underlying distribution is completely specified, that is, the parameter values are specified together with functional form of the probability density function; otherwise, it is a *composite hypothesis*. To construct a criterion for hypotheses testing, it is necessary that an alternative hypothesis be established against which hypothesis H can be tested. An example of an alternative hypothesis is simply another hypothesized distribution or, as another example, hypothesis H can be tested against the alternative hypothesis that hypothesis H is not true. In our applications, the latter choice is considered more practical and we shall in general deal with the task of either accepting or rejecting hypothesis H on the basis of a sample from the population.

X.1.1 TYPE I AND TYPE II ERRORS

As in parameter estimation, errors or risks are inherent in deciding whether a hypothesis H should be accepted or rejected on the basis of sample information. Tests for hypotheses testing are therefore generally compared in terms of the probabilities of errors that might be committed. There are basically two types of errors that are likely to be made, namely, reject H when in fact H is true or, alternatively, accept H when in fact H is false. We formalize the above with the following definition.

Definition. In testing a hypothesis H, a Type I error is committed when H is rejected when in fact H is true, and a Type II error is committed when H is accepted when in fact H is false.

In hypotheses testing, an important consideration in constructing statistical tests is thus to control, insofar as possible, the probabilities of making these errors. Let us note that, for a given test, an evaluation of Type I error can be made when hypothesis H is given, that is, when a hypothesized distribution is specified. On the other hand, the specification of an alternative hypothesis dictates Type II error probabilities. In our problem, the alternative hypothesis is simply that hypothesis H is not true. The fact that the class of alternatives is so large makes it difficult to use Type II errors as a criterion. In what follows, methods of hypotheses testing are discussed based on Type I errors only.

X.2 Chi-square Goodness-of-fit Test

As mentioned earlier, the problem to be addressed is one of testing a hypothesis H that specifies the probability distribution for a population X versus the alternative that the probability distribution of X is not of the stated type on the basis of a sample of size n from the population X. One of the most popular and most versatile tests devised for this purpose is the chi-square (χ^2) goodness-of-fit test introduced by Pearson (1900).

X.2.1 THE CASE OF KNOWN PARAMETERS

Let us first assume that the hypothesized distribution is completely specified with no unknown parameters. In order to test hypothesis H, some statistic $h(X_1, X_2, \ldots, X_n)$ of the sample is required that gives a measure of deviation of the observed distribution as constructed from the sample from the hypothesized distribution.

In the χ^2 test, the statistic used is related to, roughly speaking, the difference between the histogram constructed from the sample and a corresponding diagram constructed from the hypothesized distribution. Let the range space of X be divided into k mutually exclusive intervals $A_1, A_2, \ldots,$ and A_k, and let N_i be the number of X_j's falling into A_i, $i = 1, 2, \ldots, k$. Then the *observed* probabilities $P(A_i)$ are given by

$$\text{Observed } P(A_i) = \frac{N_i}{n} \qquad i = 1, 2, \ldots, k \qquad (10.1)$$

On the other hand, the *theoretical* probabilities $P(A_i)$ can be obtained from the hypothesized population distribution. Let us denote these by

$$\text{Theoretical } P(A_i) = p_i \qquad i = 1, 2, \ldots, k \qquad (10.2)$$

A logical choice of a statistic giving a measure of deviation is

$$\sum_{i=1}^{k} c_i \left(\frac{N_i}{n} - p_i \right)^2 \qquad (10.3)$$

which is a natural least-square type deviation measure. Pearson (1900) has shown that, if we take the coefficients $c_i = n/p_i$, the statistic defined by Equation 10.3 has particularly simple properties. Hence, we choose as our deviation measure

$$D = \sum_{i=1}^{k} \frac{n}{p_i} \left(\frac{N_i}{n} - p_i \right)^2 = \sum_{i=1}^{k} \frac{(N_i - np_i)^2}{np_i}$$

$$= \sum_{i=1}^{k} \frac{N_i^2}{np_i} - n \qquad (10.4)$$

MODEL VERIFICATION

Let us note that D is a statistic since it is a function of N_i, which are in turn functions of the sample $X_1, \ldots, X_n$. The distribution of the statistic D is given in the following theorem due to Pearson (1900).

Theorem

Assuming that hypothesis H is true, the distribution of D defined by Equation 10.4 approaches a χ^2 distribution with $(k-1)$ degrees of freedom as $n \to \infty$. Its pdf is given by (see Equation 7.67)

$$\boxed{\begin{aligned} f_D(d) &= \frac{1}{2^{(k-1)/2} \Gamma\left(\frac{k-1}{2}\right)} d^{(k-3)/2} e^{-d/2} \qquad d \geq 0 \\ &= 0 \qquad\qquad\qquad\qquad\qquad\qquad\qquad \text{elsewhere} \end{aligned}} \qquad (10.5)$$

Let us note that this distribution is independent of the hypothesized distribution.

Proof

The complete proof, which can be found in Cramér (1946) and in other advanced texts in statistics, will not be attempted here. To demonstrate its plausibility, we only sketch the proof for the $k = 2$ case.

For $k = 2$, the r.v. D is

$$D = \frac{(N_1 - np_1)^2}{np_1} + \frac{(N_2 - np_2)^2}{np_2}$$

Since $N_1 + N_2 = n$ and $p_1 + p_2 = 1$, we can write

$$D = \frac{(N_1 - np_1)^2}{np_1} + \frac{[n - N_1 - n(1 - p_1)]^2}{np_2}$$

$$= (N_1 - np_1)^2 \left(\frac{1}{np_1} + \frac{1}{np_2}\right) = \frac{(N_1 - np_1)^2}{np_1(1 - p_1)} \qquad (10.6)$$

Now, recalling that N_1 is the number of, say, successes in n trials with p_1 being the probability of success, it is a binomial r.v. with $E\{N_1\} = np_1$ and $\text{var}\{N_1\} = np_1(1 - p_1)$ if hypothesis H is true. As n increases, we have seen in Chapter VII that N_1 approaches a normal distribution by virtue of the central limit theorem. Hence, the distribution of r.v. U defined by

$$U = \frac{N_1 - np_1}{\sqrt{np_1(1 - p_1)}}$$

approaches $N(0, 1)$ as $n \to \infty$. Since

$$D = U^2$$

following Equation 10.6, the r.v. D thus approaches a χ^2 distribution with one degree of freedom and the proof is complete for $k = 2$. The proof for an arbitrary k would proceed in a similar fashion.

By means of the theorem given above, a test of hypothesis H considered above can be constructed based on the assignment of a probability of Type I error. Suppose that we wish to achieve a Type I error probability of α. The χ^2 test suggests that hypothesis H is rejected whenever

$$d = \sum_{i=1}^{k} \frac{n_i^2}{np_i} - n > \chi^2_{k-1,\alpha} \tag{10.7}$$

and is accepted otherwise, where d is the sample value of D based on the sample values x_i, $i = 1, \ldots, n$, and the number $\chi^2_{k-1,\alpha}$ takes the value such that (see Figure 10.1)

$$P(D > \chi^2_{k-1,\alpha}) = \alpha$$

Since D has a χ^2 distribution with $(k - 1)$ degrees of freedom for large n, an approximate value for $\chi^2_{k-1,\alpha}$ can be found from Table A.5 in Appendix A for the χ^2 distribution when α is specified.

The probability α of Type I error is referred to as *significance level* in this context. As seen from Figure 10.1, it represents the area under $f_D(d)$ to the right of $\chi^2_{k-1,\alpha}$. Letting $\alpha = 0.05$, for example, the criterion given by Equation 10.7 implies that we reject hypothesis H whenever the deviation measure d as calculated from a

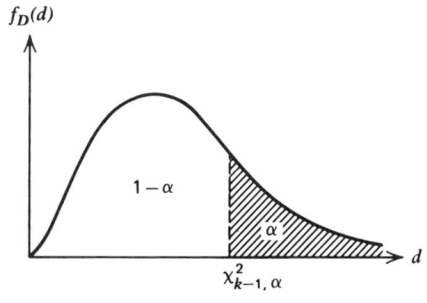

Figure 10.1 χ^2 distribution with $(k - 1)$ degrees of freedom.

MODEL VERIFICATION

given set of sample values falls within the 5% region. In other words, we expect to reject H about 5 percent of the time when in fact H is true. Which significance level should be adopted in a given situation will, of course, depend on the particular case involved. In practice, common values for α are 0.001, 0.01, and 0.05; a value of α between 5% and 1% is regarded as *almost significant*, a value between 1% and 0.1% as *significant*, and a value below 0.1% as *highly significant*.

Let us now give a step-by-step procedure for carrying out the χ^2 test when the distribution of a population X is completely specified.

1. Divide the range space of X into k mutually exclusive and numerically convenient intervals A_i, $i = 1, 2, \ldots, k$. Let n_i be the number of sample values falling into A_i. As a rule, if the number of sample values in any A_i is less than five, combine the interval A_i with either A_{i-1} or A_{i+1}.
2. Compute theoretical probabilities $P(A_i) = p_i$, $i = 1, 2, \ldots, k$, by means of the hypothesized distribution.
3. Construct d as given by Equation 10.7.
4. Choose a value of α and determine from Table A.5 for the χ^2 distribution of $(k - 1)$ degrees of freedom the value of $\chi^2_{k-1, \alpha}$.
5. Reject hypothesis H if $d > \chi^2_{k-1, \alpha}$. Otherwise, accept H.

Example 10.1. Three hundred light bulbs are tested for their burning time t (in hours) and the result is shown in Table 10.1. Suppose that the random burning time T is postulated to be exponentially distributed with mean burning time $1/\lambda = 200$ hours, that is, $\lambda = 0.005$ and

$$f_T(t) = 0.005 e^{-0.005t} \qquad t \geq 0 \qquad (10.8)$$

We wish to test this hypothesis using the χ^2 test at 5% significance level

Table 10.1. *Sample values in Example 10.1*

Burning Time t	Number
$t < 100$	121
$100 \leq t < 200$	78
$200 \leq t < 300$	43
$300 \leq t$	58
	$n = 300$

Table 10.2. *Table for χ^2 Test—Example 10.1*

Intervals A_i	Observed Number of Occurrences n_i	Theoretical $P(A_i)$ p_i	np_i	n_i^2/np_i
$t < 100$	121	0.39	117	125.1
$100 \leq t < 200$	78	0.24	72	84.5
$200 \leq t < 300$	43	0.15	45	41.1
$300 \leq t$	58	0.22	66	51.0
	300	1.00	300	301.7

The necessary steps in carrying out the χ^2 test are indicated in Table 10.2. The first column gives intervals A_i, which are chosen in this case to be the intervals of t given in Table 10.1. The theoretical probabilities $P(A_i) = p_i$ in the third column are easily calculated using Equation 10.8. For example,

$$p_1 = P(A_1) = \int_0^{100} 0.005 e^{-0.005t} \, dt = 1 - e^{-0.5} = 0.39$$

$$p_2 = P(A_2) = \int_{100}^{200} 0.005 e^{-0.005t} \, dt = 1 - e^{-1} - 0.39 = 0.24$$

For convenience, the theoretical numbers of occurrences as predicted by the model are given in the fourth column of Table 10.2, which, when compared with the values in the second column, give a measure of the goodness-of-fit of the model to the data. Column 5 is included in order to facilitate the calculation of d. Thus, from Equation 10.7 we have

$$d = \sum_{i=1}^{k} \frac{n_i^2}{np_i} - n = 301.7 - 300 = 1.7$$

Now, $k = 4$. From Table A.5 for the χ^2 distribution with three degrees of freedom, we find

$$\chi^2_{3, 0.05} = 7.815$$

Since $d < \chi^2_{3, 0.05}$, we would accept at 5% significance level the hypothesis that the observed data represent a sample from an exponential distribution with $\lambda = 0.005$.

MODEL VERIFICATION

Table 10.3. *Table for χ^2 Test—Example 10.2*

Intervals A_i	Observed Number of Occurrences n_i	Theoretical $P(A_i)$ p_i	np_i	n_i^2/np_i
$x \leq 0$	5147	0.6188	4853	5459
$0 < x \leq 1$	1859	0.2970	2329	1484
$1 < x \leq 2$	595	0.0713	559	633
$2 < x \leq 3$	167	0.0114	89	313
$3 < x \leq 4$	54	0.0013	10	292
$4 < x \leq 5$	14	0.0001	1	196
$5 < x$	6	0.0001	1	36
	7842	1.0	7842	8413

Example 10.2. A six-year accident record of 7842 California drivers was given in Table 8.2. On the basis of these sample values, we wish to test the hypothesis that X, the number of accidents in six years per driver, is Poisson distributed with mean rate $\lambda = 0.08$ per year at 1% significance level.

Since X is discrete, a natural choice of the intervals A_i is to use those centered around the discrete values as indicated in the first column of Table 10.3. Note that the interval $x > 5$ would be combined with $4 < x \leq 5$ if the number n_7 were less than five.

The hypothesized distribution for X is

$$p_X(x) = \frac{(\lambda t)^x e^{-\lambda t}}{x!} = \frac{(0.48)^x e^{-0.48}}{x!} \qquad x = 0, 1, 2, \ldots \qquad (10.9)$$

We thus have

$$P(A_i) = p_i = \frac{(0.48)^{i-1} e^{-0.48}}{(i-1)!} \qquad i = 1, 2, \ldots, 6$$

$$P(A_7) = p_7 = 1 - \sum_{i=1}^{6} p_i$$

whose values are indicated in the third column of Table 10.3.

Column 5 of Table 10.3 gives

$$d = \sum_{i=1}^{k} \frac{n_i^2}{np_i} - n = 8413 - 7842 = 571$$

With $k = 7$, the value of $\chi^2_{k-1,\alpha} = \chi^2_{6,0.01}$ is found from Table A.5 to be

$$\chi^2_{6,0.01} = 16.812$$

Since $d > \chi^2_{6,0.01}$, the hypothesis is rejected at 1 % significance level.

X.2.2 THE CASE OF ESTIMATED PARAMETERS

Let us now consider a more common situation in which parameters in the hypothesized distribution also need to be estimated from the data.

A natural procedure for a goodness-of-fit test in this case is to first estimate the parameters using one of the methods developed in Chapter IX and then follow the χ^2 test for known parameters already discussed in Section X.2.1. In doing so, however, a complication arises in that the theoretical probabilities p_i defined by Equation 10.2 are, being functions of the distribution parameters, functions of the sample. The statistic D now takes the form

$$D = \sum_{i=1}^{k} \frac{n}{\hat{P}_i} \left(\frac{N_i}{n} - \hat{P}_i \right)^2 = \sum_{i=1}^{k} \frac{N_i^2}{n\hat{P}_i} - n \qquad (10.10)$$

where $\hat{P}_i$ is an estimator for p_i and is thus a statistic. We see that D is now a much more complicated function of the sample $X_1, X_2, \ldots, X_n$. The important question to be answered is: What is the new distribution of D?

The problem of determining the limiting distribution of D in this situation was first considered by Fisher (1922, 1924) who showed that, as $n \to \infty$, the distribution of D needs to be modified and the modification obviously depends on the method of parameter estimation used. Fortunately, for a class of important methods of estimation, such as the maximum likelihood method, the modification required is a simple one, namely, the statistic D still approaches a χ^2 distribution as $n \to \infty$ but now with $(k - r - 1)$ degrees of freedom, where r is the number of parameters in the hypothesized distribution to be estimated. In other words, it is only necessary to reduce the number of degrees of freedom in the limiting distribution defined by Equation 10.5 by one for *each* parameter estimated from the sample.

We can now state a step-by-step procedure for the case in which r parameters in the distribution are to be estimated from the data.

1. Divide the range space of X into k mutually exclusive and numerically convenient intervals $A_i, i = 1, \ldots, k$. Let n_i be the number of sample values falling into A_i. As a rule, if the number of sample values in any A_i is less than five, combine the interval A_i with either A_{i-1} or A_{i+1}.
2. Estimate the r parameters by the method of maximum likelihood from the data.

MODEL VERIFICATION

3. Compute theoretical probabilities $P(A_i) = p_i$, $i = 1, \ldots, k$, by means of the hypothesized distribution with estimated parameter values.
4. Construct d as given by Equation 10.7.
5. Choose a value of α and determine from Table A.5 for the χ^2 distribution of $(k - r - 1)$ degrees of freedom the value of $\chi^2_{k-r-1,\alpha}$. It is assumed, of course, that $k - r - 1 > 0$.
6. Reject hypothesis H if $d > \chi^2_{k-r-1,\alpha}$. Otherwise, accept H.

Example 10.3. Vehicle arrivals at a toll gate on the New York Thruway were recorded. The vehicle counts at one-minute intervals were taken for 106 minutes and are given in Table 10.4. On the basis of these observations, determine whether a Poisson distribution is appropriate for X, the number of arrivals per minute, at 5% significance level.

Table 10.4 *One-minute Arrivals in Example 10.3*

Number of Vehicles per Minute	Number of Occurrences
0	0
1	0
2	1
3	3
4	5
5	7
6	13
7	12
8	8
9	9
10	13
11	10
12	5
13	6
14	4
15	5
16	4
17	0
18	1
	$n = \overline{106}$

The hypothesized distribution is

$$p_X(x) = \frac{\lambda^x e^{-\lambda}}{x!} \qquad x = 0, 1, 2, \ldots \qquad (10.11)$$

where the parameter λ needs to be estimated from the data. Thus, $r = 1$.

To proceed, we first determine the appropriate intervals A_i such that $n_i \geq 5$ for all i; these are shown in the first column of Table 10.5. Hence, $k = 11$.

The maximum likelihood estimate for λ is given by

$$\hat{\lambda} = \bar{x} = \frac{1}{n}\sum_{j=1}^{n} x_j = 9.09$$

The substitution of this value for the parameter λ in Equation 10.11 permits us to calculate the probabilities $P(A_i) = p_i$. For example,

$$p_1 = \sum_{j=0}^{4} p_X(j) = 0.052$$

$$p_2 = p_X(5) = 0.058$$

These theoretical probabilities are given in the third column of Table 10.5.

Table 10.5. *Table for χ^2 Test—Example 10.3*

Intervals A_i	Observed Number of Occurrences n_i	Theoretical $P(A_i)$ p_i	np_i	n_i^2/np_i
$0 \leq x < 5$	9	0.052	5.51	14.70
$5 \leq x < 6$	7	0.058	6.15	7.97
$6 \leq x < 7$	13	0.088	9.33	18.11
$7 \leq x < 8$	12	0.115	12.19	11.81
$8 \leq x < 9$	8	0.131	13.89	4.61
$9 \leq x < 10$	9	0.132	13.99	5.79
$10 \leq x < 11$	13	0.120	12.72	13.29
$11 \leq x < 12$	10	0.099	10.49	9.53
$12 \leq x < 13$	5	0.075	7.95	3.14
$13 \leq x < 14$	6	0.054	5.72	6.29
$14 \leq x$	14	0.076	8.06	24.32
	106	1.0	106	119.56

From column 5 of Table 10.5, we obtain

$$d = \sum_{i=1}^{k} \frac{n_i^2}{np_i} - n = 119.56 - 106 = 13.56$$

Table A.5 with $\alpha = 0.05$ and $k - r - 1 = 9$ degrees of freedom gives

$$\chi_{9,0.05}^2 = 16.92$$

Since $d < \chi_{9,0.05}^2$, the hypothesized distribution with $\lambda = 9.09$ is accepted at 5% significance level.

Example 10.4. Based upon the snowfall data given in Problem 8.8, test the hypothesis that the Buffalo yearly snowfall can be modeled by a normal distribution at 5% significance level.

For this problem, the assumed distribution for X, the Buffalo yearly snowfall measured in inches, is $N(m, \sigma^2)$ where m and σ^2 must be estimated from the data. Since the MLE for m and σ^2 are $\hat{M} = \bar{X}$ and $\widehat{\Sigma^2} = [(n-1)/n]S^2$, respectively, we have

$$\hat{m} = \bar{x} = \frac{1}{70} \sum_{j=1}^{70} x_j = 83.6$$

$$\hat{\sigma}^2 = \frac{69}{70} s^2 = \frac{1}{70} \sum_{j=1}^{70} (x_j - 83.6)^2 = 777.4$$

With intervals A_i defined as shown in the first column of Table 10.6, the theoretical probabilities $P(A_i)$ now can be calculated with the aid of Table A.3. For example, the first two of these probabilities are

$$P(A_1) = P(X \leq 56) = P\left(U \leq \frac{56 - 83.6}{\sqrt{777.4}}\right) = F_U(-0.990)$$

$$= 1 - F_U(0.990) = 1 - 0.8389 = 0.161$$

$$P(A_2) = P(56 < X \leq 72) = P(-0.990 < U \leq -0.416)$$

$$= [1 - F_U(0.416)] - [1 - F_U(0.990)]$$

$$= 0.339 - 0.161 = 0.178$$

The information given above allows us to construct Table 10.6 as shown below. Hence, we have

$$d = \sum_{i=1}^{k} \frac{n_i^2}{np_i} - n = 72.40 - 70 = 2.40$$

Table 10.6. Table for χ^2 Test—Example 10.4

Intervals A_i	Observed Number of Occurrences n_i	Theoretical $P(A_i)$ p_i	np_i	n_i^2/np_i
$x \leq 56$	13	0.161	11.27	15.00
$56 < x \leq 72$	10	0.178	12.46	8.03
$72 < x \leq 88$	20	0.224	15.68	25.51
$88 < x \leq 104$	13	0.205	14.35	11.78
$104 < x \leq 120$	8	0.136	9.52	6.72
$120 < x$	6	0.096	6.72	5.36
	70	1.0	70	72.40

The number of degrees of freedom in this case in $k - r - 1 = 6 - 2 - 1 = 3$. Table A.5 thus gives

$$\chi^2_{3, 0.05} = 7.815$$

Since $d < \chi^2_{3, 0.05}$, the normal distribution $N(83.6, 777.4)$ is acceptable at 5% significance level.

Before leaving this section, let us remark again that the statistic D in the χ^2 test is χ^2 distributed only when $n \to \infty$. It is thus a *large sample* test. As a rule, $n > 50$ is considered satisfactory for fulfilling the large sample requirement.

X.3 Kolmogorov–Smirnov Test

The so-called *Kolmogorov–Smirnov goodness-of-fit test*, referred to as the K–S test in the rest of the chapter, is based on a statistic which measures the deviation of observed *cumulative histogram* from the hypothesized *cumulative* distribution function.

Given a set of sample values $x_1, x_2, \ldots, x_n$ observed from a population X, a cumulative histogram can be constructed by (a) arranging the sample values in increasing order of magnitude, denoted here by $x_{(1)}, x_{(2)}, \ldots, x_{(n)}$, (b) determining the *observed* distribution function of X at $x_{(1)}, x_{(2)}, \ldots$, denoted by $F^0(x_{(1)})$, $F^0(x_{(2)}), \ldots$, from the relation $F^0(x_{(i)}) = i/n$, and (c) connecting the values of $F^0(x_{(i)})$ by straight-line segments.

The test statistic to be used in this case is

$$D_2 = \max_{i=1}^{n} \left[|F^0(X_{(i)}) - F_X(X_{(i)})| \right]$$
$$= \max_{i=1}^{n} \left[\left| \frac{i}{n} - F_X(X_{(i)}) \right| \right] \quad (10.12)$$

where $X_{(i)}$ is the ith order statistic of the sample. The statistic D_2 thus measures the maximum of absolute values of the n differences between observed PDF and hypothesized PDF evaluated at the observed sample values. In the case where parameters in the hypothesized distribution must be estimated, the values for $F_X(X_{(i)})$ are obtained using estimated parameter values.

While the distribution of D_2 is difficult to obtain analytically, its distribution function at various values can be computed numerically and tabulated. It can be shown that the probability distribution of D_2 is independent of the hypothesized distribution and is only a function of n, the sample size [see, for example, Massey (1951)].

The execution of the K–S test now follows that of the χ^2 test. At a specified α significance level, the form of the operating rule is to reject hypothesis H if $d_2 > c_{n,\alpha}$; otherwise, accept H. Here, d_2 is the sample value of D_2 and the value of $c_{n,\alpha}$ is defined by

$$P(D_2 > c_{n,\alpha}) = \alpha \quad (10.13)$$

The values of $c_{n,\alpha}$ for $\alpha = 0.01, 0.05$, and 0.10 are given in Table A.6 as functions of n.

It is instructive to note some of the important differences between this test and the χ^2 test. While the χ^2 test is a large sample test, the K–S test is valid for all values of n. Furthermore, the K–S test utilizes sample values in their unaltered and unaggregated form whereas data lumping is necessary in the execution of the χ^2 test. On the negative side, the K–S test is strictly valid only for continuous distributions. We also remark that the values of $c_{n,\alpha}$ given in Table A.6 are based on a completely specified hypothesized distribution. When the parameter values must be estimated, no rigorous method of adjustment is available. In these cases, it can only be stated that the values of $c_{n,\alpha}$ should be somewhat reduced.

The step-by-step procedure for executing the K–S test is now outlined as follows.

1. Rearrange sample values $x_1, x_2, \ldots, x_n$ in increasing order of magnitude and label them $x_{(1)}, x_{(2)}, \ldots, x_{(n)}$.
2. Determine the observed distribution function $F^0(x)$ at each $x_{(i)}$ using $F^0(x_{(i)}) = i/n$.

3. Determine the theoretical distribution function $F_X(x)$ at each $x_{(i)}$ using the hypothesized distribution. Parameters of the distribution are estimated from the data if necessary.
4. Form the differences $|F^0(x_{(i)}) - F_X(x_{(i)})|$ for $i = 1, 2, \ldots, n$.
5. Calculate $d_2 = \max_{i=1}^{n} [|F^0(x_{(i)}) - F_X(x_{(i)})|]$. The determination of this maximum value requires enumeration of n quantities. This labor can be somewhat reduced by plotting $F^0(x)$ and $F_X(x)$ as functions of x and noting the location of maximum by inspection.
6. Choose a value of α and determine from Table A.6 the value of $c_{n,\alpha}$.
7. Reject hypothesis H if $d_2 > c_{n,\alpha}$. Otherwise, accept H.

Example 10.5. Ten measurements of the tensile strength of one type of biological tissue are made. In dimensionless forms, they are 30.1, 30.5, 28.7, 31.6, 32.5, 29.0, 27.4, 29.1, 33.5, and 31.0. On the basis of this data set, test the hypothesis that the tensile strength follows a normal distribution at 5% significance level.

A reordering of the data yields $x_{(1)} = 27.4$, $x_{(2)} = 28.7, \ldots, x_{(10)} = 33.5$. The determination of $F^0(x_{(i)})$ is straightforward. We have, for example,

$$F^0(27.4) = 0.1 \qquad F^0(28.7) = 0.2 \cdots \qquad F^0(33.5) = 1$$

With regard to the theoretical distribution function, estimates of the mean and variance are first obtained from

$$\hat{m} = \bar{x} = \frac{1}{10} \sum_{j=1}^{10} x_j = 30.3$$

$$\hat{\sigma}^2 = \left(\frac{n-1}{n}\right) s^2 = \frac{1}{10} \sum_{j=1}^{10} (x_j - 30.3)^2 = 3.14$$

The values of $F_X(x_{(i)})$ can now be found based on the distribution $N(30.3, 3.14)$ for X. For example, with the aid of Table A.3 for the standardized normal r.v. U, we have

$$F_X(27.4) = F_U\left(\frac{27.4 - 30.3}{\sqrt{3.14}}\right) = F_U(-1.64)$$

$$= 1 - F_U(1.64) = 1 - 0.9495 = 0.0505$$

$$F_X(28.7) = F_U\left(\frac{28.7 - 30.3}{\sqrt{3.14}}\right) = F_U(-0.90)$$

$$= 1 - 0.8159 = 0.1841$$

and so on.

MODEL VERIFICATION

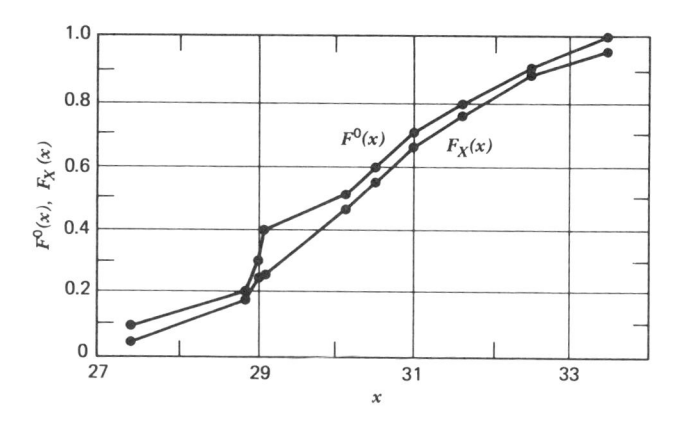

Figure 10.2 $F^0(x)$ and $F_X(x)$ in Example 10.5.

In order to determine d_2, it is constructive to plot $F^0(x)$ and $F_X(x)$ as functions of x as shown in Figure 10.2. It is clearly seen from the figure that the maximum of the differences between $F^0(x)$ and $F_X(x)$ occurs at $x = x_{(4)} = 29.1$. Hence,

$$d_2 = |F^0(29.1) - F_X(29.1)| = 0.4 - 0.2483 = 0.1517$$

With $\alpha = 0.05$ and $n = 10$, Table A.6 gives

$$c_{10, 0.05} = 0.41$$

Since $d_2 < c_{10, 0.05}$, we have no reason to reject the normal distribution $N(30.3, 3.14)$ at 5% significance level.

Let us remark that, since the parameter values were also estimated from the data, it is more appropriate to compare d_2 with a value somewhat smaller that 0.41. In view of the fact that the value of d_2 is well below 0.41, we are safe in making the conclusion given above.

REFERENCES AND COMMENTS

1. L. R. Beard, *Statistical Methods in Hydrology*, Army Corps of Engineers, Sacramento, Cal., 1962.
2. H. Cramér, *Mathematical Methods of Statistics*, Princeton University Press, Princeton, N.J., 1946.

3. R. A. Fisher, "On the Interpretation of χ^2 from Contingency Tables, and the Calculation of P," *J. Roy. Stat. Soc.*, **85**, 87–93, 1922.
4. R. A. Fisher, "The Conditions under which χ^2 Measures the Discrepancy between Observation and Hypothesis," *J. Roy. Stat. Soc.*, **87**, 442–476, 1924.
5. F. J. Massey, "The Kolmogorov Test for Goodness of Fit," *J. Am. Stat. Assoc.*, **46**, 68–78, 1951.
6. K. Pearson, "On a Criterion that a System of Deviations from the Probable in the Case of a Correlated System of Variables is such that it can be Reasonably Supposed to have Arisen in Random Sampling," *Phil. Mag.*, **50**, 157–175, 1900.

We have been rather selective in our choice of topics in this chapter. A number of important areas in hypotheses testing are not included and they can be found in more complete texts devoted to statistical inference, such as the following one.

7. E. L. Lehmann, *Testing Statistical Hypotheses*, Wiley, New York, 1959.

PROBLEMS

10.1. In the χ^2 test, is a hypothesized distribution more likely to be accepted at $\alpha = 0.05$ than at $\alpha = 0.01$? Explain.

10.2. To test whether or not a coin is fair, it is tossed 100 times with the following outcome: heads 41 times and tails 59 times. Is it fair on the basis of these tosses at 5% significance level?

10.3. Based upon telephone numbers listed on a typical page of a telephone directory, test the hypothesis that the last digit of telephone numbers is equally likely to be any number from zero to nine at 5% significance level.

10.4. The daily output of a production line is normally distributed with mean $m = 8000$ items and standard deviation $\sigma = 1000$ items. A second production line is set up and 100 daily output readings are shown in the following table. On the basis of this sample, does the second production line behave in the same statistical manner as the first? Use $\alpha = 0.01$.

MODEL VERIFICATION

Daily Output Intervals	Number of Occurrences
Below 4000	3
4000–5000	3
5000–6000	7
6000–7000	16
7000–8000	27
8000–9000	22
9000–10,000	11
10,000–11,000	8
11,000–12,000	2
Above 12,000	1
	$n = \overline{100}$

10.5. In a given plant, a sample of a given number of production items was taken from each of the five production lines and the number of defective items was recorded as follows. Test the hypothesis that the proportion of defects is constant from one line to another. Use $\alpha = 0.01$.

Production Line	Number of Defects
1	11
2	13
3	9
4	12
5	8

10.6. We have rejected in Example 10.2 the Poisson distribution with $\lambda = 0.08$ on the basis of accident data at 1% significance level. At the same α:
   (a) Would a Poisson distribution with λ estimated from the data be acceptable?
   (b) Would a negative binomial distribution be more appropriate?

10.7. Car pooling is encouraged in a city. A survey of 321 passenger vehicles coming into the city gives the following car occupancy profile. Suggest a probabilistic model for X, the number of passengers per vehicle, and test your hypothesized distribution at $\alpha = 0.05$ on the basis of this survey.

Number of Passengers per Vehicle (Driver Excluded)	Number of Vehicles
0	224
1	47
2	31
3	16
4 or more	3
	$n = 321$

10.8. Problem 8.4 gives 100 measurements of time gaps. On the basis of these data, postulate a likely distribution for X and test your hypothesis at 5% significance level.

10.9. Consider data given in Problem 8.5 for the sum of two consecutive time gaps. Postulate a likely distribution for X and test your hypothesis at 5% significance level.

10.10. Problem 8.6 gives data for one-minute vehicle arrivals. Postulate a likely distribution for X and test your hypothesis at 5% significance level.

10.11. Problem 8.9 gives a histogram for X, the peak combustion pressure. On the basis of these data, postulate a likely distribution for X and test your hypothesis at 5% significance level.

10.12. Suppose that the number of drivers sampled is 200. Based on the histogram given in Problem 8.10, postulate a likely distribution for X_1 and test your hypothesis at 1% significance level.

10.13. Problem 8.11 gives a histogram for X, the number of blemishes in television tubes. On the basis of this sample, postulate a likely distribution for X and test your hypothesis at 5% significance level.

10.14. Twenty-four readings of the annual sediment load (in 10^6 tons) in the Colorado river at the Grand Canyon are (arranged in increasing order of magnitude) 49, 50, 50, 66, 70, 75, 84, 85, 98, 118, 122, 135, 143, 146, 157, 172, 177, 190, 225, 235, 265, 270, 400, 480. Using the Kolmogorov–Smirnov test at 5% significance level, test the hypothesis that the annual sediment load follows a lognormal distribution. [Data taken from Beard (1962).]

10.15. For the snowfall data given in Problem 8.8, use the Kolmogorov–Smirnov test and test the normal distribution hypothesis on the basis of snowfall data from 1969–1979.

XI
Linear Models and Linear Regression

The tools developed in Chapters IX and X for parameter estimation and model verification are applied in this chapter to a very useful class of models encountered in science and engineering. A commonly occurring situation is one in which a random quantity, Y, is a function of one or more independent (and deterministic) variables $x_1, x_2, \ldots,$ and x_m. For example, wind load (Y) acting on a structure is a function of height (x), the intensity (Y) of strong motion earthquakes is dependent on the distance from the epicenter (x), housing price (Y) is a function of location (x_1) and age (x_2), and chemical yield (Y) may be related to temperature (x_1), pressure (x_2), and acid content (x_3).

Given a sample of Y values with their associated values of $x_i, i = 1, 2, \ldots, m,$ we are interested in estimating on the basis of this sample the relationship between Y and the independent variables $x_1, x_2, \ldots,$ and x_m. In what follows, we concentrate on some simple cases of the broadly defined problem stated above.

XI.1 Simple Linear Regression

We assume in this section that r.v. Y is a function of only one independent variable and that their relationship is linear. By a linear relationship we mean that the mean of Y, $E\{Y\}$, is known to be a linear function of x, that is,

$$E\{Y\} = \alpha + \beta x \qquad (11.1)$$

The two constants, intercept α and slope β, are unknown and are to be estimated from a sample of Y values with their associated values of x. Note that $E\{Y\}$ is a function x. In any single experiment, x will assume a certain value x_i and the mean of Y will take the value

$$E\{Y_i\} = \alpha + \beta x_i \qquad (11.2)$$

The r.v. Y is, of course, itself a function of x. If we define a random variable E by

$$E = Y - (\alpha + \beta x) \qquad (11.3)$$

we can write

$$Y = \alpha + \beta x + E \qquad (11.4)$$

where E has mean zero and variance σ^2, which is identical with the variance of Y. The value of σ^2 is not known in general but it is assumed to be a constant and not a function of x.

Equation 11.4 is a standard expression of a *simple linear regression model*. The unknown parameters α and β are called *regression coefficients* and r.v. E represents the deviation of Y about its mean. As with simple models discussed in Chapters IX and X, simple linear regression analysis is concerned with estimation of the regression parameters, quality of these estimators, and model verification on the basis of the sample. We note that, instead of a simple sample such as $Y_1, Y_2, \ldots, Y_n$ as in previous cases, our sample in the present context takes the form of pairs $(x_1, Y_1), (x_2, Y_2), \ldots, (x_n, Y_n)$. For each value x_i assigned to x, Y_i is an independent observation from the population Y defined by Equation 11.4. Hence, (x_j, Y_j), $j = 1, 2, \ldots, n$, may be considered as a sample from the r.v. Y for given values $x_1, x_2, \ldots,$ and x_n of x; these x-values need not be all distinct but, in order to estimate both α and β, we will see that we must have at least two distinct values of x represented in the sample.

XI.1.1 LEAST SQUARES METHOD OF ESTIMATION

As one approach to point estimation of regression parameters α and β, the method of least squares suggests that their estimates, $\hat{\alpha}$ and $\hat{\beta}$, be chosen so that the sum of

LINEAR MODELS AND LINEAR REGRESSION

the squared differences between observed sample values y_i and the estimated expected value of Y, $\hat{\alpha} + \hat{\beta} x_i$, is minimized. Let us write

$$e_i = y_i - (\hat{\alpha} + \hat{\beta} x_i) \tag{11.5}$$

The least-square estimates $\hat{\alpha}$ and $\hat{\beta}$, respectively, of α and β are found by minimizing

$$Q = \sum_{i=1}^{n} e_i^2 = \sum_{i=1}^{n} [y_i - (\hat{\alpha} + \hat{\beta} x_i)]^2 \tag{11.6}$$

In the above, the sample-value pairs are $(x_1, y_1), (x_2, y_2), \ldots, (x_n, y_n)$ and the e_i's are called the *residuals*. Figure 11.1 gives a graphical presentation of this procedure. We see that the residuals are the vertical distances between the observed values of Y, y_i, and the least-square estimate $\hat{\alpha} + \hat{\beta} x$ of the true regression line $\alpha + \beta x$.

The estimates α and β are easily found based on the least squares procedure. The results are stated below as a theorem.

Theorem

Consider the simple linear regression model defined by Equation 11.4. Let $(x_1, y_1), (x_2, y_2), \ldots, (x_n, y_n)$ be an observed sample of Y with associated values of x. Then the least-square estimates of α and β are

$$\hat{\alpha} = \bar{y} - \hat{\beta} \bar{x} \tag{11.7}$$

$$\hat{\beta} = \frac{\sum_{i=1}^{n} (x_i - \bar{x})(y_i - \bar{y})}{\sum_{i=1}^{n} (x_i - \bar{x})^2} \tag{11.8}$$

where

$$\bar{x} = \frac{1}{n} \sum_{i=1}^{n} x_i \quad \text{and} \quad \bar{y} = \frac{1}{n} \sum_{i=1}^{n} y_i$$

Proof

The estimates $\hat{\alpha}$ and $\hat{\beta}$ are found by taking partial derivatives of Q given by Equation 11.6 with respect to $\hat{\alpha}$ and $\hat{\beta}$, setting these derivatives to zero, and solving for $\hat{\alpha}$ and $\hat{\beta}$, which minimize the sum of squared residuals. Hence, we have

$$\frac{\partial Q}{\partial \hat{\alpha}} = \sum_{i=1}^{n} -2[y_i - (\hat{\alpha} + \hat{\beta} x_i)]$$

$$\frac{\partial Q}{\partial \hat{\beta}} = \sum_{i=1}^{n} -2x_i[y_i - (\hat{\alpha} + \hat{\beta} x_i)]$$

PARAMETER ESTIMATION AND MODEL VERIFICATION

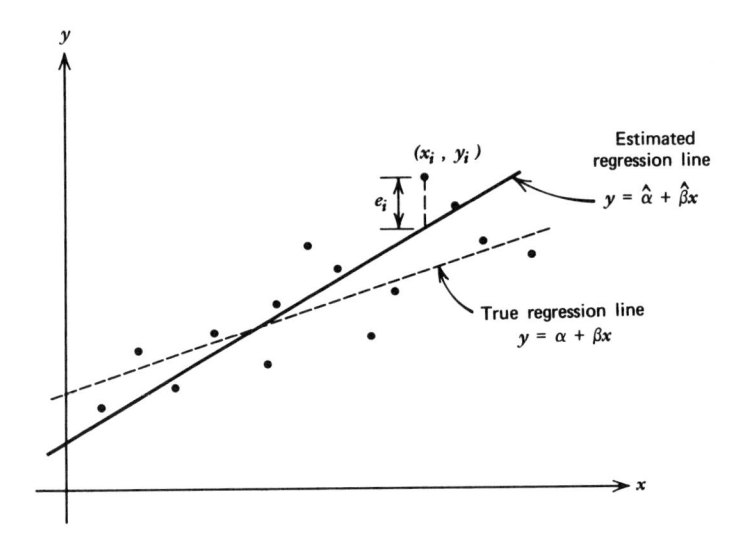

Figure 11.1 Least squares method of estimation.

Upon simplifying and setting the above equations to zero, we have the so-called *normal equations*

$$n\hat{\alpha} + n\bar{x}\hat{\beta} = n\bar{y} \tag{11.9}$$

$$n\bar{x}\hat{\alpha} + \hat{\beta}\sum_{i=1}^{n} x_i^2 = \sum_{i=1}^{n} x_i y_i \tag{11.10}$$

Their solutions are easily found to be those given by Equations 11.7 and 11.8.

To ensure that these solutions correspond to the minimum of the sum of squared residuals, we need to verify that

$$\frac{\partial^2 Q}{\partial \hat{\alpha}^2} > 0$$

and

$$D = \begin{vmatrix} \dfrac{\partial^2 Q}{\partial \hat{\alpha}^2} & \dfrac{\partial^2 Q}{\partial \hat{\alpha}\, \partial \hat{\beta}} \\ \dfrac{\partial^2 Q}{\partial \hat{\alpha}\, \partial \hat{\beta}} & \dfrac{\partial^2 Q}{\partial \hat{\beta}^2} \end{vmatrix} > 0$$

at $\hat{\alpha}$ and $\hat{\beta}$. Elementary calculations show that

$$\frac{\partial^2 Q}{\partial \hat{\alpha}^2} = 2n > 0 \quad \text{and} \quad D = 4n \sum_{i=1}^{n} (x_i - \bar{x})^2 > 0$$

The proof of this theorem is thus complete. Note that D would be zero if all x_i's take the same value. Hence, at least two distinct x_i-values are needed for the determination of $\hat{\alpha}$ and $\hat{\beta}$.

It is instructive at this point to restate the foregoing results using a more compact vector-matrix notation. As we will see, results in vector-matrix form facilitate calculations. Also, they are capable of easy generalizations when we consider more general regression models.

In terms of the observed sample values $(x_1, y_1), (x_2, y_2), \ldots, (x_n, y_n)$, we have a system of observed regression equations

$$y_i = \alpha + \beta x_i + e_i \quad i = 1, 2, \ldots, n \tag{11.11}$$

Let

$$C = \begin{bmatrix} 1 & x_1 \\ 1 & x_2 \\ \vdots & \vdots \\ 1 & x_n \end{bmatrix} \quad \mathbf{y} = \begin{bmatrix} y_1 \\ y_2 \\ \vdots \\ y_n \end{bmatrix} \quad \mathbf{e} = \begin{bmatrix} e_1 \\ e_2 \\ \vdots \\ e_n \end{bmatrix}$$

and

$$\boldsymbol{\theta} = \begin{bmatrix} \alpha \\ \beta \end{bmatrix}$$

Equations 11.11 can be represented by the vector-matrix equation

$$\mathbf{y} = C\boldsymbol{\theta} + \mathbf{e} \tag{11.12}$$

The sum of squared residuals given by Equation 11.6 is now

$$Q = \mathbf{e}^T \mathbf{e} = (\mathbf{y} - C\boldsymbol{\theta})^T (\mathbf{y} - C\boldsymbol{\theta}) \tag{11.13}$$

The least-square estimate of $\boldsymbol{\theta}$, $\hat{\boldsymbol{\theta}}$, is found by minimizing Q. Applying the variational principle discussed in Section IX.4.1, we have

$$\delta Q = -\delta \boldsymbol{\theta}^T C^T (\mathbf{y} - C\boldsymbol{\theta}) - (\mathbf{y} - C\boldsymbol{\theta})^T C \delta \boldsymbol{\theta}$$
$$= -2\delta \boldsymbol{\theta}^T C^T (\mathbf{y} - C\boldsymbol{\theta})$$

Setting $\delta Q = 0$, the solution for $\hat{\boldsymbol{\theta}}$ is obtained from the normal equation

$$C^T(\mathbf{y} - C\hat{\boldsymbol{\theta}}) = 0 \tag{11.14}$$

or

$$C^T C\hat{\boldsymbol{\theta}} = C^T \mathbf{y}$$

which gives

$$\boxed{\hat{\boldsymbol{\theta}} = (C^T C)^{-1} C^T \mathbf{y}} \tag{11.15}$$

In the above, the inverse of matrix $C^T C$ exists if there are at least two distinct values of x_i represented in the sample.

We can easily check that Equation 11.15 is identical to Equations 11.7 and 11.8 by noting that

$$C^T C = \begin{bmatrix} 1 & 1 & \cdots & 1 \\ x_1 & x_2 & \cdots & x_n \end{bmatrix} \begin{bmatrix} 1 & x_1 \\ 1 & x_2 \\ \vdots & \vdots \\ 1 & x_n \end{bmatrix} = \begin{bmatrix} n & n\bar{x} \\ n\bar{x} & \sum_{i=1}^{n} x_i^2 \end{bmatrix}$$

$$C^T \mathbf{y} = \begin{bmatrix} 1 & 1 & \cdots & 1 \\ x_1 & x_2 & \cdots & x_n \end{bmatrix} \begin{bmatrix} y_1 \\ y_2 \\ \vdots \\ y_n \end{bmatrix} = \begin{bmatrix} n\bar{y} \\ \sum_{i=1}^{n} x_i y_i \end{bmatrix}$$

and

$$\hat{\boldsymbol{\theta}} = (C^T C)^{-1} C^T \mathbf{y} = \begin{bmatrix} n & n\bar{x} \\ n\bar{x} & \sum_{i=1}^{n} x_i^2 \end{bmatrix}^{-1} \begin{bmatrix} n\bar{y} \\ \sum_{i=1}^{n} x_i y_i \end{bmatrix}$$

$$= \begin{bmatrix} \bar{y} - \left[\dfrac{\sum_{i=1}^{n} x_i y_i - n\bar{x}\bar{y}}{\sum_{i=1}^{n} x_i^2 - n\bar{x}^2} \right] \bar{x} \\ \dfrac{\sum_{i=1}^{n} x_i y_i - n\bar{x}\bar{y}}{\sum_{i=1}^{n} x_i^2 - n\bar{x}^2} \end{bmatrix}$$

$$= \begin{bmatrix} \bar{y} - \hat{\beta}\bar{x} \\ \dfrac{\sum_{i=1}^{n} (x_i - \bar{x})(y_i - \bar{y})}{\sum_{i=1}^{n} (x_i - \bar{x})^2} \end{bmatrix}$$

LINEAR MODELS AND LINEAR REGRESSION

Example 11.1. It is expected that the average percent yield, Y, from a chemical process is linearly related to the process temperature, x, in °C. Determine the least-square regression line for $E\{Y\}$ on the basis of 10 observations given below.

i	1	2	3	4	5	6	7	8	9	10
x_i (°C)	45	50	55	60	65	70	75	80	85	90
y_i (% yield)	43	45	48	51	55	57	59	63	66	68

In view of Equations 11.7 and 11.8, we need the following quantities:

$$\bar{x} = \frac{1}{n}\sum_{i=1}^{n} x_i = \tfrac{1}{10}(45 + 50 + \cdots + 90) = 67.5$$

$$\bar{y} = \frac{1}{n}\sum_{i=1}^{n} y_i = \tfrac{1}{10}(43 + 45 + \cdots + 68) = 55.5$$

$$\sum_{i=1}^{n} (x_i - \bar{x})^2 = 2062.5$$

$$\sum_{i=1}^{n} (x_i - \bar{x})(y_i - \bar{y}) = 1182.5$$

The substitution of these values into Equations 11.7 and 11.8 gives

$$\hat{\beta} = \frac{1182.5}{2062.5} = 0.57$$

$$\hat{\alpha} = 55.5 - 0.57(67.5) = 17.03$$

The estimated regression line together with observed sample values is shown in Figure 11.2.

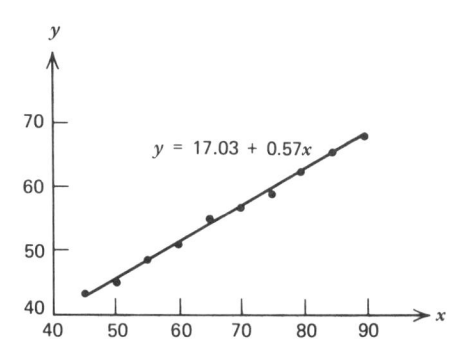

Figure 11.2 Estimated regression line and observed data in Example 11.1.

It is noteworthy that regression relationships are only valid for the range of x-values represented by the data. Thus, the estimated regression line in this example only holds for temperatures between 45 and 90°C. Extrapolation of the result beyond this range can be misleading and is not valid in general.

Another word of caution has to do with the basic linear assumption between $E\{Y\}$ and x. Linear regression analysis such as the one performed above is based on the assumption that the true relationship between $E\{Y\}$ and x is linear. Indeed, if the underlying relationship is nonlinear or nonexistent, linear regression produces meaningless results even if a straight line appears to provide a good fit to the data.

XI.1.2 PROPERTIES OF LEAST-SQUARE ESTIMATORS

The properties of estimators for regression coefficients α and β can be determined in a straightforward fashion following the vector-matrix expression (11.15). Let $\hat{A}$ and $\hat{B}$ denote, respectively, the estimators for α and β following the method of least squares and let

$$\hat{\Theta} = \begin{bmatrix} \hat{A} \\ \hat{B} \end{bmatrix} \tag{11.16}$$

We see from Equation 11.15 that

$$\hat{\Theta} = (C^T C)^{-1} C^T Y \tag{11.17}$$

where

$$Y = \begin{bmatrix} Y_1 \\ \vdots \\ Y_n \end{bmatrix} \tag{11.18}$$

and $Y_j, j = 1, 2, \ldots, n$, are independent and identically distributed according to Equation 11.4. Thus, if we write

$$Y = C\Theta + E \tag{11.19}$$

then E is a zero-mean random vector with covariance matrix $\Lambda = \sigma^2 I$, I being an $n \times n$ identity matrix.

The mean and variance of the estimator $\hat{\Theta}$ are now easily determined. In view of Equations 11.17 and 11.19, we have

$$\begin{aligned} E\{\hat{\Theta}\} &= (C^TC)^{-1}C^TE\{\mathbf{Y}\} \\ &= (C^TC)^{-1}C^T[C\theta + E\{\mathbf{E}\}] \\ &= (C^TC)^{-1}(C^TC)\theta = \theta \end{aligned} \qquad (11.20)$$

Hence, the estimators $\hat{A}$ and $\hat{B}$ for α and β, respectively, are unbiased.

The covariance matrix associated with $\hat{\Theta}$ is given by, as seen from Equation 11.17,

$$\begin{aligned} \operatorname{cov}\{\hat{\Theta}\} &= E\{(\hat{\Theta} - \theta)(\hat{\Theta} - \theta)^T\} \\ &= (C^TC)^{-1}C^T \operatorname{cov}\{\mathbf{Y}\}C(C^TC)^{-1} \end{aligned}$$

But $\operatorname{cov}\{\mathbf{Y}\} = \sigma^2 I$, we thus have

$$\operatorname{cov}\{\hat{\Theta}\} = \sigma^2(C^TC)^{-1}C^TC(C^TC)^{-1} = \sigma^2(C^TC)^{-1} \qquad (11.21)$$

The diagonal elements of the matrix in Equation 11.21 give the variances of $\hat{A}$ and $\hat{B}$. In terms of the elements of C, we can write

$$\operatorname{var}\{\hat{A}\} = \frac{\sigma^2 \sum_{i=1}^{n} x_i^2}{n \sum_{i=1}^{n} (x_i - \bar{x})^2} \qquad (11.22)$$

$$\operatorname{var}\{\hat{B}\} = \frac{\sigma^2}{\sum_{i=1}^{n} (x_i - \bar{x})^2} \qquad (11.23)$$

It is seen that these variances decrease as sample size n increases according to $1/n$. Thus, it follows from our discussion in Chapter IX that these estimators are consistent, a desirable property. We further note that, for a fixed n, the variance of $\hat{B}$ can be reduced by selecting the x_i's in such a way that the denominator of Equation 11.23 is maximized; this can be accomplished by spreading the x_i's as far apart as possible. In Example 11.1, for example, assuming that we are free to choose the values of x_i, the quality of $\hat{\beta}$ is improved if one-half of the x readings are taken at one extreme of the temperature range and the other half at the other extreme. On the other hand, the sampling strategy for minimizing $\operatorname{var}(\hat{A})$ for a fixed n is to make $\bar{x}$ as close to zero as possible.

Are the variances given by Equations 11.22 and 11.23 minimum variances associated with any unbiased estimators for α and β? An answer to this important question can be found by comparing the results given by Equations 11.22 and 11.23 with the Cramér–Rao lower bounds defined in Chapter IX. In order to evaluate these lower bounds, a probability distribution of Y must be made available. Without this knowledge, however, we can still show in the following theorem that the least squares technique leads to *linear unbiased minimum-variance estimators* for α and β, that is, among all unbiased estimators which are *linear* in $\mathbf{Y}$, least-square estimators have minimum variance.

Theorem

Let r.v. Y be defined by Equation 11.4. Given a sample $(x_1, Y_1), (x_2, Y_2), \ldots, (x_n, Y_n)$ of Y with its associated x values, the least-square estimators $\hat{A}$ and $\hat{B}$ given by Equation 11.17 are, respectively, *minimum-variance linear unbiased estimators* for α and β.

Proof

The proof of this important theorem is sketched below using the vector-matrix notation.

Consider a linear unbiased estimator of the form

$$\mathbf{\Theta}^* = [(C^T C)^{-1} C^T + G]\mathbf{Y} \qquad (11.24)$$

We thus wish to prove that $G = 0$ if $\mathbf{\Theta}^*$ is to be minimum variance.

The unbiasedness requirement leads to, in view of Equation 11.19,

$$GC = 0 \qquad (11.25)$$

Consider now the covariance matrix

$$\operatorname{cov}\{\mathbf{\Theta}^*\} = E\{(\mathbf{\Theta}^* - \mathbf{\theta})(\mathbf{\Theta}^* - \mathbf{\theta})^T\} \qquad (11.26)$$

Upon using Equations 11.19, 11.24, and 11.25 and expanding the covariance, we have

$$\operatorname{cov}\{\mathbf{\Theta}^*\} = \sigma^2[(C^T C)^{-1} + GG^T]$$

Now, in order to minimize the variances associated with the components of $\mathbf{\Theta}^*$, we must minimize each diagonal element of GG^T. Since the iith diagonal element of GG^T is given by

$$(GG^T)_{ii} = \sum_{j=1}^{n} g_{ij}^2$$

where g_{ij} is the ijth element of G, we must have

$$g_{ij} = 0 \qquad \text{for all } i \text{ and } j$$

LINEAR MODELS AND LINEAR REGRESSION

and we obtain

$$G = 0 \tag{11.27}$$

This completes the proof. The theorem stated above is a special case of the *Gauss–Markov theorem*.

Another interesting comparison is that between the least-square estimators for α and β and their maximum likelihood estimators with an assigned distribution for the r.v. Y. It is left as an exercise to show that the maximum likelihood estimators for α and β are identical with their least-square counterparts under the added assumption that Y is normally distributed.

XI.1.3 AN UNBIASED ESTIMATOR FOR σ^2

As we have shown, the method of least squares does not lead to an estimator for the variance σ^2 of Y, which is also an unknown quantity in general in linear regression models. In order to propose an estimator for σ^2, an intuitive choice is

$$\widehat{\Sigma^2} = k \sum_{i=1}^{n} [Y_i - (\hat{A} + \hat{B}x_i)]^2 \tag{11.28}$$

where the coefficient k is to be chosen so that $\widehat{\Sigma^2}$ is unbiased. In order to carry out the expectation of $\widehat{\Sigma^2}$, we note that (see Equation 11.7)

$$Y_i - \hat{A} - \hat{B}x_i = Y_i - (\bar{Y} - \hat{B}\bar{x}) - \hat{B}x_i$$
$$= (Y_i - \bar{Y}) - \hat{B}(x_i - \bar{x}) \tag{11.29}$$

Hence, it follows that

$$\sum_{i=1}^{n} (Y_i - \hat{A} - \hat{B}x_i)^2 = \sum_{i=1}^{n} (Y_i - \bar{Y})^2 - \hat{B}^2 \sum_{i=1}^{n} (x_i - \bar{x})^2 \tag{11.30}$$

since (see Equation 11.8)

$$\sum_{i=1}^{n} (x_i - \bar{x})(Y_i - \bar{Y}) = \hat{B} \sum_{i=1}^{n} (x_i - \bar{x})^2 \tag{11.31}$$

Upon taking expectations terms by term, we can show that

$$E\{\widehat{\Sigma^2}\} = kE\left\{\sum_{i=1}^{n}(Y_i - \bar{Y})^2 - \hat{B}^2 \sum_{i=1}^{n}(x_i - \bar{x})^2\right\}$$
$$= k(n-2)\sigma^2$$

Hence, $\widehat{\Sigma^2}$ is unbiased with $k = 1/(n-2)$, giving

$$\widehat{\Sigma^2} = \frac{1}{n-2} \sum_{i=1}^{n} [Y_i - (\hat{A} + \hat{B}x_i)]^2 \qquad (11.32)$$

or, in view of Equation 11.30,

$$\widehat{\Sigma^2} = \frac{1}{n-2} \left[\sum_{i=1}^{n} (Y_i - \bar{Y})^2 - \hat{B}^2 \sum_{i=1}^{n} (x_i - \bar{x})^2 \right] \qquad (11.33)$$

Example 11.2. In Example 11.1, use the results given above and determine an unbiased estimate for σ^2.

We have found in Example 11.1 that

$$\sum_{i=1}^{n} (x_i - \bar{x})^2 = 2062.5$$

$$\hat{\beta} = 0.57$$

In addition, we easily obtain

$$\sum_{i=1}^{n} (y_i - \bar{y})^2 = 680.5$$

Equation 11.33 thus gives

$$\widehat{\sigma^2} = \tfrac{1}{8}[680.5 - (0.57)^2(2062.5)]$$
$$= 1.30$$

Example 11.3. An experiment on lung tissue elasticity as a function of lung expansion properties is performed and the measurements given below are those of the tissue's Young's modulus in g/cm^2(Y) at varying values of lung expansion in terms of stress in g/cm^2(x). Assuming that $E\{Y\}$ is linearly related to x and that $\sigma_Y^2 = \sigma^2$ (a constant), determine the least-square estimates of the regression coefficients and an unbiased estimate of σ^2.

x	2	2.5	3	5	7	9	10	12	15	16	17	18	19	20
y	9.1	19.2	18.0	31.3	40.9	32.0	54.3	49.1	73.0	91.0	79.0	68.0	110.5	130.8

In this case, we have $n = 14$. The quantities of interest are

$$\bar{x} = \frac{1}{n}\sum_{i=1}^{n} x_i = \tfrac{1}{14}(2 + 2.5 + \cdots + 20) = 11.11$$

$$\bar{y} = \frac{1}{n}\sum_{i=1}^{n} y_i = \tfrac{1}{14}(9.1 + 19.2 + \cdots + 130.8) = 57.59$$

$$\sum_{i=1}^{n}(x_i - \bar{x})^2 = 546.09$$

$$\sum_{i=1}^{n}(y_i - \bar{y})^2 = 17{,}179.54$$

$$\sum_{i=1}^{n}(x_i - \bar{x})(y_i - \bar{y}) = 2862.12$$

The substitution of these values into Equations 11.7, 11.8, and 11.33 gives

$$\hat{\beta} = 2862.12/546.09 = 5.24$$

$$\hat{\alpha} = 57.59 - 5.24(11.11) = -0.63$$

$$\widehat{\sigma^2} = \tfrac{1}{12}[17{,}179.54 - (5.24)^2(546.09)] = 182.10$$

The estimated regression line together with the data is shown in Figure 11.3. The estimated standard deviation is $\hat{\sigma} = \sqrt{182.10} = 13.49$ g/cm^2 and the 1σ – band is also shown in the figure.

XI.1.4 CONFIDENCE INTERVALS FOR REGRESSION COEFFICIENTS

In addition to point estimators for the slope and intercept in linear regression, it is also easy to construct confidence intervals for them and for $\alpha + \beta x$, the mean of Y, under certain distributional assumptions. In what follows, let us assume that Y is normally distributed according to $N(\alpha + \beta x, \sigma^2)$. Since the estimators $\hat{A}$, $\hat{B}$, and $\hat{A} + \hat{B}x$ are linear functions of the sample of Y, they are also normal random variables. Let us note that, when sample size n is large, $\hat{A}$, $\hat{B}$, and $\hat{A} + \hat{B}x$ are expected to follow normal distributions as a consequence of the central limit theorem, no matter how Y is distributed.

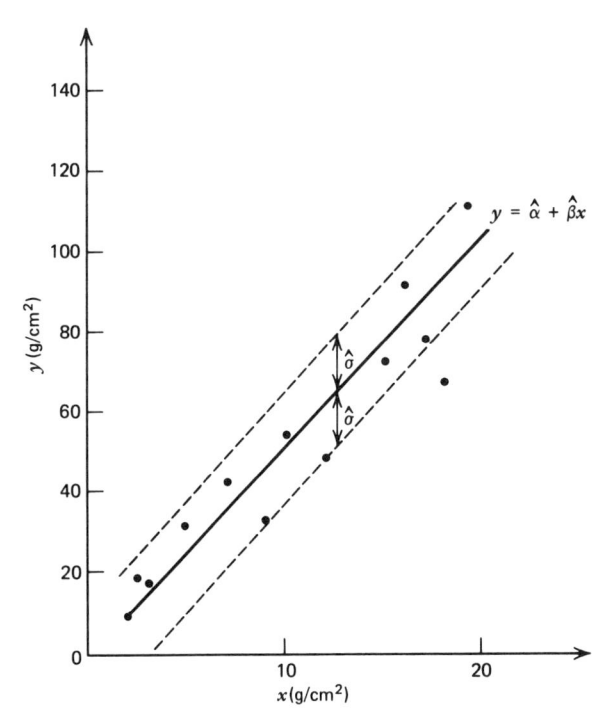

Figure 11.3 Estimated regression line and observed data in Example 11.3.

We follow our development in Section IX.5 in establishing the desired confidence limits. Based on our experience in Section IX.5, the following are not difficult to verify.

1. Let $\widehat{\Sigma}^2$ be the unbiased estimator for σ^2 as defined by Equation 11.33 and let

$$D = \frac{(n-2)\widehat{\Sigma}^2}{\sigma^2} \tag{11.34}$$

It follows from the results given in Section IX.5.3 that D is a χ^2 distributed random variable with $(n-2)$ degrees of freedom.

2. Consider the random variables

$$\frac{\hat{A} - \alpha}{\sqrt{\widehat{\Sigma}^2 \sum_{i=1}^{n} x_i^2 / [n \sum_{i=1}^{n} (x_i - \bar{x})^2]}} \tag{11.35}$$

and

$$\frac{\hat{B} - \beta}{\sqrt{\widehat{\Sigma^2}/[\sum_{i=1}^{n} (x_i - \bar{x})^2]}} \tag{11.36}$$

where, as seen from Equations 11.20, 11.22, and 11.23, α and β are, respectively, the means of $\hat{A}$ and $\hat{B}$ and the denominators are, respectively, the standard deviations of $\hat{A}$ and $\hat{B}$ with σ^2 estimated by $\widehat{\Sigma^2}$. The derivation given in Section IX.5.2 shows that each of these random variables has a t-distribution with $(n - 2)$ degrees of freedom.

3. The estimator $\widehat{E\{Y\}}$ for the mean of Y is normally distributed with mean $\alpha + \beta x$ and variance

$$\begin{aligned} \text{var}\{\widehat{E\{Y\}}\} &= \text{var}\{\hat{A} + \hat{B}x\} \\ &= \text{var}\{\hat{A}\} + x^2 \, \text{var}\{\hat{B}\} + 2x \, \text{cov}\{\hat{A}, \hat{B}\} \\ &= \frac{\sigma^2}{\sum_{i=1}^{n} (x_i - \bar{x})^2} \left(\frac{1}{n} \sum_{i=1}^{n} x_i^2 + x^2 - 2x\bar{x} \right) \\ &= \sigma^2 \left[\frac{1}{n} + \frac{(x - \bar{x})^2}{\sum_{i=1}^{n} (x_i - \bar{x})^2} \right] \end{aligned} \tag{11.37}$$

Hence, again following the derivation given in Section IX.5.2, the random variable

$$\frac{\widehat{E\{Y\}} - (\alpha + \beta x)}{\sqrt{\widehat{\Sigma^2} \left[\frac{1}{n} + \frac{(x - \bar{x})^2}{\sum_{i=1}^{n} (x_i - \bar{x})^2} \right]}} \tag{11.38}$$

is also t-distributed with $(n - 2)$ degrees of freedom.

Based on the results presented above, we can now easily establish confidence limits for all the parameters of interest. The results given below are a direct consequence of the development in Section IX.5.

1. A $100(1 - \gamma)\%$ confidence interval for α is determined by (see Equation 9.141)

$$\boxed{L_{1,2} = \hat{A} \mp t_{n-2, \gamma/2} \sqrt{\frac{\widehat{\Sigma^2} \sum_{i=1}^{n} x_i^2}{n \sum_{i=1}^{n} (x_i - \bar{x})^2}}} \tag{11.39}$$

2. A $100(1-\gamma)\%$ confidence interval for β is determined by (see Equation 9.141)

$$L_{1,2} = \hat{B} \mp t_{n-2,\gamma/2} \sqrt{\frac{\widehat{\Sigma^2}}{\sum_{i=1}^{n}(x_i - \bar{x})^2}} \qquad (11.40)$$

3. A $100(1-\gamma)\%$ confidence interval for $E\{Y\} = \alpha + \beta x$ is determined by (see Equation 9.141).

$$L_{1,2} = \widehat{E\{Y\}} \mp t_{n-2,\gamma/2} \sqrt{\widehat{\Sigma^2}\left[\frac{1}{n} + \frac{(x-\bar{x})^2}{\sum_{i=1}^{n}(x_i - \bar{x})^2}\right]} \qquad (11.41)$$

4. A two-sided $100(1-\gamma)\%$ confidence interval for σ^2 is determined by (see Equation 9.144)

$$L_1 = \frac{(n-2)\widehat{\Sigma^2}}{\chi^2_{n-2,\gamma/2}}$$
$$L_2 = \frac{(n-2)\widehat{\Sigma^2}}{\chi^2_{n-2,1-\gamma/2}} \qquad (11.42)$$

If a one-sided confidence interval for σ^2 is desired, it is given by (see Equation 9.145)

$$L_1 = \frac{(n-2)\widehat{\Sigma^2}}{\chi^2_{n-2,\gamma}} \qquad (11.43)$$

A number of observations can be made regarding these confidence intervals. In each case, both position and width of the interval will vary from sample to sample. In addition, the confidence interval for $\alpha + \beta x$ is shown to be a function of x. If one is to plot the observed values of L_1 and L_2, they will form a *confidence band* about the estimated regression line as shown in Figure 11.4. Equation 11.41 clearly shows that the narrowest point of the band occurs at $x = \bar{x}$; it becomes broader as x moves away from $\bar{x}$ in either direction.

Example 11.4. In Example 11.3, assuming that Y is normally distributed, determine a 95% confidence band for the mean $\alpha + \beta x$.

LINEAR MODELS AND LINEAR REGRESSION

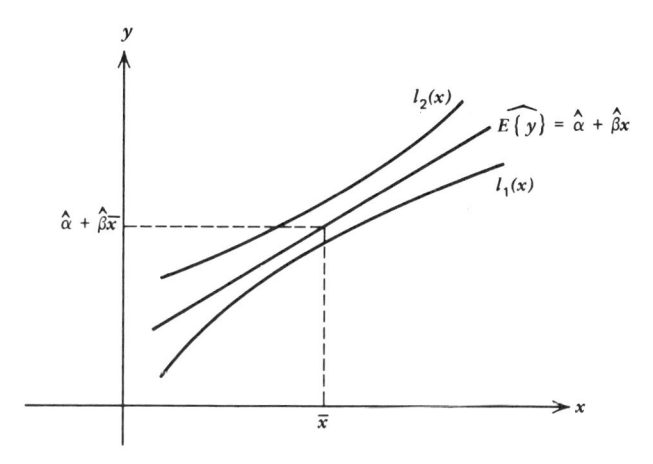

Figure 11.4 Confidence band for $E\{Y\} = \alpha + \beta x$.

Equation 11.41 gives the desired confidence limits with $n = 14$, $\gamma = 0.05$, and

$$\widehat{E\{y\}} = \hat{\alpha} + \hat{\beta}x = -0.63 + 5.24x$$

$$t_{n-2,\gamma/2} = t_{12,0.025} = 2.179 \text{ (from Table A.4)}$$

$$\bar{x} = 11.11$$

$$\sum_{i=1}^{n}(x_i - \bar{x})^2 = 546.09$$

$$\widehat{\sigma^2} = 182.10$$

The observed confidence limits are thus given by

$$l_{1,2} = (-0.63 + 5.24x) \mp 2.179 \sqrt{182.10 \left[\frac{1}{14} + \frac{(x - 11.11)^2}{546.09}\right]}$$

This result is graphically shown in Figure 11.5.

XI.1.5 SIGNIFICANCE TESTS

Following the results given above, tests of hypotheses about the values of α and β can be carried out based upon the approach discussed in Chapter X. Let us demonstrate the underlying ideas by testing the hypothesis $H_0: \beta = \beta_0$ versus $H_1: \beta \neq \beta_0$, where β_0 is some specified value.

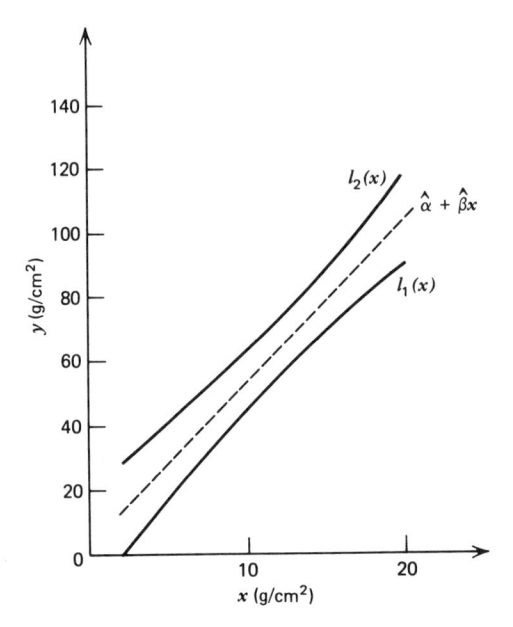

Figure 11.5 95% confidence band for $E\{Y\}$ in Example 11.4.

Using $\hat{B}$ as the test statistic, we have shown in Section XI.1.4 that the random variable defined by Equation 11.36 has a t-distribution with $(n-2)$ degrees of freedom. Suppose that we wish to achieve a Type I error probability of γ. We would reject H_0 if $|\hat{\beta} - \beta_0|$ exceeds (see Figure 11.6)

$$t_{n-2,\gamma/2}\sqrt{\frac{\widehat{\sigma^2}}{\sum_{i=1}^{n}(x_i - \bar{x})^2}} \tag{11.44}$$

Similarly, significance tests about the value of α can be easily carried out using $\hat{A}$ as the test statistic.

An important special case of the above is the test of $H_0: \beta = 0$ versus $H_1: \beta \neq 0$. This particular situation corresponds essentially to the significance test of linear regression. Accepting H_0 is equivalent to concluding that there is no reason to accept a linear relationship between $E\{Y\}$ and x at a specified significant level γ. In many cases, this may indicate the lack of a causal relationship between $E\{Y\}$ and the independent variable x.

Example 11.5. It is speculated that the starting salary of a college graduate in a given field is a function of the graduate's height. Assuming that salary (Y) is

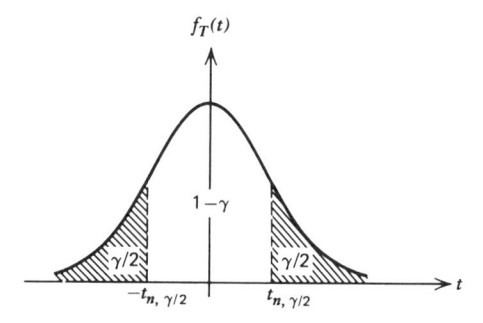

Figure 11.6 Probability density function of $T = (\hat{\beta} - \beta_0)/\sqrt{\widehat{\Sigma^2}/[\sum_{i=1}^{n}(x_i - \bar{x})^2]}$.

normally distributed and its mean is linearly related to height (x), we wish to use the data given below to test the assumption that $E\{Y\}$ and x are linearly related at 5% significance level.

x(ft)	5.7	5.7	5.7	5.7	6.1	6.1	6.1	6.1
y($10,000)	2.25	2.10	1.90	1.95	2.40	1.95	2.10	2.25

In this case, we wish to test $H_0: \beta = 0$ versus $H_1: \beta \neq 0$ with $\gamma = 0.05$. From the foregoing data, we have

$$\hat{\beta} = \frac{\sum_{i=1}^{n}(x_i - \bar{x})(y_i - \bar{y})}{\sum_{i=1}^{n}(x_i - \bar{x})^2} = 0.31$$

$$t_{n-2, \gamma/2} = t_{6, 0.025} = 2.447 \text{ (from Table A.4)}$$

$$\widehat{\sigma^2} = \frac{1}{n-2}\left[\sum_{i=1}^{n}(y_i - \bar{y})^2 - \hat{\beta}^2 \sum_{i=1}^{n}(x_i - \bar{x})^2\right] = 0.02$$

$$\sum_{i=1}^{n}(x_i - \bar{x})^2 = 0.32$$

According to Equation 11.44, we have

$$t_{6, 0.025}\sqrt{\frac{\widehat{\sigma^2}}{\sum_{i=1}^{n}(x_i - \bar{x})^2}} = 0.61$$

Since $\hat{\beta} = 0.31 < 0.61$, we accept H_0. That is, we conclude that the data do not indicate a linear relationship between $E\{Y\}$ and x; the probability that we are wrong in accepting H_0 is 0.05.

In closing, let us remark that we are often called on to perform tests of *simultaneous hypotheses*. For example, one may wish to test $H_0: \alpha = 0$ and $\beta = 1$ versus $H_1: \alpha \neq 0$ or $\beta \neq 1$ or both. Such tests involve both estimators $\hat{A}$ and $\hat{B}$ and hence require their joint distribution. This is also often the case in multiple linear regression to be discussed in the next section. Such tests customarily involve F-distributed test statistics and we will not pursue them here. A general treatment of simultaneous hypotheses testing can be found in Rao (1965), for example.

XI.2 Multiple Linear Regression

The vector-matrix approach proposed in the preceding section provides a smooth transition from simple linear regression to linear regression involving more than one independent variable. In multiple linear regression, the model takes the form

$$E\{Y\} = \beta_0 + \beta_1 x_1 + \beta_2 x_2 + \cdots + \beta_m x_m \qquad (11.45)$$

Again, we assume that the variance of Y is σ^2 and is independent of $x_1, x_2, \ldots,$ and x_m. As in simple linear regression, we are interested in estimating the $(m + 1)$ regression coefficients $\beta_0, \beta_1, \ldots,$ and β_m, obtaining certain interval estimates, and testing hypotheses about these parameters on the basis of a sample of Y values with their associated values of $(x_1, x_2, \ldots, x_m)$. Let us note that our sample of size n in this case takes the form of arrays $(x_{11}, x_{21}, \ldots, x_{m1}, Y_1), (x_{12}, x_{22}, \ldots, x_{m2}, Y_2), \ldots, (x_{1n}, x_{2n}, \ldots, x_{mn}, Y_n)$. For each set of values $x_{ki}, k = 1, 2, \ldots, m,$ of x_k, Y_i is an independent observation from the population Y defined by

$$Y = \beta_0 + \beta_1 x_1 + \cdots + \beta_m x_m + E \qquad (11.46)$$

As before, E is the random error with mean zero and variance σ^2.

XI.2.1 LEAST SQUARES METHOD OF ESTIMATION

To estimate the regression coefficients, the method of least squares will again be employed. Given observed sample-value sets $(x_{1i}, x_{2i}, \ldots, x_{mi}, y_i), i = 1, 2, \ldots, n,$ the system of observed regression equations in this case takes the form

$$y_i = \beta_0 + \beta_1 x_{1i} + \cdots + \beta_m x_{mi} + e_i \qquad i = 1, 2, \ldots, n \qquad (11.47)$$

LINEAR MODELS AND LINEAR REGRESSION

If we let

$$C = \begin{bmatrix} 1 & x_{11} & x_{21} & \cdots & x_{m1} \\ 1 & x_{12} & x_{22} & \cdots & x_{m2} \\ \vdots & \vdots & \vdots & & \vdots \\ 1 & x_{1n} & x_{2n} & \cdots & x_{mn} \end{bmatrix} \quad \mathbf{y} = \begin{bmatrix} y_1 \\ y_2 \\ \vdots \\ y_n \end{bmatrix} \quad \mathbf{e} = \begin{bmatrix} e_1 \\ e_2 \\ \vdots \\ e_n \end{bmatrix}$$

and

$$\boldsymbol{\theta} = \begin{bmatrix} \beta_0 \\ \beta_1 \\ \vdots \\ \beta_m \end{bmatrix}$$

equations 11.47 can again be represented by the vector-matrix equation

$$\mathbf{y} = C\boldsymbol{\theta} + \mathbf{e} \tag{11.48}$$

Comparing Equation 11.48 with Equation 11.12 in simple linear regression, we see that observed regression equations in both cases are identical except that the C matrix is now $n \times (m + 1)$ and $\boldsymbol{\theta}$ is a $(m + 1)$-dimensional vector. Keeping this dimension difference in mind, the results obtained in the case of simple linear regression based on Equation 11.12 again hold in the multiple linear regression case. Thus, without further derivation, we have for solution of the least-square estimates $\hat{\boldsymbol{\theta}}$ of $\boldsymbol{\theta}$ (see Equation 11.15)

$$\boxed{\hat{\boldsymbol{\theta}} = (C^T C)^{-1} C^T \mathbf{y}} \tag{11.49}$$

The existence of the matrix inverse $(C^T C)^{-1}$ requires that there are at least $(m + 1)$ distinct sets of values of $(x_{1i}, x_{2i}, \ldots, x_{mi})$ represented in the sample. It is noted that $C^T C$ is always a $(m + 1) \times (m + 1)$ symmetric matrix.

Example 11.6. The average monthly electric power consumption (Y) at a certain manufacturing plant is considered to be linearly dependent on the average ambient temperature (x_1) and the number of working days in a month (x_2). Consider the one-year monthly data given below. Determine the least-square estimates of the associated linear regression coefficients.

$x_1(°F)$	20	26	41	55	60	67	75	79	70	55	45	33
$x_2(days)$	23	21	24	25	24	26	25	25	24	25	25	23
$y(1000\ kW\text{-}h)$	210	206	260	244	271	285	270	265	234	241	258	230

In this case, C is a 12×3 matrix and

$$C^TC = \begin{bmatrix} 12 & 626 & 290 \\ 626 & 36{,}776 & 15{,}336 \\ 290 & 15{,}336 & 7{,}028 \end{bmatrix}$$

$$C^T\mathbf{y} = \begin{bmatrix} 2{,}974 \\ 159{,}011 \\ 72{,}166 \end{bmatrix}$$

We thus have, upon finding the inverse of C^TC using either matrix inversion formulas or readily available matrix inversion computer programs,

$$\hat{\boldsymbol{\theta}} = (C^TC)^{-1}C^T\mathbf{y} = \begin{bmatrix} -33.84 \\ 0.39 \\ 10.80 \end{bmatrix}$$

or

$$\hat{\beta}_0 = -33.84 \qquad \hat{\beta}_1 = 0.39 \qquad \hat{\beta}_2 = 10.80$$

The estimated regression equation based on the data is thus

$$\widehat{E\{y\}} = \hat{\beta}_0 + \hat{\beta}_1 x_1 + \hat{\beta}_2 x_2$$
$$= -33.84 + 0.39 x_1 + 10.80 x_2$$

Since Equation 11.48 is identical to its counterpart in the case of simple linear regression, much of the results obtained therein concerning properties of least-square estimators, confidence intervals, and hypotheses testing can be duplicated here with, of course, due regard to the new definitions for matrix C and vector $\boldsymbol{\theta}$.

Let us write estimator $\hat{\boldsymbol{\Theta}}$ for $\boldsymbol{\theta}$ in the form

$$\hat{\boldsymbol{\Theta}} = (C^TC)^{-1}C^T\mathbf{Y} \qquad (11.50)$$

We see immediately that

$$E\{\hat{\boldsymbol{\Theta}}\} = (C^TC)^{-1}C^T E\{\mathbf{Y}\} = \boldsymbol{\theta} \qquad (11.51)$$

Hence, the least-square estimator $\hat{\boldsymbol{\Theta}}$ is again unbiased. It also follows from Equation 11.21 that the covariance matrix for $\hat{\boldsymbol{\Theta}}$ is given by

$$\text{cov}\{\hat{\boldsymbol{\Theta}}\} = \sigma^2 (C^TC)^{-1} \qquad (11.52)$$

Confidence intervals for the regression parameters in this case can also be established following similar procedures employed in the case of simple linear regression. Concerning hypotheses testing, it was mentioned in Section XI.1.5 that testing of simultaneous hypotheses is more appropriate in multiple linear regression and we will not pursue it here.

XI.3 Other Regression Models

In science and engineering, one often finds it necessary to consider regression models that are nonlinear in the independent variables. Common examples of this class of models include

$$Y = \beta_0 + \beta_1 x + \beta_2 x^2 + E \tag{11.53}$$

$$Y = \beta_0 \exp(\beta_1 x + E) \tag{11.54}$$

$$Y = \beta_0 + \beta_1 x_1 + \beta_2 x_2 + \beta_{11} x_1^2 + \beta_{22} x_2^2 + \beta_{12} x_1 x_2 + E \tag{11.55}$$

$$Y = \beta_0 \beta_1^x + E \tag{11.56}$$

Polynomial models such as Equation 11.53 or Equation 11.55 are still linear regression models in that they are linear in the unknown parameters $\beta_0, \beta_1, \beta_2, \ldots$, etc. Hence, they can be estimated using multiple linear regression techniques. Indeed, let $x_1 = x$ and $x_2 = x^2$ in Equation 11.53, it takes the form of that of a multiple linear regression model with two independent variables and can thus be analyzed as such. Similar equivalence can be established between Equation 11.55 and a multiple linear regression model with five independent variables.

Consider the exponential model given by Equation 11.54. Taking logarithms of both sides, we have

$$\ln Y = \ln \beta_0 + \beta_1 x + E \tag{11.57}$$

In terms of r.v. $\ln Y$, Equation 11.57 represents a linear regression equation with regression coefficients $\ln \beta_0$ and β_1. Linear regression techniques again apply in this case. Equation 11.56, on the other hand, cannot be conveniently put into linear regression form

Example 11.7. On the average, the rate of population increase (Y) associated with a given city varies with x, the number of years after 1970. Assuming that

$$E\{Y\} = \beta_0 + \beta_1 x + \beta_2 x^2$$

compute the least-square estimates for β_0, β_1, and β_2 based on the data presented below.

x	0	1	2	3	4	5
y(%)	1.03	1.32	1.57	1.75	1.83	2.33

Let $x_1 = x$ and $x_2 = x^2$ and let

$$\theta = \begin{bmatrix} \beta_0 \\ \beta_1 \\ \beta_2 \end{bmatrix}$$

The least-square estimate for θ, $\hat{\theta}$, is given by Equation 11.49 with

$$C = \begin{bmatrix} 1 & 0 & 0 \\ 1 & 1 & 1 \\ 1 & 2 & 4 \\ 1 & 3 & 9 \\ 1 & 4 & 16 \\ 1 & 5 & 25 \end{bmatrix} \quad \text{and} \quad y = \begin{bmatrix} 1.03 \\ 1.32 \\ 1.57 \\ 1.75 \\ 1.83 \\ 2.33 \end{bmatrix}$$

Thus

$$\hat{\theta} = (C^T C)^{-1} C^T y = \begin{bmatrix} 6 & 15 & 55 \\ 15 & 55 & 225 \\ 55 & 225 & 979 \end{bmatrix}^{-1} \begin{bmatrix} 9.83 \\ 28.68 \\ 110.88 \end{bmatrix}$$

$$= \begin{bmatrix} 1.07 \\ 0.20 \\ 0.01 \end{bmatrix}$$

or

$$\hat{\beta}_0 = 1.07 \quad \hat{\beta}_1 = 0.20 \quad \text{and} \quad \hat{\beta}_2 = 0.01$$

Let us note in this example that, since $x_2 = x_1^2$, the matrix C is constrained in that its elements in the third column are the squared values of their corresponding elements in the second column. It needs to be cautioned that, for high-order polynomial regression models, constraints of this type may render the matrix $C^T C$ ill conditioned and lead to matrix inversion difficulties.

REFERENCES AND COMMENTS

1. C. R. Rao, *Linear Statistical Inference and Its Applications*, Wiley, New York, 1965.

Some additional useful references on regression analysis are given below.

2. R. L. Anderson and T. A. Bancroft, *Statistical Theory in Research*, McGraw-Hill, New York, 1952.
3. J. S. Bendat and A. G. Piersol, *Measurement and Analysis of Random Data*, Wiley, New York, 1966.
4. N. Draper and H. Smith, *Applied Regression Analysis*, Wiley, New York, 1966.
5. F. A. Graybill, *An Introduction to Linear Statistical Models*, Vol. 1, McGraw-Hill, New York, 1961.

PROBLEMS

11.1. A special case of simple linear regression is given by

$$Y = \beta x + E$$

Determine:
(a) The least-square estimator $\hat{B}$ for β.
(b) The mean and variance of $\hat{B}$.
(c) An unbiased estimator for σ^2.

11.2. In simple linear regression, show that the maximum likelihood estimators for α and β are identical with their least-square estimators when Y is normally distributed.

11.3. Determine the maximum likelihood estimator for variance σ^2 of Y in simple linear regression assuming that Y is normally distributed. Is it a biased estimator?

11.4. Since data quality is generally not uniform among data points, it is sometimes desirable to estimate the regression coefficients by minimizing the sum of *weighted* squared residuals, that is, $\hat{\alpha}$ and $\hat{\beta}$ in simple linear regression are found by minimizing

$$\sum_{i=1}^{n} w_i e_i^2$$

where w_i are assigned weights. In vector-matrix notation, show that the estimates $\hat{\alpha}$ and $\hat{\beta}$ now take the form

$$\hat{\theta} = \begin{bmatrix} \hat{\alpha} \\ \hat{\beta} \end{bmatrix} = (C^T W C)^{-1} C^T W y$$

where

$$W = \begin{bmatrix} w_1 & & & 0 \\ & w_2 & & \\ & & \ddots & \\ 0 & & & w_n \end{bmatrix}$$

11.5. (a) In simple linear regression (Equation 11.4), use vector-matrix notation and show that the unbiased estimator for σ^2 given by Equation 11.33 can be written in the form

$$\widehat{\Sigma^2} = \frac{1}{n-2}[(Y - C\hat{\Theta})^T (Y - C\hat{\Theta})]$$

(b) In multiple linear regression (Equation 11.46), show that an unbiased estimator for σ^2 is given by

$$\widehat{\Sigma^2} = \frac{1}{n-m-1}[(Y - C\hat{\Theta})^T (Y - C\hat{\Theta})]$$

11.6. Given the data below

x	0	1	2	3	4	5	6	7	8	9
y	3.2	3.1	3.9	4.7	4.3	4.4	4.8	5.3	5.9	6.0

(a) Determine the least-square estimates of α and β in the linear regression equation

$$Y = \alpha + \beta x + E$$

(b) Determine an unbiased estimate of σ^2.
  (c) Estimate $E\{Y\}$ at $x = 5$.
  (d) Determine a 95% confidence interval for β.
  (e) Determine a 95% confidence band for $\alpha + \beta x$.

11.7. In transportation studies, it is assumed that, on the average, peak vehicle noise level (Y) is linearly related to the logarithm of vehicle speed (v). Some measurements taken for a class of light vehicles are given below. Assuming that

$$Y = \alpha + \beta \log_{10} v + E$$

determine the estimated regression line for Y as a function of $\log_{10} v$.

v(km/h)	20	30	40	50	60	70	80	90	100
y(Db)	55	63	68	70	72	78	74	76	79

11.8. An experimental study of nasal deposition of particles was carried out and showed a linear relationship between $E\{Y\}$ and $\ln d^2 f$, where Y is the fraction of particles of aerodynamic diameter, d, which is deposited in the nose during an inhalation of f (l/min). Consider the data given below (four readings are taken at each value of $\ln d^2 f$). Estimate the regression parameters in the linear regression equation

$$E\{Y\} = \alpha + \beta \ln d^2 f$$

and estimate σ^2.

$\ln d^2 f$	1.6	1.7	2.0	2.2	2.8	3.0	3.6
y	0.39	0.41	0.42	0.61	0.83	0.79	0.98
	0.30	0.28	0.34	0.51	0.79	0.69	0.88
	0.21	0.20	0.22	0.47	0.70	0.63	0.87
	0.12	0.10	0.18	0.39	0.61	0.59	0.83

11.9. In a study of stress-strain history of soft biological tissues, experimental results relating dynamic moduli of aorta (D) to stress frequency (ω) are given below.
  (a) Assuming that $E\{D\} = \alpha + \beta \omega$ and $\sigma_D^2 = \sigma^2$, estimate the regression coefficients α and β.
  (b) Determine a one-sided 95% confidence interval for the variance of D.

(c) Test if the slope estimate is significantly different from zero at 5% significance level.

$\omega(Hz)$	1	2	3	4	5	6	7	8	9	10
d(normalized)	1.60	1.51	1.40	1.57	1.60	1.59	1.80	1.59	1.82	1.59

11.10 Given the following data:

x_1	−1	−1	1	1	2	2	3	3
x_2	1	2	3	4	5	6	7	8
y	2.0	3.1	4.8	4.9	5.4	6.8	6.9	7.5

(a) Determine the least-square estimates of β_0, β_1, and β_2 assuming that

$$E\{Y\} = \beta_0 + \beta_1 x_1 + \beta_2 x_2$$

(b) Estimate $E\{Y\}$ at $x_1 = x_2 = 2$.

11.11. In Problem 11.7, when vehicle weight is taken into account, we have the multiple linear regression equation

$$Y = \beta_0 + \beta_1 \log_{10} v + \beta_2 \log_{10} w + E$$

where w is vehicle unladen weight in Mg. Use the data given below and estimate the regression parameters in this case.

v	20	40	60	80	100	120
w	1.0	1.0	1.7	3.0	1.0	0.7
y	54	59	78	91	78	67

11.12. Given the following data:

x	0	1	2	3	4	5	6	7
y	3.2	2.8	5.1	7.3	7.6	5.9	4.1	1.8

(a) Determine the least-square estimates of β_0, β_1, and β_2 assuming that

$$E\{Y\} = \beta_0 + \beta_1 x + \beta_2 x^2$$

(b) Estimate $E\{Y\}$ at $x = 3$.

11.13. A large number of socioeconomic variables are important to account for mortality rate. Assuming a multiple linear regression model, one version of the model for mortality rate (Y) is expressed by

$$Y = \beta_0 + \beta_1 x_1 + \beta_2 x_2 + \beta_3 x_3 + \beta_4 x_4 + E$$

where

x_1 = mean annual precipitation in inches
x_2 = education in terms of median school years completed for those over 25
x_3 = percent of area population that is nonwhite
x_4 = relative pollution potential of SO_2

Some available data are presented below. Determine the least-square estimates of the regression parameters.

x_1	13	11	21	30	35	27	27	40
x_2	9	10.5	11	10	9	12.3	9	9
x_3	1.5	7	21	27	30	6	27	33
x_4	4	21	64	67	17	28	82	101
y	795	841	820	1050	1010	970	980	1090

APPENDIX A
Tables

Table A.1. Binomial Mass Function. A Table of $p_X(k) = \binom{n}{k} p^k (1-p)^{n-k}$ to $n = 2$ to 10 $p = 0.01$ to 0.50

n	k	p = 0.01	0.05	0.10	0.15	0.20	0.25	0.30	$\frac{1}{3}$	0.35	0.40	0.45	0.49	0.50
2	0	0.9801	0.9025	0.8100	0.7225	0.6400	0.5625	0.4900	0.4444	0.4225	0.3600	0.3025	0.2601	0.2500
	1	0.0198	0.0950	0.1800	0.2550	0.3200	0.3750	0.4200	0.4444	0.4550	0.4800	0.4950	0.4998	0.5000
	2	0.0001	0.0025	0.0100	0.0225	0.0400	0.0625	0.0900	0.1111	0.1225	0.1600	0.2025	0.2401	0.2500
3	0	0.9703	0.8574	0.7290	0.6141	0.5120	0.4219	0.3430	0.2963	0.2746	0.2160	0.1664	0.1327	0.1250
	1	0.0294	0.1354	0.2430	0.3251	0.3840	0.4219	0.4410	0.4444	0.4436	0.4320	0.4084	0.3823	0.3750
	2	0.0003	0.0071	0.0270	0.0574	0.0960	0.1406	0.1890	0.2222	0.2389	0.2880	0.3341	0.3674	0.3750
	3	0.0000	0.0001	0.0010	0.0034	0.0080	0.0156	0.0270	0.0370	0.0429	0.0640	0.0911	0.1176	0.1250
4	0	0.9606	0.8145	0.6561	0.5220	0.4096	0.3164	0.2401	0.1975	0.1785	0.1296	0.0915	0.0677	0.0625
	1	0.0388	0.1715	0.2916	0.3685	0.4096	0.4219	0.4116	0.3951	0.3845	0.3456	0.2995	0.2600	0.2500
	2	0.0006	0.0135	0.0486	0.0975	0.1536	0.2109	0.2646	0.2963	0.3105	0.3456	0.3675	0.3747	0.3750
	3	0.0000	0.0005	0.0036	0.0115	0.0256	0.0469	0.0756	0.0988	0.1115	0.1536	0.2005	0.2400	0.2500
	4	0.0000	0.0000	0.0001	0.0005	0.0016	0.0039	0.0081	0.0123	0.0150	0.0256	0.0410	0.0576	0.0625
5	0	0.9510	0.7738	0.5905	0.4437	0.3277	0.2373	0.1681	0.1317	0.1160	0.0778	0.0503	0.0345	0.0312
	1	0.0480	0.2036	0.3280	0.3915	0.4096	0.3955	0.3602	0.3292	0.3124	0.2592	0.2059	0.1657	0.1562
	2	0.0010	0.0214	0.0729	0.1382	0.2048	0.2637	0.3087	0.3292	0.3364	0.3456	0.3369	0.3185	0.3125
	3	0.0000	0.0011	0.0081	0.0244	0.0512	0.0879	0.1323	0.1646	0.1811	0.2304	0.2757	0.3060	0.3125
	4	0.0000	0.0000	0.0004	0.0022	0.0064	0.0146	0.0284	0.0412	0.0488	0.0768	0.1128	0.1470	0.1562
	5	0.0000	0.0000	0.0000	0.0001	0.0003	0.0010	0.0024	0.0041	0.0053	0.0102	0.0185	0.0283	0.0312

n	k													
6	0	0.9415	0.7351	0.5314	0.3771	0.2621	0.1780	0.1176	0.0878	0.0754	0.0467	0.0277	0.0176	0.0156
	1	0.0571	0.2321	0.3543	0.3993	0.3932	0.3560	0.3025	0.2634	0.2437	0.1866	0.1359	0.1014	0.0938
	2	0.0014	0.0305	0.0984	0.1762	0.2458	0.2966	0.3241	0.3292	0.3280	0.3110	0.2780	0.2437	0.2344
	3	0.0000	0.0021	0.0146	0.0415	0.0819	0.1318	0.1852	0.2195	0.2355	0.2765	0.3032	0.3121	0.3125
	4	0.0000	0.0001	0.0012	0.0055	0.0154	0.0330	0.0595	0.0823	0.0951	0.1382	0.1861	0.2249	0.2344
	5	0.0000	0.0000	0.0001	0.0004	0.0015	0.0044	0.0102	0.0165	0.0205	0.0369	0.0609	0.0864	0.0938
	6	0.0000	0.0000	0.0000	0.0000	0.0001	0.0002	0.0007	0.0014	0.0018	0.0041	0.0083	0.0139	0.0156
7	0	0.9321	0.6983	0.4783	0.3206	0.2097	0.1335	0.0824	0.0585	0.0490	0.0280	0.0152	0.0090	0.0078
	1	0.0659	0.2573	0.3720	0.3960	0.3670	0.3115	0.2471	0.2048	0.1848	0.1306	0.0872	0.0603	0.0547
	2	0.0020	0.0406	0.1240	0.2097	0.2753	0.3115	0.3177	0.3073	0.2985	0.2613	0.2140	0.1740	0.1641
	3	0.0000	0.0036	0.0230	0.0617	0.1147	0.1730	0.2269	0.2561	0.2679	0.2903	0.2918	0.2786	0.2734
	4	0.0000	0.0002	0.0026	0.0109	0.0287	0.0577	0.0972	0.1280	0.1442	0.1935	0.2388	0.2676	0.2734
	5	0.0000	0.0000	0.0002	0.0012	0.0043	0.0115	0.0250	0.0384	0.0466	0.0774	0.1172	0.1543	0.1641
	6	0.0000	0.0000	0.0000	0.0001	0.0004	0.0013	0.0036	0.0064	0.0084	0.0172	0.0320	0.0494	0.0547
	7	0.0000	0.0000	0.0000	0.0000	0.0000	0.0001	0.0002	0.0005	0.0006	0.0016	0.0037	0.0068	0.0078
8	0	0.9227	0.6634	0.4305	0.2725	0.1678	0.1001	0.0576	0.0390	0.0319	0.0168	0.0084	0.0046	0.0039
	1	0.0746	0.2793	0.3826	0.3847	0.3355	0.2670	0.1977	0.1561	0.1373	0.0896	0.0548	0.0352	0.0312
	2	0.0026	0.0515	0.1488	0.2376	0.2936	0.3115	0.2965	0.2731	0.2587	0.2090	0.1569	0.1183	0.1094
	3	0.0001	0.0054	0.0331	0.0839	0.1468	0.2076	0.2541	0.2731	0.2786	0.2787	0.2568	0.2273	0.2188
	4	0.0000	0.0004	0.0046	0.0185	0.0459	0.0865	0.1361	0.1707	0.1875	0.2322	0.2627	0.2730	0.2734
	5	0.0000	0.0000	0.0004	0.0026	0.0092	0.0231	0.0467	0.0683	0.0808	0.1239	0.1719	0.2098	0.2188
	6	0.0000	0.0000	0.0000	0.0002	0.0011	0.0038	0.0100	0.0171	0.0217	0.0413	0.0703	0.1008	0.1094
	7	0.0000	0.0000	0.0000	0.0000	0.0001	0.0004	0.0012	0.0024	0.0033	0.0079	0.0164	0.0277	0.0312
	8	0.0000	0.0000	0.0000	0.0000	0.0000	0.0000	0.0001	0.0002	0.0002	0.0007	0.0017	0.0033	0.0039

Continued

Table A.1. (continued)

n	k	$p = 0.01$	0.05	0.10	0.15	0.20	0.25	0.30	$\frac{1}{3}$	0.35	0.40	0.45	0.49	0.50
9	0	0.9135	0.6302	0.3874	0.2316	0.1342	0.0751	0.0404	0.0260	0.0207	0.0101	0.0046	0.0023	0.0020
	1	0.0830	0.2985	0.3874	0.3679	0.3020	0.2253	0.1556	0.1171	0.1004	0.0605	0.0339	0.0202	0.0176
	2	0.0034	0.0629	0.1722	0.2597	0.3020	0.3003	0.2668	0.2341	0.2162	0.1612	0.1110	0.0776	0.0703
	3	0.0001	0.0077	0.0446	0.1069	0.1762	0.2336	0.2668	0.2731	0.2716	0.2508	0.2119	0.1739	0.1641
	4	0.0000	0.0006	0.0074	0.0283	0.0661	0.1168	0.1715	0.2048	0.2194	0.2508	0.2600	0.2506	0.2461
	5	0.0000	0.0000	0.0008	0.0050	0.0165	0.0389	0.0735	0.1024	0.1181	0.1672	0.2128	0.2408	0.2461
	6	0.0000	0.0000	0.0001	0.0006	0.0028	0.0087	0.0210	0.0341	0.0424	0.0743	0.1160	0.1542	0.1641
	7	0.0000	0.0000	0.0000	0.0000	0.0003	0.0012	0.0039	0.0073	0.0098	0.0212	0.0407	0.0635	0.0703
	8	0.0000	0.0000	0.0000	0.0000	0.0000	0.0001	0.0004	0.0009	0.0013	0.0035	0.0083	0.0153	0.0176
	9	0.0000	0.0000	0.0000	0.0000	0.0000	0.0000	0.0000	0.0001	0.0001	0.0003	0.0008	0.0016	0.0020
10	0	0.9044	0.5987	0.3487	0.1969	0.1074	0.0563	0.0282	0.0173	0.0135	0.0060	0.0025	0.0012	0.0010
	1	0.0914	0.3151	0.3874	0.3474	0.2684	0.1877	0.1211	0.0867	0.0725	0.0403	0.0207	0.0114	0.0098
	2	0.0042	0.0746	0.1937	0.2759	0.3020	0.2816	0.2335	0.1951	0.1757	0.1209	0.0736	0.0495	0.0439
	3	0.0001	0.0105	0.0574	0.1298	0.2013	0.2503	0.2668	0.2601	0.2522	0.2150	0.1665	0.1267	0.1172
	4	0.0000	0.0010	0.0112	0.0401	0.0881	0.1460	0.2001	0.2276	0.2377	0.2508	0.2384	0.2130	0.2051
	5	0.0000	0.0001	0.0015	0.0085	0.0264	0.0584	0.1029	0.1366	0.1536	0.2007	0.2340	0.2456	0.2461
	6	0.0000	0.0000	0.0001	0.0012	0.0055	0.0162	0.0368	0.0569	0.0689	0.1115	0.1596	0.1966	0.2051
	7	0.0000	0.0000	0.0000	0.0001	0.0008	0.0031	0.0090	0.0163	0.0212	0.0425	0.0746	0.1080	0.1172
	8	0.0000	0.0000	0.0000	0.0000	0.0001	0.0004	0.0014	0.0030	0.0043	0.0106	0.0229	0.0389	0.0439
	9	0.0000	0.0000	0.0000	0.0000	0.0000	0.0000	0.0001	0.0003	0.0005	0.0016	0.0042	0.0083	0.0098
	10	0.0000	0.0000	0.0000	0.0000	0.0000	0.0000	0.0000	0.0000	0.0000	0.0001	0.0003	0.0008	0.0010

Taken from Parzen (1960). Reprinted with permission.

Table A.2. Poisson Mass Function. A Table of $p_k(0, t) = (\lambda t)^k e^{-\lambda t}/k!$ $k = 0 \text{ to } 24$ $\lambda t = 0.1 \text{ to } 10$

λt \ k	0	1	2	3	4	5	6	7	8	9	10	11	12
0.1	0.9048	0.0905	0.0045	0.0002	0.0000								
0.2	0.8187	0.1637	0.0164	0.0011	0.0001	0.0000							
0.3	0.7408	0.2222	0.0333	0.0033	0.0002	0.0000							
0.4	0.6703	0.2681	0.0536	0.0072	0.0007	0.0001	0.0000						
0.5	0.6065	0.3033	0.0758	0.0126	0.0016	0.0002	0.0000						
0.6	0.5488	0.3293	0.0988	0.0198	0.0030	0.0004	0.0000						
0.7	0.4966	0.3476	0.1217	0.0284	0.0050	0.0007	0.0001	0.0000					
0.8	0.4493	0.3595	0.1438	0.0383	0.0077	0.0012	0.0002	0.0000					
0.9	0.4066	0.3659	0.1647	0.0494	0.0111	0.0020	0.0003	0.0000					
1.0	0.3679	0.3679	0.1839	0.0613	0.0153	0.0031	0.0005	0.0001	0.0000				
1.1	0.3329	0.3662	0.2014	0.0738	0.0203	0.0045	0.0008	0.0001	0.0000				
1.2	0.3012	0.3614	0.2169	0.0867	0.0260	0.0062	0.0012	0.0002	0.0000				
1.3	0.2725	0.3543	0.2303	0.0998	0.0324	0.0084	0.0018	0.0003	0.0001	0.0000			
1.4	0.2466	0.3452	0.2417	0.1128	0.0395	0.0111	0.0026	0.0005	0.0001	0.0000			
1.5	0.2231	0.3347	0.2510	0.1255	0.0471	0.0141	0.0035	0.0008	0.0001	0.0000			

Continued

Table A.2. (*continued*)

λt \ k	0	1	2	3	4	5	6	7	8	9	10	11	12
1.6	0.2019	0.3230	0.2584	0.1378	0.0551	0.0176	0.0047	0.0011	0.0002	0.0000			
1.7	0.1827	0.3106	0.2640	0.1496	0.0636	0.0216	0.0061	0.0015	0.0003	0.0001	0.0000		
1.8	0.1653	0.2975	0.2678	0.1607	0.0723	0.0260	0.0078	0.0020	0.0005	0.0001	0.0000		
1.9	0.1496	0.2842	0.2700	0.1710	0.0812	0.0309	0.0098	0.0027	0.0006	0.0001	0.0000		
2.0	0.1353	0.2707	0.2707	0.1804	0.0902	0.0361	0.0120	0.0034	0.0009	0.0002	0.0000		
2.2	0.1108	0.2438	0.2681	0.1966	0.1082	0.0476	0.0174	0.0055	0.0015	0.0004	0.0001	0.0000	
2.4	0.0907	0.2177	0.2613	0.2090	0.1254	0.0602	0.0241	0.0083	0.0025	0.0007	0.0002	0.0000	
2.6	0.0743	0.1931	0.2510	0.2176	0.1414	0.0735	0.0319	0.0118	0.0038	0.0011	0.0003	0.0001	0.0000
2.8	0.0608	0.1703	0.2384	0.2225	0.1557	0.0872	0.0407	0.0163	0.0057	0.0018	0.0005	0.0001	0.0000
3.0	0.0498	0.1494	0.2240	0.2240	0.1680	0.1008	0.0504	0.0216	0.0081	0.0027	0.0008	0.0002	0.0001
3.2	0.0408	0.1304	0.2087	0.2226	0.1781	0.1140	0.0608	0.0278	0.0111	0.0040	0.0013	0.0004	0.0001
3.4	0.0334	0.1135	0.1929	0.2186	0.1858	0.1264	0.0716	0.0348	0.0148	0.0056	0.0019	0.0006	0.0002
3.6	0.0273	0.0984	0.1771	0.2125	0.1912	0.1377	0.0826	0.0425	0.0191	0.0076	0.0028	0.0009	0.0003
3.8	0.0224	0.0850	0.1615	0.2046	0.1944	0.1477	0.0936	0.0508	0.0241	0.0102	0.0039	0.0013	0.0004
4.0	0.0183	0.0733	0.1465	0.1954	0.1954	0.1563	0.1042	0.0595	0.0298	0.0132	0.0053	0.0019	0.0006

λt								
5.0	0.0067	0.0337	0.0842	0.1404	0.1755	0.1755	0.1462	0.1044	0.0653	0.0363	0.0181	0.0082	0.0034
6.0	0.0025	0.0149	0.0446	0.0892	0.1339	0.1606	0.1606	0.1377	0.1033	0.0688	0.0413	0.0225	0.0113
7.0	0.0009	0.0064	0.0223	0.0521	0.0912	0.1277	0.1490	0.1490	0.1304	0.1014	0.0710	0.0452	0.0264
8.0	0.0003	0.0027	0.0107	0.0286	0.0573	0.0916	0.1221	0.1396	0.1396	0.1241	0.0993	0.0722	0.0481
9.0	0.0001	0.0011	0.0050	0.0150	0.0337	0.0607	0.0911	0.1171	0.1318	0.1318	0.1186	0.0970	0.0728
10.0	0.0000	0.0005	0.0023	0.0076	0.0189	0.0378	0.0631	0.0901	0.1126	0.1251	0.1251	0.1137	0.0948

λt \ k	13	14	15	16	17	18	19	20	21	22	23	24
5.0	0.0013	0.0005	0.0002									
6.0	0.0052	0.0022	0.0009	0.0003	0.0001							
7.0	0.0142	0.0071	0.0033	0.0014	0.0006	0.0002	0.0001					
8.0	0.0296	0.0169	0.0090	0.0045	0.0021	0.0009	0.0004	0.0002	0.0001			
9.0	0.0504	0.0324	0.0194	0.0109	0.0058	0.0029	0.0014	0.0006	0.0003	0.0001		
10.0	0.0729	0.0521	0.0347	0.0217	0.0128	0.0071	0.0037	0.0019	0.0009	0.0004	0.0002	0.0001

Taken from Parzen (1960). Reprinted with permission.

Table A.3. *Standardized Normal Distribution Function. A Table of* $F_U(u) = (1/\sqrt{2\pi}) \int_{-\infty}^{u} e^{-x^2/2} \, dx$ $u = 0.0$ to 3.69

u	0.00	0.01	0.02	0.03	0.04	0.05	0.06	0.07	0.08	0.09
0.0	0.5000	0.5040	0.5080	0.5120	0.5160	0.5199	0.5239	0.5279	0.5319	0.5359
0.1	0.5398	0.5438	0.5478	0.5517	0.5557	0.5596	0.5636	0.5675	0.5714	0.5753
0.2	0.5793	0.5832	0.5871	0.5910	0.5948	0.5987	0.6026	0.6064	0.6103	0.6141
0.3	0.6179	0.6217	0.6255	0.6293	0.6331	0.6368	0.6406	0.6443	0.6480	0.6517
0.4	0.6554	0.6591	0.6628	0.6664	0.6700	0.6736	0.6772	0.6808	0.6844	0.6879
0.5	0.6915	0.6950	0.6985	0.7019	0.7054	0.7088	0.7123	0.7157	0.7190	0.7224
0.6	0.7257	0.7291	0.7324	0.7357	0.7389	0.7422	0.7454	0.7486	0.7517	0.7549
0.7	0.7580	0.7611	0.7642	0.7673	0.7704	0.7734	0.7764	0.7794	0.7823	0.7852
0.8	0.7881	0.7910	0.7939	0.7967	0.7995	0.8023	0.8051	0.8078	0.8106	0.8133
0.9	0.8159	0.8186	0.8212	0.8238	0.8264	0.8289	0.8315	0.8340	0.8365	0.8389
1.0	0.8413	0.8438	0.8461	0.8485	0.8508	0.8531	0.8554	0.8577	0.8599	0.8621
1.1	0.8643	0.8665	0.8686	0.8708	0.8729	0.8749	0.8770	0.8790	0.8810	0.8830
1.2	0.8849	0.8869	0.8888	0.8907	0.8925	0.8944	0.8962	0.8980	0.8997	0.9015
1.3	0.9032	0.9049	0.9066	0.9082	0.9099	0.9115	0.9131	0.9147	0.9162	0.9177
1.4	0.9192	0.9207	0.9222	0.9236	0.9251	0.9265	0.9279	0.9292	0.9306	0.9319
1.5	0.9332	0.9345	0.9357	0.9370	0.9382	0.9394	0.9406	0.9418	0.9429	0.9441
1.6	0.9452	0.9463	0.9474	0.9484	0.9495	0.9505	0.9515	0.9525	0.9535	0.9545
1.7	0.9554	0.9564	0.9573	0.9582	0.9591	0.9599	0.9608	0.9616	0.9625	0.9633
1.8	0.9641	0.9649	0.9656	0.9664	0.9671	0.9678	0.9686	0.9693	0.9699	0.9706
1.9	0.9713	0.9719	0.9726	0.9732	0.9738	0.9744	0.9750	0.9756	0.9761	0.9767
2.0	0.9772	0.9778	0.9783	0.9788	0.9793	0.9798	0.9803	0.9808	0.9812	0.9817
2.1	0.9821	0.9826	0.9830	0.9834	0.9838	0.9842	0.9846	0.9850	0.9854	0.9857
2.2	0.9861	0.9864	0.9868	0.9871	0.9875	0.9878	0.9881	0.9884	0.9887	0.9890
2.3	0.9893	0.9896	0.9898	0.9901	0.9904	0.9906	0.9909	0.9911	0.9913	0.9916
2.4	0.9918	0.9920	0.9922	0.9925	0.9927	0.9929	0.9931	0.9932	0.9934	0.9936
2.5	0.9938	0.9940	0.9941	0.9943	0.9945	0.9946	0.9948	0.9949	0.9951	0.9952
2.6	0.9953	0.9955	0.9956	0.9957	0.9959	0.9960	0.9961	0.9962	0.9963	0.9964
2.7	0.9965	0.9966	0.9967	0.9968	0.9969	0.9970	0.9971	0.9972	0.9973	0.9974
2.8	0.9974	0.9975	0.9976	0.9977	0.9977	0.9978	0.9979	0.9979	0.9980	0.9981
2.9	0.9981	0.9982	0.9982	0.9983	0.9984	0.9984	0.9985	0.9985	0.9986	0.9986
3.0	0.9987	0.9987	0.9987	0.9988	0.9988	0.9989	0.9989	0.9989	0.9990	0.9990
3.1	0.9990	0.9991	0.9991	0.9991	0.9992	0.9992	0.9992	0.9992	0.9993	0.9993
3.2	0.9993	0.9993	0.9994	0.9994	0.9994	0.9994	0.9994	0.9995	0.9995	0.9995
3.3	0.9995	0.9995	0.9995	0.9996	0.9996	0.9996	0.9996	0.9996	0.9996	0.9997
3.4	0.9997	0.9997	0.9997	0.9997	0.9997	0.9997	0.9997	0.9997	0.9997	0.9998
3.6	0.9998	0.9998	0.9999	0.9999	0.9999	0.9999	0.9999	0.9999	0.9999	0.9999

Taken from Parzen (1960). Reprinted with permission.

Table A.4. *Student's t Distribution with n Degrees of Freedom. A Table of* $t_{n,\alpha}$ *in* $P(T > t_{n,\alpha}) = \alpha$
$\alpha = 0.005$ *to* 0.10 $n = 1, 2, \ldots$

n \ α	0.10	0.05	0.025	0.01	0.005
1	3.078	6.314	12.706	31.821	63.657
2	1.886	2.920	4.303	6.965	9.925
3	1.638	2.353	3.182	4.541	5.841
4	1.533	2.132	2.776	3.747	4.604
5	1.476	2.015	2.571	3.365	4.032
6	1.440	1.943	2.447	3.143	3.707
7	1.415	1.895	2.365	2.998	3.499
8	1.397	1.860	2.306	2.896	3.355
9	1.383	1.833	2.262	2.821	3.250
10	1.372	1.812	2.228	2.764	3.169
11	1.363	1.796	2.201	2.718	3.106
12	1.356	1.782	2.179	2.681	3.055
13	1.350	1.771	2.160	2.650	3.012
14	1.345	1.761	2.145	2.624	2.977
15	1.341	1.753	2.131	2.602	2.947
16	1.337	1.746	2.120	2.583	2.921
17	1.333	1.740	2.110	2.567	2.898
18	1.330	1.734	2.101	2.552	2.878
19	1.328	1.729	2.093	2.539	2.861
20	1.325	1.725	2.086	2.528	2.845
21	1.323	1.721	2.080	2.518	2.831
22	1.321	1.717	2.074	2.508	2.819
23	1.319	1.714	2.069	2.500	2.807
24	1.318	1.711	2.064	2.492	2.797
25	1.316	1.708	2.060	2.485	2.787
26	1.315	1.706	2.056	2.479	2.779
27	1.314	1.703	2.052	2.473	2.771
28	1.313	1.701	2.048	2.467	2.763
29	1.311	1.699	2.045	2.462	2.756
Inf.	1.282	1.645	1.960	2.326	2.576

Taken from Fisher (1925). Copyright © 1970 University of Adelaide. Reprinted with permission.

Table A.5. *Chi Square Distribution with* n *Degrees of Freedom. A Table of* $\chi^2_{n,\alpha}$ *in* $P(D > \chi^2_{n,\alpha}) = \alpha$ $\alpha = 0.005$ *to* 0.995 $n = 1$ *to* 30

n \ α	0.995	0.99	0.975	0.95	0.05	0.025	0.01	0.005
1	$0.0^4 393$	$0.0^3 157$	$0.0^3 982$	$0.0^2 393$	3.841	5.024	6.635	7.879
2	0.0100	0.0201	0.0506	0.103	5.991	7.378	9.210	10.597
3	0.0717	0.115	0.216	0.352	7.815	9.348	11.345	12.838
4	0.207	0.297	0.484	0.711	9.488	11.143	13.277	14.860
5	0.412	0.554	0.831	1.145	11.070	12.832	15.086	16.750
6	0.676	0.872	1.237	1.635	12.592	14.449	16.812	18.548
7	0.989	1.239	1.690	2.167	14.067	16.013	18.475	20.278
8	1.344	1.646	2.180	2.733	15.507	17.535	20.090	21.955
9	1.735	2.088	2.700	3.325	16.919	19.023	21.666	23.589
10	2.156	2.558	3.247	3.940	18.307	20.483	23.209	25.188
11	2.603	3.053	3.816	4.575	19.675	21.920	24.725	26.757
12	3.074	3.571	4.404	5.226	21.026	23.337	26.217	28.300
13	3.565	4.107	5.009	5.892	22.362	24.736	27.688	29.819
14	4.075	4.660	5.629	6.571	23.685	26.119	29.141	31.319
15	4.601	5.229	6.262	7.261	24.996	27.488	30.578	32.801
16	5.142	5.812	6.908	7.962	26.296	28.845	32.000	34.267
17	5.697	6.408	7.564	8.672	27.587	30.191	33.409	35.718
18	6.265	7.015	8.231	9.390	28.869	31.526	34.805	37.156
19	6.844	7.633	8.907	10.117	30.144	32.852	36.191	38.582
20	7.434	8.260	9.591	10.851	31.410	34.170	37.566	39.997
21	8.034	8.897	10.283	11.591	32.671	35.479	38.932	41.401
22	8.643	9.542	10.982	12.338	33.924	36.781	40.289	42.796
23	9.260	10.196	11.689	13.091	35.172	38.076	41.638	44.181
24	9.886	10.856	12.401	13.848	36.415	39.364	42.980	45.558
25	10.520	11.524	13.120	14.611	37.652	40.646	44.314	46.928
26	11.160	12.198	13.844	15.379	38.885	41.923	45.642	48.290
27	11.808	12.879	14.573	16.151	40.113	43.194	46.963	49.645
28	12.461	13.565	15.308	16.928	41.337	44.461	48.278	50.993
29	13.121	14.256	16.047	17.708	42.557	45.722	49.588	52.336
30	13.787	14.953	16.791	18.493	43.773	46.979	50.892	53.672

Taken from Pearson and Hartley (1954). Reprinted with permission of Biometrika Trustees.

Table A.6. D_2 Distribution with Sample Size n.
A Table of $c_{n,\alpha}$ in $P(D_2 > c_{n,\alpha}) = \alpha$ $\alpha = 0.01$ to 0.10 $n = 5, 10, \ldots$

n	0.10	0.05	0.01
5	0.51	0.56	0.67
10	0.37	0.41	0.49
15	0.30	0.34	0.40
20	0.26	0.29	0.35
25	0.24	0.26	0.32
30	0.22	0.24	0.29
40	0.19	0.21	0.25
Large n	$1.22/\sqrt{n}$	$1.36/\sqrt{n}$	$1.63/\sqrt{n}$

Taken from Lindgren (1962). Reprinted with permission.

REFERENCES

1. E. Parzen, *Modern Probability Theory and Its Applications*, Wiley, New York, 1960.
2. R. A. Fisher, *Statistical Methods for Research Workers*, 14th Ed., Hafner Press, New York, 1970 (First Edition, 1925).
3. E. S. Pearson and H. O. Hartley, *Biometrika Tables for Statisticians*, Vol. 1, Cambridge University Press, Cambridge, 1954.
4. B. W. Lindgren, *Statistical Theory*, Macmillan, New York, 1962.

APPENDIX B
Answers to Selected Problems

Chapter II

2.1 (a) Incorrect, (b) Correct, (c) Incorrect, (d) Correct, (e) Correct, (f) Correct, (g) Correct, (h) Correct.

2.4 (a) $\{1, 2, \ldots, 10\}$, (b) $\{1, 3, 4, 5, 6\}$, (c) $\{2, 7\}$, (d) $\{2, 4, 6, 7, 8, 9, 10\}$, (e) $\{1, 2, \ldots, 10\}$, (f) $\{1, 3, 4, 5, 6\}$, (g) $\{1, 5\}$.

2.6 (a) $\bar{A}\bar{B}\bar{C}$, (b) $A\bar{B}\bar{C}$, (c) $(A\bar{B}\bar{C}) \cup (\bar{A}B\bar{C}) \cup (\bar{A}\bar{B}C)$, (d) $A \cup B \cup C$, (e) $(AB\bar{C}) \cup (A\bar{B}C)$, (f) $\bar{A}BC$, (g) $(AB) \cup (BC) \cup (CA)$, (h) $\overline{ABC}$, (i) ABC.

2.9 (a) 0.553, (b) 0.053, (c) 0.395.

2.11 0.9999.

2.13 (a) 0.8865, (b) $[1 - (1 - p_A)(1 - p_C)][1 - (1 - p_B)(1 - p_D)]$.

2.15 No.

2.17 No, (a) $P(A) = P(B) = 0.5$, (b) Impossible.

APPENDIX B ANSWERS TO SELECTED PROBLEMS   371

2.18 Under condition of mutual exclusiveness: (a) F,   (b) T,   (c) F,
      (d) T,   (e) F.
      Under condition of independence: (a) T,   (b) F,   (c) F,
      (d) F,   (e) T.

2.19 (a) Approximately 10^{-5},   (b) Yes,   (c) 0.00499.

2.21 (a) $\dfrac{t_1 - t_0}{t}$,   (b) $\dfrac{t_1 - t_0}{t - t_0}$.

2.23 (a) 0.35,   (b) 0.1225,   (c) 0.65.

2.25 (a) 0.08,   (b) 0.375.

2.27 (a) 0.351,   (b) 0.917,   (c) 0.25.

2.29 (a) 0.002,   (b) 0.086,   (c) 0.4904.

Chapter III

3.1 (a) $a = 1$, $p(x) = 1$ $x = 5$
                   $= 0$ elsewhere

     (c) $a = 2$, $p(x) = \dfrac{1}{2^x}$ $x = 1, 2, \ldots$

     (e) $a > 0$, $f(x) = ax^{a-1}$ $0 \le x \le 1$
                   $= 0$ elsewhere
         or
         $a = 0$, $p(x) = 1$ $x = 0$
                   $= 0$ elsewhere

     (g) $a = \frac{1}{2}$,   neither pdf nor pmf exists.

3.2 (a) $1, \frac{1}{3}, \frac{63}{64}, 1 - e^{-6a}, 1, 1, \dfrac{2 - e^{-1/3}}{2}$;

     (b) $1, 1, \frac{127}{128}, e^{-a/2} - e^{-7a}, 1 - (\frac{1}{2})^a, \frac{1}{2}, \dfrac{e^{-1/4} - e^{-7/2}}{2}$.

3.4 (a) $F_X(x) = 0$ $x < 90$
              $= 0.1x - 9$ $90 \le x < 100$
              $= 1$ $x \ge 100$
     (b) $F_X(x) = 0$ $x < 0$
              $= 2x - x^2$ $0 \le x \le 1$
              $= 1$ $x > 1$
     (c) $F_X(x) = \dfrac{1}{\pi} \tan^{-1} x + \frac{1}{2}$ $-\infty < x < \infty$

3.5 $\frac{2}{3}$.

3.7 $F_X(x) = 0 \quad x < 0$
$ = \dfrac{x}{b} \quad 0 \le x \le b$
$ = 1 \quad x > b$
$f_X(x) = \dfrac{1}{b} \quad 0 \le x \le b$
$ = 0 \quad \text{elsewhere}$

3.9 $\dfrac{3}{a}$.

3.11 (b) $\tfrac{1}{6}$.

3.12 (a): (i) $p_X(x) = 0.6 \quad x = 1 \quad p_Y(y) = 0.6 \quad y = 1$
$ = 0.4 \quad = 2 \quad = 0.4 \quad = 2$
(iii) $f_X(x) = e^{-x} \quad x \ge 0$
$ = 0 \quad\quad \text{elsewhere}$
$ f_Y(y) = e^{-y} \quad y \ge 0$
$ = 0 \quad\quad \text{elsewhere}$
(b): (i) No,   (iii) Yes.

3.15 $\dfrac{F_X(x) - F_X(100)}{1 - F_X(100)} \quad x \ge 100.$

3.17 (a) 0.087,   (b) 0.3174,   (c) 0.274.

3.19 0.0039.

3.23 (a) $p_{X_3}(x) = 0.016 \quad x = 1$
$\phantom{(a) \quad p_{X_3}(x)} = 0.035 \quad = 2$
$\phantom{(a) \quad p_{X_3}(x)} = 0.080 \quad = 3$
$\phantom{(a) \quad p_{X_3}(x)} = 0.125 \quad = 4$
$\phantom{(a) \quad p_{X_3}(x)} = 0.415 \quad = 5$
$\phantom{(a) \quad p_{X_3}(x)} = 0.192 \quad = 6$
$\phantom{(a) \quad p_{X_3}(x)} = 0.137 \quad = 7$

(b) **Table of** $p_{X_4 X_3}(i, j)$

i \ j	1	2	3	4	5	6	7
1	0.006	0.004	0.003	0.003	0.004	0.000	0.000
2	0.002	0.009	0.008	0.005	0.010	0.002	0.001
3	0.003	0.008	0.015	0.014	0.031	0.008	0.005
4	0.001	0.004	0.015	0.027	0.051	0.017	0.011
5	0.002	0.007	0.029	0.054	0.196	0.075	0.050
6	0.001	0.002	0.005	0.015	0.071	0.060	0.032
7	0.000	0.001	0.005	0.008	0.052	0.030	0.038

APPENDIX B ANSWERS TO SELECTED PROBLEMS

Chapter IV

4.1 (a) 5, 0; (c) 2, 2; (e) $\dfrac{a}{a+1}, \dfrac{a}{(a+1)^2(a+2)}$; (g) 1, 3.

4.3 2.44 *min.*

4.5 (a) $\frac{1}{2}$, (b) 2, 4; (c) 0, 1.

4.9 (a) $\dfrac{1-p}{\lambda}$, (b) $\dfrac{1}{\lambda}$.

4.11 24 min.

4.13 $P(|X-1| \le 0.75) \ge 0.41$ by the Chebyshev inequality, $P(|X-1| \le 0.75) = 0.75$.

4.16 (a) $P(55 \le X \le 85) \ge 0$,
(b) $P(55 \le X \le 85) \ge \frac{5}{9}$, much more improved bound.

4.17 $\frac{3}{4}$.

4.19 3.05.

4.21 $\dfrac{\sigma_{X_2}^2}{(\sigma_{X_1}^2 + \sigma_{X_2}^2)^{1/2}(\sigma_{X_2}^2 + \sigma_{X_3}^2)^{1/2}}$

4.23 (a) $m_{X_1} + m_{X_2}, \sigma_{X_1}^2 + \sigma_{X_2}^2$;

(b) $\dfrac{\sigma_{X_2}}{(\sigma_{X_1}^2 + \sigma_{X_2}^2)^{1/2}}$, it approaches one if $\sigma_{X_2}^2 \gg \sigma_{X_1}^2$.

4.24 $n, 2n$.

4.26 (a) $\phi_X(t) = e^{5jt}$, 5, 0;

(c) $\phi_X(t) = \sum_{k=1}^{\infty} \dfrac{1}{2^k} e^{jtk}$, 2, 2;

(e) $\phi_X(t) = \dfrac{2e^{jt}}{jt}\left(1 - \dfrac{1}{jt}\right) + \dfrac{2}{(jt)^2}, \frac{2}{3}, \frac{1}{18}$.

Chapter V

5.1 (a) $F_Y(y) = 0 \quad y < 8$
$ = \frac{1}{3} \quad 8 \le y \le 17$
$ = 1 \quad y > 17$
(b) $F_Y(y) = 0 \quad y < 8$
$ = \dfrac{y-8}{9} \quad 8 \le y \le 17$
$ = 1 \quad y > 17$

5.3 $f_Y(y) = 0 \qquad y < -1$

$\qquad = \dfrac{y+1}{9} \qquad -1 \le y < 2$

$\qquad = \dfrac{5-y}{9} \qquad 2 \le y < 5$

$\qquad = 0 \qquad y \ge 5$

5.5 $f_Y(y) = \dfrac{1}{y\sqrt{2\pi}} e^{-\ln^2 y/2} \qquad y \ge 0$

$\qquad = 0 \qquad \text{elsewhere}$

5.8 $f_X(x) = \dfrac{2}{\pi\sqrt{a^2 - x^2}} \qquad 0 \le x \le a$

$\qquad = 0 \qquad \text{elsewhere}$

5.9 (a) $f_W(w) = \dfrac{0.19}{2a\sqrt{w/a}} \left(\dfrac{\sqrt{w/a}}{36.6}\right)^{-7.96} \exp\left[-\left(\dfrac{\sqrt{w/a}}{36.6}\right)^{-6.96}\right] \qquad w > 0$

$\qquad = 0 \qquad \text{elsewhere}$

$m_W = 1.71a \times 10^3 \qquad \sigma_W^2 = 8.05a^2 \times 10^5$

(b) Same as (a).

5.11 Y is discrete and

$p_Y(y) = \displaystyle\int_0^\infty f_X(x)\,dx \qquad y = 1$

$\qquad = \displaystyle\int_{-\infty}^0 f_X(x)\,dx \qquad = 0$

5.13 (a) $f_A(a) = \dfrac{1}{0.04 r_0 \sqrt{a/\pi}} \qquad \pi(0.99 r_0)^2 \le a \le \pi(1.01 r_0)^2$

$\qquad = 0 \qquad \text{elsewhere}$

(b) $f_V(v) = \dfrac{1}{0.08\pi r_0}\left(\dfrac{3v}{4\pi}\right)^{-2/3} \qquad \tfrac{4}{3}\pi(0.99 r_0)^3 \le v \le \tfrac{4}{3}\pi(1.01 r_0)^3$

$\qquad = 0 \qquad \text{elsewhere}$

5.15 (a) $f_Y(y) = \dfrac{2+y}{4} \qquad -2 < y \le 0$

$\qquad = \dfrac{2-y}{4} \qquad 0 < y \le 2$

$\qquad = 0 \qquad \text{elsewhere}$

(b) Same as (a).

APPENDIX B   ANSWERS TO SELECTED PROBLEMS

5.19 $f_T(t) = (a_1 + a_2 + \cdots + a_n)e^{-(a_1 + a_2 + \cdots + a_n)t}$ $t > 0$
            $= 0$ elsewhere

5.21 $f_Y(y) = \int_{-\infty}^{\infty} f_{X_2}(x_2)[f_{X_1}(x_2 + y) + f_{X_1}(x_2 - y)]dx_2$ $-\infty < y < \infty$

5.23 $f_Y(y) = ye^{-y^2/2}$ $y \geq 0$
            $= 0$ elsewhere

5.25 $f_Y(y) = \dfrac{3y^2}{(1 + y)^4}$ $y \geq 0$
            $= 0$ elsewhere

5.27 $f_{R\Phi}(r, \phi) = \dfrac{r}{2\pi\sigma^2} e^{-r^2/2\sigma^2}$ $r \geq 0$ $-\pi \leq \phi \leq \pi$
            $= 0$ elsewhere

      $f_R(r) = \dfrac{r}{\sigma^2} e^{-r^2/2\sigma^2}$ $r \geq 0$
            $= 0$ elsewhere

      $f_\Phi(\phi) = \dfrac{1}{2\pi}$ $-\pi \leq \phi \leq \pi$
            $= 0$ elsewhere

      R and Φ are independent.

Chapter VI

6.3   (a) 0.237,   (b) 3.75.
6.5   0.611, 4.2
6.7   (a) $1 - \sum_{k=0}^{m} \binom{n}{k} p^k (1 - p)^{n-k}$
      (b) $\sum_{k=m+1}^{n} (k - m) \binom{n}{k} p^k (1 - p)^{n-k}$
6.9   0.584.
6.11 $\sum_{k=r}^{s+r-1} \binom{k-1}{r-1} p^r (1 - p)^{k-r}$.
6.13 0.096.

6.15 (a) 0.349, (b) 5.
6.21 0.93.
6.23 1.4×10^{-24}.
6.25 $p_X(k) = \dfrac{22.5^k e^{-22.5}}{k!}$ $\quad k = 0, 1, 2, \ldots$
6.27 $p_k(0, t) = \left(\dfrac{t^v}{w}\right)^k \dfrac{\exp(-t^v/w)}{k!}$ $\quad k = 0, 1, 2, \ldots$

Chapter VII

7.1 0.847.
7.3 (a) 0.9, (b) 0.775.
7.6 (a) 4.566×10^{-3}, (b) 0.8944, (c) 0.383, (d) 0.385.
7.9 X_2 is preferred in both cases.
7.13 0.0062.
7.18 (a) 0.221
(b) $f_Y(y) = \dfrac{1}{(0.294)\sqrt{2\pi}(y-a)} \exp\left[-\dfrac{1}{0.172}\ln^2\left(\dfrac{y-a}{0.958b}\right)\right]$ $\quad y \geq a$
$ = 0$ $\hfill$ elsewhere
$m_Y = a + b$
7.19 0.153.
7.27 (a) 0.056, (b) 0.989.
7.31 0.125, 0, 0, 0.875. No partial failure is possible.
7.33 $f_S(s) = n(n-1)\displaystyle\int_{-\infty}^{\infty} [F_X(y) - F_X(y-s)]^{n-2} f_X(y-s) f_X(y) dy \quad s \geq 0$
$ = 0$ $\hfill$ elsewhere

Chapter VIII

8.2 (a) Type I asymptotic maximum-value distribution is suggested,
(b) $\alpha \cong 0.025$, $\quad u \cong 46.92$.
8.5 (a) Gamma is suggested, (b) $\lambda \cong 0.317$.
8.7 (a) Poisson is suggested, (b) $v \cong 45.81$.

APPENDIX B ANSWERS TO SELECTED PROBLEMS

8.9 (a) Normal is suggested, (b) $m \cong 2860, \sigma \cong 202.9$.
8.11 (a) Poisson is suggested, (b) $v \cong 7.0$.
8.13 (a) Lognormal is suggested, (b) $\theta_X \cong 76.2, \sigma^2_{\ln X} \cong 0.203$.

Chapter IX

9.1 1.75, 27.96.

9.5 (a) $f_Y(y) = 10y^9 \quad 0 \le y \le 1$
$ = 0 \quad$ elsewhere
$f_Z(z) = 10(1-z)^9 \quad 0 \le z \le 1$
$ = 0 \quad$ elsewhere

(c) 0.091, 0.91.

9.7 $\hat{\Theta}_1$ is better.

9.9 It is biased, but unbiased as $n \to \infty$.

9.12 (a) $\dfrac{\theta^2}{n}$, (b) $\dfrac{\theta^2}{n}$, (c) $\dfrac{\theta(1-\theta)}{n}$, (d) $\dfrac{\theta}{n}$.

9.13 $\dfrac{\sigma^2}{n}, \dfrac{2\sigma^4}{n}$.

9.15 (a) $1 - \dfrac{1}{\overline{X}}$,

(b) $1 - \sum_{k=1}^{9} \dfrac{1}{\overline{X}}\left(1 - \dfrac{1}{\overline{X}}\right)^{k-1}$

9.17 $\hat{A}_{ML} = X_{(1)}, \hat{A}_{ME} = \overline{X} - 1$.

9.19 $\hat{\lambda} = 0.13 \text{ sec}^{-1}$.

9.21 (a) $\hat{\Lambda}_{ML} = \hat{\Lambda}_{ME} = \dfrac{1}{\overline{T} - t_0}$

(b) $\hat{T}_{0_{ML}} = T_{(1)}, \hat{T}_{0_{ME}} = \overline{T} - \dfrac{1}{\lambda}$

(c) $\hat{\Lambda}_{ML} = \dfrac{1}{\overline{T} - T_{(1)}}, \hat{T}_{0_{ML}} = T_{(1)}$

$\hat{\Lambda}_{ME} = \dfrac{1}{\sqrt{M_2 - \overline{T}^2}}, \hat{T}_{0_{ME}} = \overline{T} - \sqrt{M_2 - \overline{T}^2}$

9.23 (a) $\hat{m}_{ML} = \hat{m}_{ME} = 83.6$, $\widehat{\sigma^2_{ML}} = \widehat{\sigma^2_{ME}} = 777.4$
(b) 2×10^{-5}

9.25 $\dfrac{1}{\sqrt{2\alpha}}$.

9.27 (a) $l_{1,2} = 63.65, 81.55$; (b) $l_{1,2} = 70.57, 84.43$;
(c) $l_{1,2} = 77.74, 89.46$.

9.29 (a) 9.16, (b) $l_{1,2} = 8.46, 9.86$.

9.31 (a) $l_{1,2} = 1072, 1128$; (b) $l_{1,2} = 1340, 6218$ and $l_1 = 1478$.

9.33 384.

Chapter X

10.1 More likely to be accepted at $\alpha = 0.01$.
10.3 Hypothesis is accepted.
10.5 Hypothesis is accepted.
10.7 Poisson hypothesis is rejected.
10.9 Gamma hypothesis is accepted.
10.11 Normal hypothesis is accepted.
10.13 Poisson hypothesis is accepted.
10.15 Hypothesis is accepted.

Chapter XI

11.1 (a) $\hat{B} = \dfrac{\sum_{i=1}^{n} x_i Y_i}{\sum_{i=1}^{n} x_i^2}$, (b) $E\{\hat{B}\} = \beta$ and $\text{var}\{\hat{B}\} = \dfrac{\sigma^2}{\sum_{i=1}^{n} x_i^2}$,
(c) $\widehat{\Sigma^2} = \dfrac{1}{n-1} \sum_{i=1}^{n} (Y_i - \hat{B}x_i)^2$.

11.3 $\widehat{\Sigma^2} = \dfrac{1}{n} \sum_{i=1}^{n} (Y_i - \hat{A} - \hat{B}x_i)^2$, $E\{\widehat{\Sigma^2}\} = \dfrac{n-2}{n} \sigma^2$, hence biased.

11.7 $14.98 + 32.14 \log_{10} v$.

11.9 (a) $\hat{\alpha} = 1.486$ and $\hat{\beta} = 0.022$, (b) $l_1 = 0.006$,
(c) Not significantly different from zero.

11.11 66.18, 0.42, 46.10.

11.13 717.18, 10.84, -3.78, -1.57, 0.38.

Index

Average value, *see* Mean
Axioms of probability, 15

Bayes' theorem, 26
Bernoulli trials, 156
Beta distribution, 214-219, 230
  generalized, 219
  mean, 216, 230
  variance, 216, 230
Bias, 260
Binomial distribution, 43, 157-162, 178
  characteristic function, 159
  mean, 159, 178
  Poisson approximation, 176
  table, 360-362
  variance, 159, 178
Boole's inequality, 31
Borel field (σ-field), 12
Brownian motion, 102

Cauchy distribution, 120
Central limit theorem, 193
Characteristic function, 93
  joint, 104
Chebyshev inequality, 82
Chi-square distribution, 126, 213-214, 230
  mean, 214, 230
  table, 368
  variance, 214, 230
Chi-square test, 312
Coefficient of excess, 79
Coefficient of skewness, 79
Coefficient of variation, 77
Confidence coefficient, 291
Confidence interval, 291, 292, 294, 297, 299
Confidence limits, 291
Consistency, 269
Correlation, 84
  perfect, 86
  zero, 85
Correlation coefficient, 84
Covariance, 84
  matrix, 88
Cramér-Rao inequality, 262
  lower bound (CRLB), 264
Cumulant, 97

Cumulative distribution function, see Probability distribution function

D_2 distribution, 323
  table, 369
De Morgan's laws, 12
Density function, see Probability density function
Distribution function, see Probability distribution function

Efficiency, 265
  asymptotic, 267
Error, 311
  type I, 311
  type II, 311
Estimate, 260
Estimator, 260
  consistent, 269
  efficient, 265
  sufficient, 270
  unbiased minimum-variance, 261
Event, 13
Expectation, 71
  conditional, 79
  mathematical, 71
  operator, 71
Exponential distribution, 45, 209-213, 230
  mean, 209, 230
  variance, 209, 230
Exponential failure law, 211
Extreme value distribution, 219-229, 230
  type I, 221, 230
  type II, 226, 230
  type III, 228, 230

Failure rate, see Hazard function
Fisher-Neyman factorization criterion, 271
Frequency diagram, 242
Function of random variables, 113-148
  moments, 129
  probability distributions, 114, 131, 139, 142

Gamma distribution, 206-213, 230
  mean, 208, 230
  variance, 208, 230
Gaus-Markov theorem, 339
Gaussian distribution, see Normal distribution
Geometric distribution, 162-164, 178
  mean, 163, 178
  variance, 163, 178
Gumbel's extreme value distribution, 221

Hazard function, 212
Histogram, 242
  cumulative, 322
Hypergeometric distribution, 162, 178
  mean, 178
  variance, 178
Hypothesis testing, see Test of hypothesis

Independence, 19
  mutual, 20
Interarrival time, 209

Jacobian, 143

Kolmogorov-Smirnov test, 322

Law of large numbers, 92
Least-square estimate, 336
Least-square estimator, 336
  covariance, 337
  linear unbiased minimum variance, 338
  mean, 337
  variance, 337
Likelihood equation, 284
Likelihood function, 283
Linear regression, 329-353
  multiple, 348
  other models, 351
  simple, 329
  variance, 339, 344
Lognormal distribution, 202-206, 230
  mean, 205, 230
  variance, 205, 230

MacLaurin series, 94
Markovian property, 29
Markov's inequality, 110
Mass function, see Probability mass function
Maximum likelihood estimate, 283
Maximum likelihood estimator (MLE), 284
  consistency, 285
  efficiency, 285
  invariance property, 285
Mean, 73
  conditional, 74, 79
Median, 74
Mode, 75
Moment, 72
  central, 75
  joint, 83
  joint central, 83

INDEX

Moment estimate, 274
Moment estimator (ME), 274
   combined, 280
   consistency, 274
Moment generating function, 108, 112
Multinomial distribution, 167-168, 178
   covariance, 168, 178
   mean, 168, 178
   variance, 168, 178
Mutual exclusiveness, 14

Negative binomial distribution, 164-167, 178
   mean, 166, 178
   variance, 166, 178
Normal distribution, 190-202, 230
   bivariate, 107, 197
   characteristic function, 192
   mean, 192, 230
   multivariate, 197
   standardized, 195
   table, 366
   variance, 192, 230
Normal equation, 332
Nuisance parameter, 279

Parameter estimation, 254-301
   interval estimation, 287
   maximum likelihood method, 283
   moment method, 272
   point estimation, 273
Pascal distribution, *see* Negative binomial distribution
Poisson distribution, 168-179
   mean, 171, 178
   table, 363-365
   variance, 171, 178
Population, 255
Probability, 15
   assignment, 18
   conditional, 22
   function, 15
   measure, 15
Probability density function (pdf), 43
   conditional, 59
   joint (jpdf), 53
   marginal, 55
Probability distribution function (PDF), 39
   bivariate, 47
   conditional, 58
   joint (JPDF), 47
   marginal, 48

   mixed-type, 46
Probability mass function (pmf), 42
   conditional, 58
   joint (jpmf), 49
   marginal, 50

Random experiment, 13
Random sample, *see* Sample
Random variable (r.v.), 38
   continuous, 39
   discrete, 38
   function of, 113-148
   sum of, 89-91, 100, 139, 201
Random vector, 39
Random walk, 50
Range space, 114
Regression coefficient, 330
   confidence interval, 341
   least-square estimate, 330, 348
   test of hypothesis, 345
Relative frequency, 18
Relative likelihood, 18
Reliability, 22, 211
Residual, 331
Return period, 164

Sample, 241, 254
   size, 255
   value, 255
Sample mean, 256-257
   mean, 256
   variance, 256
Sample moment, 259
Sample point, 13
Sample space, 13
Sample variance, 257-259
   mean, 257
   variance, 257
Schwarz inequality, 87
Set, 8
   complement of, 9
   countable (enumerable), 9
   disjoint, 10
   element of, 8
   empty, 9
   equal, 9
   finite, 9
   infinite, 9
   noncountable (nonenumerable), 9
   subset of, 9

Set operation, 10
   difference, 12
   intersection (product), 10
   union (sum), 10
Significance level, 314
Standard deviation, 77
Statistic, 256
   sufficient, 270
Statistical independence, *see* Independence
Sterling's formula, 103
Student's t-distribution, 294
   table, 367
Sum of random variables, 139, 201
   characteristic function, 100
   moment, 89-91
   probability distribution, 139, 201

Test of hypothesis, 345
Total probability theorem, 25
Tree diagram, 29

Unbiasedness, 260
Uniform distribution, 55, 185-190, 230
   bivariate, 187
   mean, 186, 230
   variance, 186, 230
Unimodal distribution, 75

Variance, 76
Venn diagram, 10

Weibull distribution, 229